Fundamentals of Industrial Instrumentation (Second Edition)

Online at: https://doi.org/10.1088/978-0-7503-3755-7

Fundamentals of Industrial Instrumentation (Second Edition)

Alok Barua

Former Professor, Indian Institute of Technology, Kharagpur, India

IOP Publishing, Bristol, UK

ISBN   978-0-7503-3755-7 (ebook)
ISBN   978-0-7503-3753-3 (print)
ISBN   978-0-7503-3756-4 (myPrint)
ISBN   978-0-7503-3754-0 (mobi)

DOI   10.1088/978-0-7503-3755-7

Version: 20240701

IOP ebooks

British Library Cataloguing-in-Publication Data: A catalogue record for this book is available from the British Library.

Published by IOP Publishing, wholly owned by The Institute of Physics, London

IOP Publishing, No.2 The Distillery, Glassfields, Avon Street, Bristol, BS2 0GR, UK

US Office: IOP Publishing, Inc., 190 North Independence Mall West, Suite 601, Philadelphia, PA 19106, USA

To

Mom

Yours

Ramananda

Contents

Preface

The main goals of an industrial instrumentation course for engineering students are shaped by a variety of applications including control, quality assurance, performance testing, design and research. In this book I have adopted two main objectives: (1) to provide a fundamental grounding in the theory of industrial instrumentation and measurement system performance and (2) to establish the physical principles and practical techniques used to measure those quantities most important for instrumentation applications. This book is structured such that the chapters are short and each deals with a specific topic, either measured variables or a specific device. It is designed for a one-semester course in industrial instrumentation or instrumentation devices.

Chapters 1 and 2 provide the fundamental background to the static and dynamic characteristics of measurement systems. Chapter 1 also includes a general introduction to industrial instrumentation. Chapter 3 describes strain gauges and different load and torque cells. Both the balanced and unbalanced bridge circuits are analyzed here. The goals of these chapters are to provide an understanding of the physical principles used to measure these variables and to provide sufficient practical information for an engineer to design their own measurement system.

Chapters 4–7 describe the sensors used for the measurement of temperature, displacement, pressure and flow rate. Chapter 4 focuses on the techniques used to measure temperature. Here, three different sensors are discussed namely, the resistance temperature detector (RTD), the thermocouple, and the thermistor. The associated signal conditioning circuits are also presented in this chapter.

Chapter 5 deals with displacement measurement. Here, the linear variable differential transformer (LVDT), capacitance sensors, and their measuring circuits are discussed. Chapter 6 covers the entire spectrum of pressure measurement, from high pressure to very low pressure. Chapter 7 addresses flow transducers. There are often bewildering numbers of alternative methods for solving any given flow rate measurement, but by understanding the advantages and limitations of the various flow meters, users can choose the most appropriate method and sensor and can be attuned to their limitations in terms of sensitivity, range, precision and accuracy. The pneumatic flapper nozzle system is discussed in chapter 8. Chapter 9 is devoted to signal conditioning circuits which are frequently used in instrumentation. Here, active filters, sample and hold circuits, logarithmic and antilogarithmic amplifiers, and multiplexing and de-multiplexing are discussed. Transducers based on peizo-electric crystals and ultrasonic sensors are covered in chapters 10 and 11 respectively. Chapter 12 covers the measurement techniques used with magnetic fields. The details of the Hall effect transducer are discussed in this chapter. Chapter 13 treats optoelectronic sensors including the photoresistor, the photodiode, and fibre optic sensors. The basics of light transmission in fibre optic cables are also included here. Analytical instrumentation is discussed in chapters 14–16. The techniques used to measure pH and viscosity are covered in chapter 14. Chapter 15 is devoted to different classes of process parameter measurement. Dissolved oxygen sensing is very important for any bioreaction. The basic operational principle of the probe, its constructional details, and calibration are discussed in this chapter. Chapter 16 focuses on the methodology used to measure the compositions of gases by gas chromatography. In today's world, the air is polluted by different harmful gases produced by industries that experienced exponential growth in the last century. Moreover, the exhausts of motorized vehicles pollute the air heavily. Therefore, quantitative analysis is necessary in order to quantify the presence of these gases in the air. Chapter 17 covers pollution measurement. The last chapter (18) discusses smart sensors and their classifications.

Problems which are mathematical in nature are given at the end of the some of the chapters. The solutions to the questions are given at the end of the book. For the convenience of the student, I have written the question once again before each solution. Problems on vortex shedding and the Coriolis flowmeter have been included here, and brief introductions to these flow meters are given before the problem definition. One hundred multiple choice questions are included in this book. A third-year student should be able to answer the objective test questions, which cover not only instrumentation but also control theory and mathematics. Thermocouple charts are provided in the appendix of the book.

The major part of the material contained here is essentially that presented in a one-semester course in instrumentation devices which I have taught for a number of years at the Indian Institute of Technology, Kharagpur. My deepest gratitude must go to my wife and daughter, who were more than understanding of all of the stolen hours. Finally, my best wishes go to the students and teachers who are the ultimate users of this book.

August 2010
Alok Barua
Kharagpur

Preface to the 2nd edition

Since the first edition of this book was published in 2011, many changes have taken place in instrumentation. Nearly all students of instrumentation engineering should have more exposure to mathematics. A good knowledge of artificial intelligence or expert systems is necessary to apply many AI-based tools in the domains of instrumentation or sensor systems as a whole.

In this edition, one more set of objective questions and answers has been added. The reader should solve more numerical problems to understand the subject better. A large number of problems have been added in this edition. The solutions to these new problems have also been added as usual.

A new chapter has been added. Chapter 19 contains a basic introduction to artificial intelligence and expert systems. Its application to instrumentation systems is also covered. An expert system-based method that selects a sensor for given process variables is presented in detail here.

I wish to thank my family for their support during the process of this revision. I especially thank my wife, Mausumi for managing my computer desk and other supporting documents.

January 2024
Kolkata
Alok Barua

Acknowledgements

Author would like to express his appreciation and gratitude to the many individuals who have contributed to the development of the second edition. These include colleagues, both undergraduate and graduate students who have used the first edition and suggested comments. I apologize to my wife, Mausumi, for not having been available often during the preparation of second edition of the book. I would like to thank my daughter, Arpita, who is the continuous source of inspiration for doing any academic work. I would finally like to thank my parents for encouraging me to study and work through all my life.

Alok Barua
New Town, Kolkata

Author biography

Alok Barua

Alok Barua received his Bachelor of Technology in Instrumentation and Electronics Engineering, Master of Electronics and Telecommunication Engineering, and PhD in Electrical Engineering from Jadavpur University and the Indian Institute of Technology, Kharagpur in 1977, 1980, and 1992, respectively. With more than thirty-three years of teaching experience in the Department of Electrical Engineering, Indian Institute of Technology, Kharagpur and the Indian Institute of Technology, Jammu, he has published many papers in his teaching and research areas—instrumentation, bioreactor design and control, testing and fault diagnosis of analog and mixed-signal circuits, and image processing. He supervises several MS theses and one PhD thesis on VLSI and mixed-signal circuits. He also holds a patent for the design of the 'See Saw Bioreactor.' He had delivered invited lectures in many different universities in the USA, Europe, the Mediterranean and the Far East. He worked as a visiting professor, guest professor, or research professor at the University of Arkansas, USA, the University of Karlsruhe, Frankfurt University, Yonsei University, Korea University, and other institutions around the world.

Professor Barua is the author or coauthor of several books: '*Computer Aided Analysis, Synthesis and Expertise of Active Filters*' (1995); '*Fault Diagnosis of Analog Integrated Circuit*' (2005); '*Fundamentals of Industrial Instrumentation*' (2011); '*Analog Signal Processing: Analysis and Synthesis*' (2014); '*Bioreactors: Animal Cell Culture Control for Bioprocess Engineering*' (2015); and '*Pipelined Analog to Digital Converter and Fault Diagnosis*' *(2020)*. He has also coauthored a research monogram entitled '*3D Reconstruction with Feature Level Fusion*' published by Lambert Academic Publishing, Germany in 2010.

IOP Publishing

Fundamentals of Industrial Instrumentation (Second Edition)

Alok Barua

Chapter 1

Introduction

1.1 Introduction

The primary object of this textbook is to introduce instrumentation devices in a sufficiently complete way that the student will acquire an ability to make meaningful measurements of three important process parameters, namely temperature, pressure, and flow as well as other mechanical and chemical parameters. Any instrumentation system is incomplete without signal conditioning circuits. Various signal conditioning circuits exist, and a full chapter is devoted to this topic.

1.2 Process instrumentation systems

A typical flow control and measurement instrumentation scheme is shown in figure 1.1. The flow of the liquid in the horizontal pipe is to be controlled. An orifice plate is used to measure the flow. The upstream and downstream pressure taps are connected to a differential pressure transmitter which has a pneumatic output. The range of the pneumatic signal is 3–15 PSI, which is converted to an electrical signal in the range of 4–20 mA by a pneumatic-to-electrical signal converter. This signal goes to a controller. A predefined flow rate is given to the controller as a set point. If the measured flow rate is above or below the set point, then the controller gives the appropriate current signal output to the current-to-pressure converter and the valve closes or opens accordingly. The purpose of all the measurement and instrumentation is to maintain the predetermined flow rate of liquid through the pipe.

1.3 Instrument characteristics

The static characteristics of an instrument are only relevant to steady-state readings. A sensor converts an input quantity into an output quantity that can be monitored, recorded, and transmitted. The final goal of an instrumentation system is to provide an exact indication of the value of the input that is the measured quantity. However,

doi:10.1088/978-0-7503-3755-7ch1

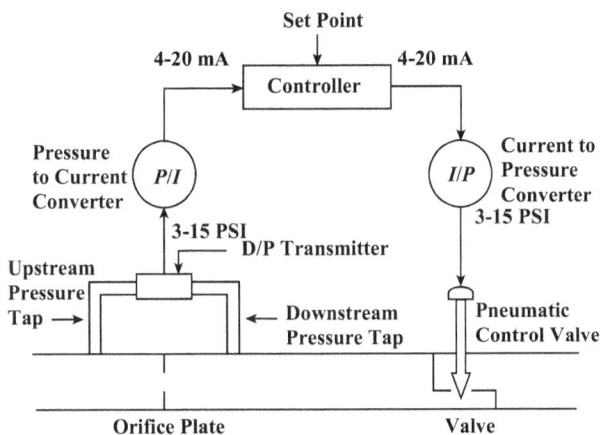

Figure 1.1. A typical process instrumentation system.

no sensor measures the exact value of a specific input variable. There is always an error in the measurement, however small it may be. If we have a room thermometer which indicates a temperature of 30 °C, it really does not matter if the true temperature is 29 °C or 31 °C. However, if a temperature sensor is installed to measure the temperature of a bioprocess, a variation of more than 0.5 °C affects the growth of the cells; therefore, a measurement accuracy of better than ±0.5 °C is required for such a process. Accuracy of measurement should be considered in the choice of sensor for a particular application. Other parameters, such as sensitivity, repeatability, linearity, etc. should also be considered in the design of an instrument. These attributes are collectively known as the 'static characteristics' of the sensor or instrument. The various static characteristics are defined in the following subsections.

1.3.1 An instrument's span

If, in a measuring instrument, the highest point of calibration is x_2 units and the lowest point is x_1 units, then the instrument range is x_2 units.

The instrument span is given by

$$\text{span} = (x_2 - x_1) \text{ units.}$$

1.3.2 The mean and standard deviation of measurements

For a set of n measurements $x_1, x_2, \ldots, x_n$, the mean value is given by

$$x_{\text{mean}} = \frac{x_1 + x_2 + \cdots + x_n}{n}.$$

The spread of any measured value (x_i) can be expressed as a deviation (d_i) as follows:

$$d_i = x_i - x_{\text{mean}}.$$

The extent to which n measured values are spread about the mean value can now be expressed by the standard deviation σ, such that

$$\sigma = \left(\frac{d_1^2 + d_2^2 + \cdots + d_n^2}{n-1} \right)^{\frac{1}{2}}.$$

1.3.3 Accuracy and precision

Accuracy is usually expressed using a phrase such as 'accurate to within x percent.' This means 'accurate to within $\pm x$ percent of the instrument span at all calibration points of the scale, unless otherwise stated.' When a temperature transducer with an error of $\pm 1\%$ indicates 100 °C, the true temperature is somewhere between 99 °C and 101 °C. Thus, the measurement accuracy of $\pm 1\%$ defines how close the measurement is to the actual measured quantity.

1.3.4 Linearity

If the calibration curve of an instrument is not a straight line, then it is not a linear instrument even though it may be very accurate. It is normally desirable that the output reading of an instrument should be linearly proportional to the quantity being measured. Figure 1.2 shows a calibration curve that is a plot of the typical output reading (cross marked) of an instrument when a sequence of input quantities is applied to it. A straight line that fits well gives the calibration curve of the instrument. The nonlinearity is thus defined as the maximum deviation of any of the output readings marked with a cross from this straight line; it is usually expressed as a percentage of the full-scale reading.

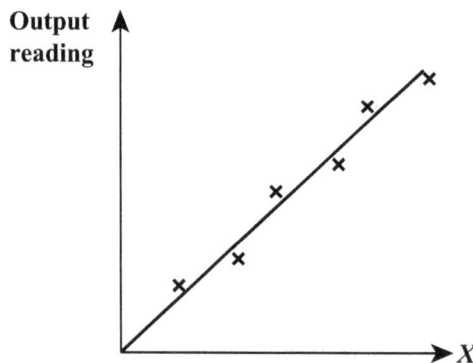

Figure 1.2. A plot of the measured quantities for a set of input values with a best-fit line.

1.3.5 Tolerance

The term 'tolerance' is closely related to accuracy; it defines the maximum error which is to be expected for some value. It is not exactly a static characteristic; however, some instruments have a quoted tolerance figure. It describes the maximum deviation of a manufactured component from some specified value. For an example, a resistor that has a nominal value of 100 ohms and a tolerance of 5% might have an actual value anywhere between 95 and 105 ohms.

1.3.6 Static error

Static error is the difference between the true value of a time-invariant measurable quantity and the value indicated by the instrument. The static error is expressed as $+y$ units or $-y$ units. For a static error given in units, the true value + the static error = the instrument reading; alternatively, the true value = the instrument reading + a static correction.

1.3.7 Repeatability

The repeatability of an instrument is the degree of closeness with which a measurable quantity may be repeatedly measured. Mathematically, it is defined as the measure of the variation in the measured data known as the standard deviation, σ. It is expressed in terms of maximum repeatability error as a percentage of the full-scale output range (FSOR).

$$\% R_e = \frac{2(\sigma)}{\text{FSOR}} \times 100.$$

1.3.8 Static sensitivity

The slope of a static calibration curve evaluated at the input value yields the static sensitivity. As shown in figure 1.3, the static sensitivity at any particular input value, say x_1, is expressed as

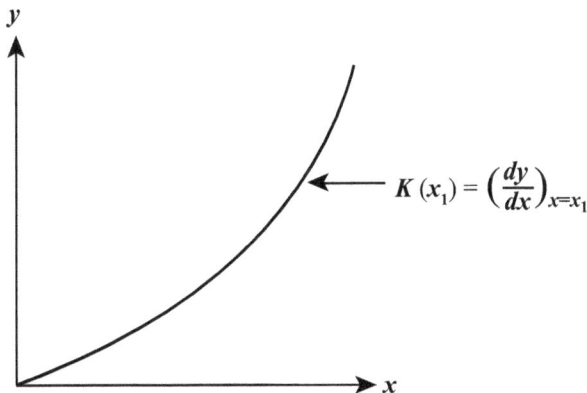

$$K(x_1) = \left(\frac{dy}{dx}\right)_{x=x_1}$$

Figure 1.3. Static sensitivity for a particular input value.

$$K = K(x_1) = \left(\frac{dy}{dx}\right)_{x=x_1}$$

1.3.9 Calibration

This is a procedure that involves the comparison of a particular instrument with either (i) a primary standard, (ii) a secondary standard with a higher accuracy than the instrument to be calibrated, or a (iii) a known output source. For example, a flow meter can be calibrated with a standard flow meter available in a national laboratory, it can be calibrated with another flow meter of known accuracy, or it can be directly calibrated using a primary measurement, for example, by weighing a certain amount of water in a tank and recording the time required for this quantity to flow through the meter.

1.3.10 Dead zone or dead space

The dead zone is the largest value of a measured variable for which the instrument does not respond. A dead zone usually occurs as a result of friction in a mechanical measurement system.

1.3.11 Hysteresis

Hysteresis error refers to the difference between the upscale sequence of calibration and the downscale sequence of calibration. The hysteresis error of an instrument is given by

$$h_e = (\{y\}\text{upscale} - \{y\}\text{downscale})_{x=x_1}.$$

Hysteresis error is depicted in figure 1.4. It is usually expressed in terms of the maximum hysteresis error as a percentage of the FSOR.

$$\% \text{ hysteresis error} = \frac{h_{e(\text{max})}}{\text{FSOR}} \times 100$$

1.3.12 Input impedance

At the input of each component in a measuring system, there exists a variable x_{i1}. The same point is associated with another variable, x_{i2}, such that their product has the dimension of power. When these two signals are identified, we can define the input impedance as follows:

$$z_i \equiv \frac{x_{i1}}{x_{i2}}.$$

Here, the power drain is $p_d = \frac{x_{i_1}^2}{z_i}$; thus, a large input impedance is needed to keep the power drain small.

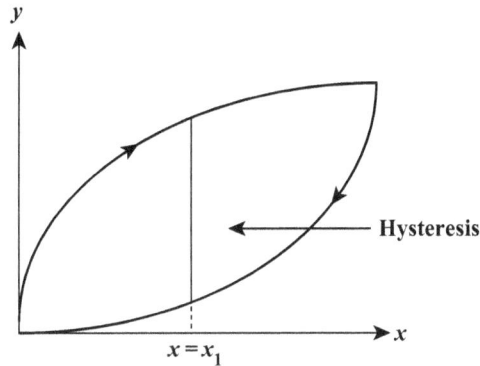

Figure 1.4. The hysteresis of a system.

1.3.12.1 Resolution
The measurement precision of an instrument defines the smallest change in the measured quantity that can be observed. For example, if 0.2 °C is the smallest temperature change that is observed using a temperature transducer, then the measurement resolution is 0.2 °C.

1.3.12.2 Bias
'Bias' means that a constant error exists over the full range of measurement of an instrument. This error can easily be removed by calibration. For example, a voltmeter may show a reading of 1 V when no input voltage is present at its terminal. If a known voltage of 30 V were now applied to the voltmeter, the reading would be 31 V. This constant bias of 1 V can be removed by calibration or by mechanical/electronic adjustment of the voltmeter itself.

1.3.12.3 Drift
The calibration of an instrument is usually performed under controlled conditions of temperature, pressure, etc. As variations occur in the ambient temperature, etc. some static characteristics may change, namely zero drift and sensitivity drift.

1.3.12.4 Zero drift
Zero drift denotes the effect by which the zero reading of an instrument is modified by a change in the ambient conditions. For example, the zero drift of a voltmeter for an ambient temperature change is expressed in $V\ °C^{-1}$. This is sometimes called the zero drift coefficient of temperature change. If there are several environmental parameters, then there will be several zero drift coefficients. The effect of zero drift is to impose bias on the instrument.

1.3.12.5 Sensitivity drift
Sensitivity drift defines the amount by which an instrument's sensitivity of measurement varies as ambient conditions change.

IOP Publishing

Fundamentals of Industrial Instrumentation (Second Edition)

Alok Barua

Chapter 2

Dynamic characteristics

Learning objectives:

- The characteristic equation of an instrument.
- Zero-order instruments and an example of such an instrument.
- First-order and second-order instruments and their responses to step, ramp, and sinusoidal inputs.
- Errors in measurement.
- Physical parameters that influence error.

2.1 Introduction

Instruments rarely respond instantaneously to changes in the measured variable. The dynamic characteristics of an instrument describe its behaviour between the time at which a measured variable changes value and the time at which the instrument output attains a steady-state value in response. The dynamic characteristic of an instrument is determined by giving it a predetermined signal. Three most common input variables are: (1) step input, (2) ramp input, and (3) sinusoidal input.

The dynamic response of an instrument to a signal input may be described by an nth-order differential equation, such as the following:

$$a_n\frac{d^n y}{dt^n} + a_{n-1}\frac{d^{n-1} y}{dt^{n-1}} + \cdots + a_1\frac{dy}{dt} + a_0 y = b_0 x, \tag{2.1}$$

where

- y = the measured quantity or value indicated by the instrument,
- x = the input quantity,
- t = time, and
- a_0, a_1, a_2, etc and b_0 are constants which represent a combination of system physical parameters.

A measurement system is shown in figure 2.1.

doi:10.1088/978-0-7503-3755-7ch2

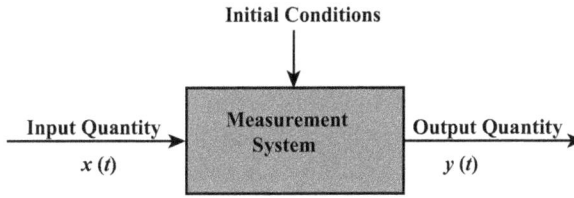

Figure 2.1. A schematic of a measurement system.

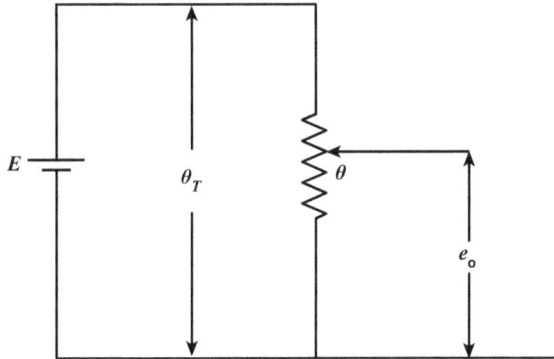

Figure 2.2. A potentiometer.

2.2 Zero-order instruments

The simplest model for a measurement system is a zero-order differential equation:

$$a_0 y = b_0 x$$
$$\text{or, } y = \frac{b_0}{a_0} x$$
$$= Kx,$$

where $K = \frac{b_0}{a_0}$ and K is called the static sensitivity.

Remarks

- It is obvious that x may vary with time; however, the instrument output follows it perfectly with no distortion or time lag.
- In the case of zero-order behaviour, the system output is considered to respond to the input signal instantly.
- The static sensitivity is found from the static calibration curve of measurement system. It is the slope of the calibration curve.
- A potentiometer is an example of a zero-order instrument (figure 2.2).

$$e_\text{o} = \frac{\theta}{\theta_T} E = K\theta$$

where $K = \frac{E}{\theta_T}$ is expressed in V rad^{-1}.

2.3 First-order instruments

A transducer that contains a storage element cannot respond instantaneously to changes in input. The mercury in a glass thermometer is an example of a first-order instrument. The bulb takes energy from the environment until the two are at the same temperature or steady-state conditions have been reached. The temperature of the bulb changes over time until an equilibrium is reached. The rate at which the temperature changes over time can be modeled using the first-order derivative, and the thermometer's behaviour is modeled by a first-order differential equation.

Therefore, the dynamic characteristics of a first-order instrument are given by

$$a_1\frac{dy}{dt} + a_0y = b_0x$$

$$\frac{a_1}{a_0}\dot{y}+y = \frac{b_0}{a_0}x$$

$$\tau\dot{y}+y = \frac{b_0}{a_0}x$$

$$\tau\dot{y}+y = Kx, \tag{2.2}$$

where $\tau = \frac{a_1}{a_0}$ and $K = \frac{b_0}{a_0}$.

τ is called the time constant of the system and it always has the dimension of time.

2.3.1 Step inputs

- The step input function is defined as

$$\left.\begin{array}{l} x = 0 \text{ for } t < 0 \\ x = x_s \text{ for } t \geqslant 0 \end{array}\right\}. \tag{2.3}$$

- Substituting equation (2.3) into equation (2.2) for $t \geqslant 0$, we get

$$\tau\dot{y}+y = Kx_s. \tag{2.4}$$

- The solution of this differential equation (2.4) gives the following for $t \geqslant 0$:

$$y = Kx_s[1-e^{-\frac{t}{\tau}}].$$

- The error in measurement at any instant of time is defined as

$$e_m = x-\frac{y}{K}= x_s-x_s[1-e^{-\frac{t}{\tau}}] \text{ (for a step input)}$$

$$= x_se^{-\frac{t}{\tau}}.$$

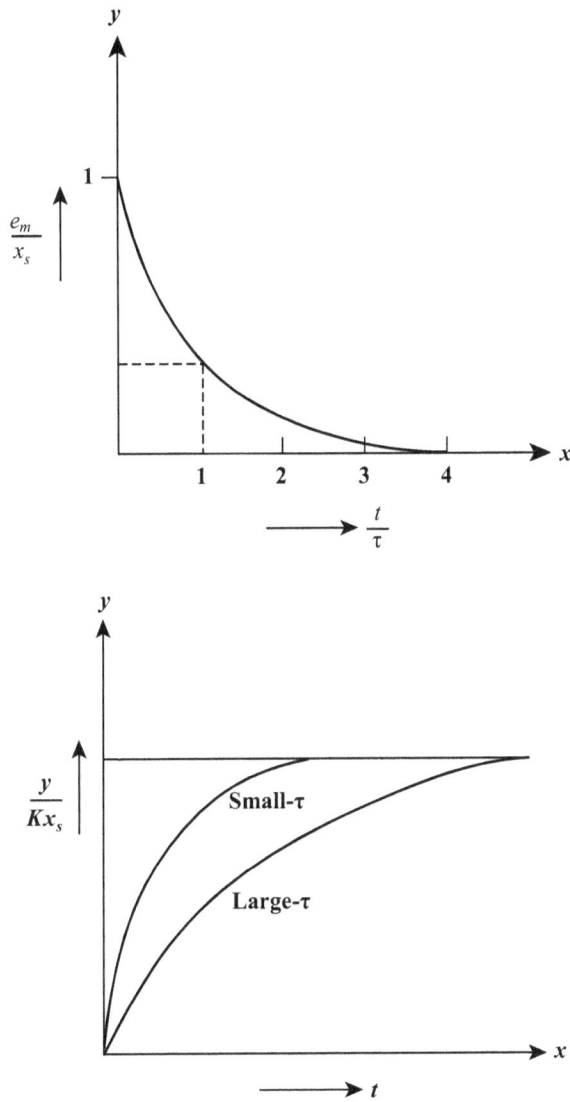

Figure 2.3. The response of a first-order system to a step input.

The response of a first-order system to a step input is shown figure 2.3.

- Normalized error $= \frac{e_m}{x_s} = e^{-\frac{t}{\tau}}$.

2.3.2 Ramp inputs

The ramp function input is defined as

$$x = \begin{Bmatrix} 0 \text{ for } t < 0 \\ \dot{x}_s t \text{ for } t \geqslant 0 \end{Bmatrix}.$$

Recalling the characteristic equation (2.2) of a first-order system, we can write

$$\tau\dot{y}+y = \dot{x}_s t.$$

The initial conditions are

$$x = y = 0, \text{ for } t = 0$$

$$y = Kx_s[\tau e^{-\frac{t}{\tau}} + t{-}\tau].$$

The error in measurement at any instant of time is given by

$$e_m = x-\frac{y}{K} = \dot{x}_s t-\dot{x}_s\tau e^{-\frac{t}{\tau}}-\dot{x}_s t + \dot{x}_s\tau$$

$$= -\dot{x}_s\tau e^{-\frac{t}{\tau}} + \dot{x}_s\tau.$$

A ramp response is shown in figure 2.4.

Remarks

- It is apparent that smaller the value of τ, the faster the disappearance of transient error.
- Moreover, once the transient has disappeared, the instrument lags behind by a constant value which is again the time constant of the instrument.
- Therefore, it is obvious that the lag of the instrument is directly dependent on the time constant.
- For example, a temperature sensor that has a time constant of 5 s will ultimately lag behind a ramp input by 5 s.
- The measurement error is directly proportional to the ramp input and the time constant.
- If low measurement error is required, the instrument must have a low time-constant value.

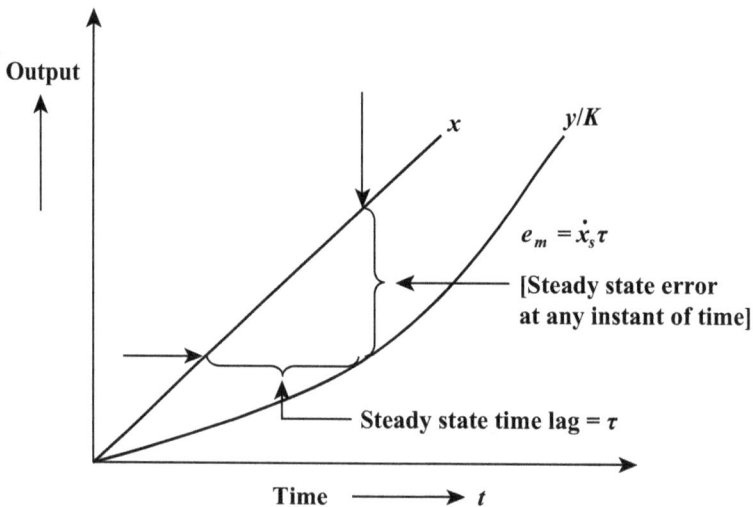

Figure 2.4. The response of a first-order system to a ramp input.

2.3.3 Sinusoidal inputs

Periodic signals are encountered in many process, such as vibration analysis, ambient temperature variation, etc When a periodic signal such as a sinusoidal input is applied to a first-order system or instrument, the frequency of the input signal influences the response of the measurement system.

The sinusoidal function is defined as

$$x(t) = x_S \sin \omega t \text{ for } t \geqslant 0.$$

The characteristic equation is as follows:

$$\tau \dot{y} + y = K x_S \sin \omega t.$$

The solution to this differential equation yields (ignoring the initial conditions)

$$y[0] = y_0$$

$$y = \frac{K x_S}{\sqrt{1 + \omega^2 \tau^2}} \sin(\omega t - \phi) \text{ where, } \phi = \tan^{-1} \omega \tau$$

$$= A \sin(\omega t - \phi),$$

where $A = \dfrac{K x_S}{\sqrt{1 + \omega^2 \tau^2}}$.

Here, A represents the amplitude of the steady-state response and ϕ is the phase shift of the output response with respect to the sinusoidal input. It is apparent that the amplitude of the output response of a first-order system depends on the frequency of the input periodic signal.

The delay in measurement is given by

$$D = \frac{\phi}{\omega},$$

where
- D is expressed in seconds,
- ϕ is in radians, and
- ω is in radian/s.

The amplitude ratio is given by

$$\frac{A}{K x_S} = \frac{1}{\sqrt{1 + \omega^2 \tau^2}}.$$

The frequency response of a first-order system is shown in figures 2.5(a) and (b).

Remarks
- It is obvious that both the time constant of the system and the input signal frequency influence the system response.

2-6

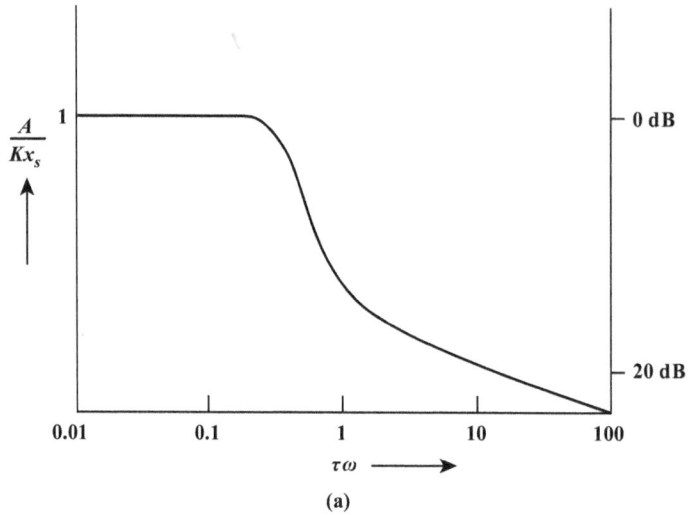

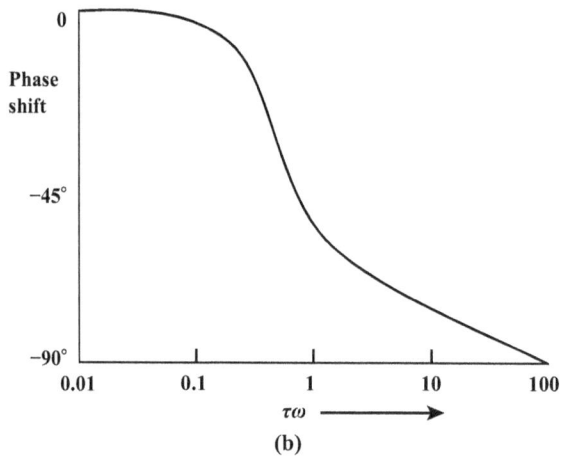

Figure 2.5. (a) An amplitude plot and (b) a phase plot of a first-order system.

- For those values of $\omega\tau$ for which $\frac{A}{Kx_S} \approx 1$, the measurement system has almost no attenuation of the input signal amplitude, and there is very little time delay.
- If the measurement of high-frequency signals is desired, then it will be necessary for the system to having small τ.
- A large time-constant system will result in the removal of the high-frequency component from the output signal.

2.4 Second-order systems

A sensor that is modeled by a second-order differential equation is called a second-order system. Accelerometers, diaphragm pressure transducers, and mercury-in-glass manometers are a few examples of second-order systems. The behaviour of a second-order measurement system can be described by

$$a_2\ddot{y}+a_1\dot{y}+a_0y = b_0x. \tag{2.5}$$

This equation can be rewritten as

$$\frac{\ddot{y}}{\omega_n^2} + \frac{2\xi}{\omega_n}\dot{y} + y = \frac{b_0}{a_0}x$$

or

$$\frac{\ddot{y}}{\omega_n^2} + \frac{2\xi}{\omega_n}\dot{y} + y = Kx,$$

where

- $\omega_n = \sqrt{\frac{a_0}{a_2}}$ = the undamped natural frequency of the system,
- $\xi = \frac{a_1}{2\sqrt{a_0a_2}}$ = the damping ratio of the system, and
- $K = \frac{b_0}{a_0}$ = the static sensitivity of the system.

The homogeneous solution of equation (2.5) describes the natural or intrinsic response of any system. The output response depends on the roots of the characteristic equation.

2.4.1 Step inputs

Equation (2.5) can be rewritten as

$$\left(\frac{D^2}{\omega_n^2} + 2\xi\frac{D}{\omega_n} + 1\right)y = Kx_s, \tag{2.6}$$

where

- D is the differential operator and $D \equiv \frac{d}{dt}$
- $y = 0$ at $t = 0$, and

$$\frac{dy}{dt} = 0 \text{ at } t = 0.$$

The particular integral of equation (2.6) is $y_p = Kx_s$. The homogenous solution of the equation is given by:

homogeneous solution = particular integral + complementary function solution.
The complementary function has three possible forms.

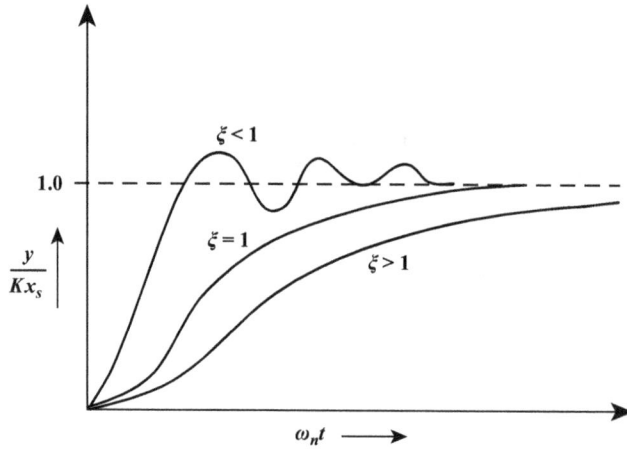

Figure 2.6. Responses to a step input for a second-order system.

Case 1: $\xi > 1$, an over-damped system (real unrepeated roots)
The normalized output can be expressed as

$$\frac{y}{Kx_s} = 1 - \frac{\xi + \sqrt{\xi^2 - 1}}{2\sqrt{\xi^2 - 1}} e^{\left(-\xi + \sqrt{\xi^2 - 1}\right)\omega_n t} + \frac{\xi - \sqrt{\xi^2 - 1}}{2\sqrt{\xi^2 - 1}} e^{\left(-\xi - \sqrt{\xi^2 - 1}\right)\omega_n t}.$$

Case 2: $\xi = 1$, a critically damped system (real repeated roots)

$$\frac{y}{Kx_s} = 1 - (1 + \omega_n t)e^{-\omega_n t}.$$

Case 3: $0 < \xi < 1$, an under-damped system (complex conjugate roots)

$$\frac{y}{Kx_s} = 1 - \frac{e^{-\xi \omega_n t}}{\sqrt{1 - \xi^2}} \sin\left(\sqrt{1 - \xi^2}\,\omega_n t + \varphi\right),$$

where $\phi = sin^{-1}\sqrt{1 - \xi^2}$.
The responses of a second-order system for different values of the damping co-efficient are shown in figure 2.6.

2.4.2 Ramp inputs

The characteristic equation for a second-order system with a ramp input can be expressed as follows:

$$\left(\frac{D^2}{\omega_n^2} + \frac{2\xi D}{\omega_n} + 1\right)y = K\dot{x}_s t$$

$$y = \frac{dy}{dt} = 0 \text{ at } t = 0.$$

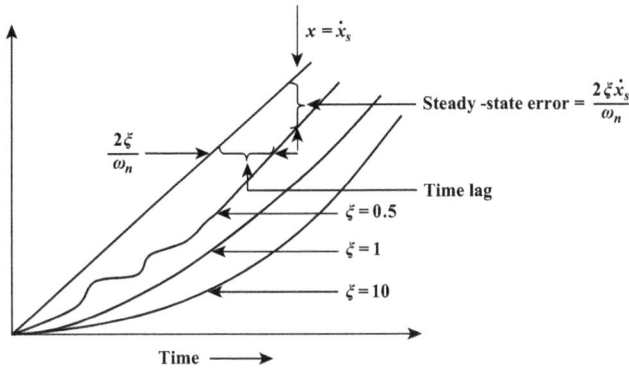

Figure 2.7. The ramp response of a second-order instrument.

Solutions can be found as described below.

• Case 1: for $\xi > 1$, an over-damped system,

$$\frac{y}{K} = \dot{x}_s t - \frac{2\xi \dot{x}_s}{\omega_n}\left[1 + \frac{2\xi^2 - 1 - 2\xi\sqrt{\xi^2-1}}{4\xi\sqrt{\xi^2-1}}e^{\left(-\xi+\sqrt{\xi^2-1}\right)\omega_n t} + \frac{-2\xi^2 + 1 - 2\xi\sqrt{\xi^2-1}}{4\xi\sqrt{\xi^2-1}}e^{\left(-\xi-\sqrt{\xi^2-1}\right)\omega_n t}\right].$$

Case 2: for $\xi = 1$, a critically damped system,

$$\frac{y}{K} = \dot{x}_s t - \frac{2\dot{x}_s}{\omega_n}\left[1 - e^{-\omega_n t\left(1+\frac{\omega_n t}{2}\right)}\right].$$

Case 3: for $0 < \xi < 1$, an under-damped system,

$$\frac{y}{K} = \dot{x}_s s t - \frac{2\xi \dot{x}_s s}{\omega_n}\left[1 - \frac{e^{-\xi\omega_n t}}{2\xi\sqrt{1-\xi^2}}\sin\left(\sqrt{1-\xi^2}\,\omega_n t + \phi\right)\right],$$

where $\phi = \tan^{-1}\frac{2\xi\sqrt{1-\xi^2}}{2\xi^2 - 1}$.

The responses of the second order system with ramp input for various damping coefficients are shown in figure 2.7.

2.4.3 Sinusoidal inputs

$$x = x_s \sin \omega t$$

$$\frac{y/K}{x_s} = \frac{\sin(\omega t + \varphi)}{\sqrt{\left[1-\left(\frac{\omega}{\omega_n}\right)^2\right]^2 + \frac{4\xi^2\omega^2}{\omega_n^2}}}$$

Phase shift $\phi = \tan^{-1}\frac{2\xi}{\frac{\omega}{\omega_n} - \frac{\omega_n}{\omega}}$.

A frequency response plot and a phase plot for a second-order system are shown in figures 2.8(a) and (b), respectively.

(a) Magnitude response

(b) Phase response

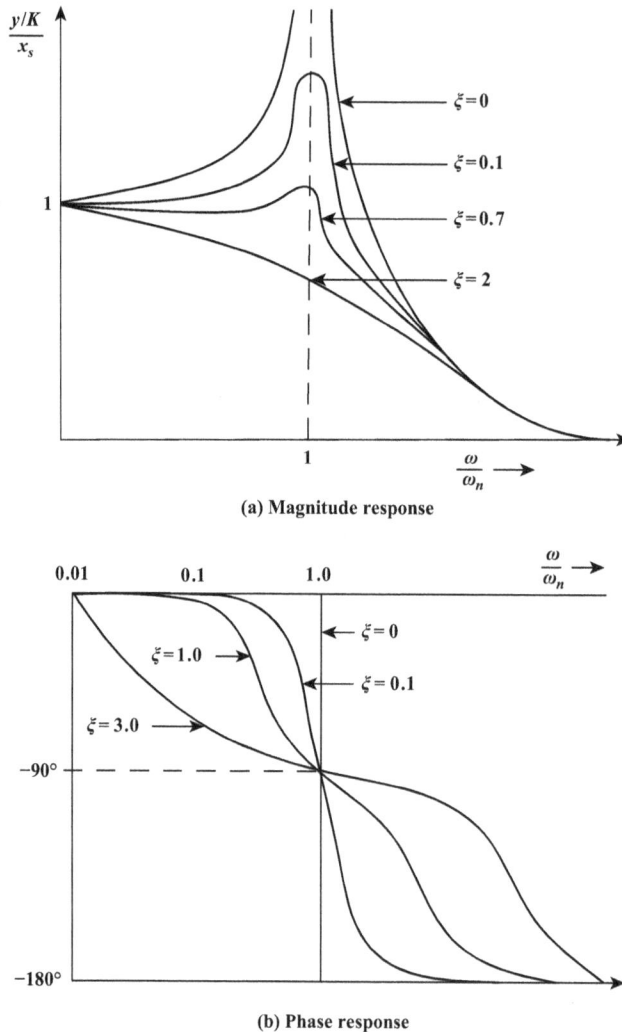

Figure 2.8. (a) Frequency response and (b) phase plot of a second order system.

2.4.3.1 Dynamic error

The dynamic error, d_e, of a system can be defined as $d_e = \frac{y/K}{x} - 1$; it represents a measure of the inability of a system to adequately reproduce the amplitude of the input signal for a particular input frequency. A measurement system with a magnitude ratio close to unity over the anticipated frequency band of the input signal is preferred in order to minimize d_e.

2.4.3.2 Settling time

A characteristic that is useful in characterizing the speed of response of any instrument is the settling time. This is the time (after the application of a step input) taken for the instrument to reach and stay within a tolerance band around its final value.

Problem 2.1. A first-order temperature transducer is used to measure the temperature of an oil bath. If the temperature exceeds 100 °C, the heat supplied to the oil bath should be stopped within 5 s of reaching 100 °C. Determine the maximum allowable time constant of the sensor if a measurement error of 5% is allowed.

Problem 2.2. A first-order instrument has a time constant of 0.5 s. It is measuring a process parameter that is sinusoidal in nature and has a frequency of 3 Hz. Determine the dynamic error of the system.

Problem 2.3. A first-order temperature sensor is suddenly immersed in a liquid that has a temperature of 100 °C. If, after 3 s, the sensor shows a temperature of 80 °C, calculate the instrument time constant. Also calculate the error in the temperature reading after 2 s.

Problem 2.4. A pressure transducer is to be selected to measure the pressure of a vessel. The pressure variation can be considered to be a sinusoidal signal whose frequency lies between 1 and 4 Hz. Several sensors are available, each with a known time constant. Select the appropriate sensor if a dynamic error of ±1% is acceptable.

Problem 2.5. A periodic signal is to be measured by a first-order instrument that has a time constant of 3 s. If a dynamic error of ±5% can be tolerated, find the highest-frequency input signal that can be measured by the instrument.

Problem 2.6. An accelerometer that is second order in nature is to be selected to measure a sinusoidal signal whose frequency is less than 100 Hz. If a dynamic error of ±6% is allowed, choose a sensor for a damping ratio of 0.6.

Problem 2.7. A first-order temperature alarm unit with a time constant of 2 min is subjected to a sudden 100 °C rise because of a fire. If an increase of 50 °C is required to activate the alarm, what is the delay in signaling the temperature change?

IOP Publishing

Fundamentals of Industrial Instrumentation (Second Edition)

Alok Barua

Chapter 3

Strain, load, and torque measurement

Learning objectives:

- Derivation of the gauge factor of a strain gauge.
- Strain gauge composition.
- Temperature compensation.
- Installation of strain gauges on a load cell.
- The sensitivity of the load cell.
- The range of the load cell.
- Different torque measurement techniques.
- Details of strain gauge based torque measurement.
- Torque cell data transmission.

3.1 Introduction

The design of load-carrying components for equipment and structures requires information about the distribution of forces within each particular component. Effective designs for mechanical shafts, pressure vessels, and a variety of support structures must consider their load-carrying capacities and allowable deflections. All mechanical designs are based on the safe level of stress within a material. The experimental analysis of stress is accomplished by measuring the deformation of a part under load. The strain gauge is a suitable device for stress analysis. However, it is also used in the construction of force, torque, pressure, flow, and acceleration transducers. Strain gauges are constructed by bending a conductor so that several lengths of wire are oriented along the axis of the gauge. The wire should be long enough to give a reasonable total resistance.

3.2 The strain gauge

Let us take a wire of length L and cross-sectional area A. If this wire is stretched or compressed, its resistance changes due to the dimensional change and because of the

doi:10.1088/978-0-7503-3755-7ch3

property of materials called piezo resistance, which is a dependence of resistivity (ρ) on strain.

The resistance R is given by

$$R = \frac{\rho L}{A}, \tag{3.1}$$

where ρ = the resistivity of the wire material;

$$\therefore dR = \frac{A(\rho dL + L d\rho) - \rho L dA}{A^2}. \tag{3.2}$$

The axial strain, $\epsilon_a = \epsilon = \frac{dL}{L}$.

The transverse strain, $\epsilon_t = \frac{dD}{D} = \frac{1}{2}\frac{dA}{A}$.

Poisson's ratio, $\nu = -\frac{\epsilon_t}{\epsilon_a}$.

Again, the volume of wire, $V = AL$.

Differentiating yields the following:

$$dV = AdL + LdA. \tag{3.3}$$

Moreover,

$$\begin{aligned} \Delta V &\cong AL \in (1 - 2\nu) \\ &= AdL(1-2\nu). \end{aligned} \tag{3.4}$$

Again, $\Delta V = AdL + LdA$.

Equating equations (3.3) and (3.4), we get

$$AdL(1-2\nu) = AdL + LdA$$

$$\therefore -2\nu AdL = LdA. \tag{3.5}$$

Substituting equation (3.5) into equation (3.2) gives

$$dR = \frac{\rho AdL + ALd\rho + 2\nu\rho AdL}{A^2}$$

$$\therefore dR = \frac{\rho dL(1 + 2\nu)}{A} + \frac{Ld\rho}{A}. \tag{3.6}$$

Dividing equation (3.6) by equation (3.1) yields

$$\frac{dR}{R} = \frac{dL(1 + 2\nu)}{L} + \frac{d\rho}{\rho}.$$

The gauge factor, $\lambda = \frac{dR/R}{dL/L} = 1 + 2\nu + \frac{d\rho/\rho}{dL/L}$.

1 = the resistance change due to the change of length.

2ν = the resistance change due to the change of area (0–0.5 for all materials).

$\frac{d\rho/\rho}{dL/L}$ = the resistance change due to the piezo resistance effect.

$$\frac{d\rho/\rho}{dL/L} = \prod_1 E,$$

where E = the modulus of elasticity and $\prod_1$ = the longitudinal piezo resistance coefficient (positive or negative).

3.2.1 Implementations of strain gauges

Strain gauges are implemented in several ways. Some types of gauge are as follows:

1. The unbonded metal wire gauge.
2. The bonded metal wire gauge.
3. The bonded metal foil gauge.
4. The bonded semiconductor gauge.
5. The diffused semiconductor gauge.

3.2.2 The compositions of strain gauges materials

Strain gauges material	Composition	Gauge factor
Constantan or advance	55% Cu, 45% Ni	$\lambda = 2$
Isoelastic 1	36% Ni, 8% Cr, 4% Mn, Si, Molybdenum, 52% Fe	$\lambda = 3.5$
Isoelastic 2	36% Ni, 8% Cr, 05% Molybdenum, 55.5% Fe	$\lambda = 3.6$
Karma	74% Ni, 20% Cr, 3% Al, 3% Fe	$\lambda = 2$
Armour D	70% Fe, 20% Cr, 10% Al	$\lambda = 2$
Platinum–tungsten	92% Pt, 8% W	$\lambda = 4$
Nichrome V	80% Ni, 20% Cr	$\lambda = 2.1$
Semiconductor strain gauge		$\lambda \cong 130$

3.2.3 The bonded metal foil gauge

The bonded metal foil gauge is the most widely used transducer for stress analysis. It uses similar materials to the materials used in wire gauges. It consists of a metallic foil pattern similar to the patterns used to produce printed circuit boards. The photo-etched metal foil pattern is installed on a backing material. A bonded metal foil gauge is shown in figure 3.1. The gauges are built with thick end turns. This local increase in area reduces resistance and hence it also reduces transverse sensitivity. In metal wire gauges, these end turns would have same cross-sectional area as that of the longitudinal cross-section; these elements increase the transverse sensitivity.

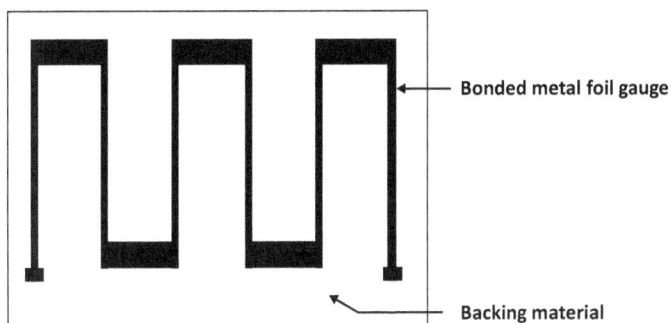

Figure 3.1. A bonded metal foil gauge.

3.2.4 The gauge length

The correct choice of gauge length is an important factor for some specific applications. The strain should measured at the location where stress is at a maximum. The strain gauge averages the measured strain over the gauge length. An an improper choice of gauge length can lead to erroneous measurements.

3.2.5 The backing material

The backing material electrically isolates the metallic gauge from the test specimen. It transmits the applied strain to the sensor and provides the surface that is used to bond the gauge to the specimen surface where the strain is to be measured. The backing material should be able to tolerate a wide temperature range. Polyimide and glass-reinforced phenolic are two commonly used backing materials.

3.2.6 The adhesive

The bond created by the adhesive serves as a mechanical and thermal coupling between the strain gauge and the test specimen. The adhesive should accurately transmit the strain applied to the test specimen. It should have thermal conduction and expansion characteristics. The adhesive should not shrink or expand during the curing process, otherwise a pseudo strain will be developed in the gauge. Epoxies, cellulose nitrate cement, and ceramic-based cements are some of the adhesives used with strain gauges.

3.2.7 Semiconductor strain gauges

Silicon is the basic material used to make semiconductor strain gauge. Usually, silicon is doped with boron to make a p-type strain gauge. Alternatively, an n-type strain gauge can be produced by doping silicon with arsenic. The resistance of the p-type gauges increases with the applied tensile strain, whereas that of the n-type gauge decreases.

The resistivity of a semiconductor material is given by

$$\rho = \frac{1}{qn\mu},$$

where

q = the charge of the electrons;

n = the number of charge carriers, which depends on the doping; and

μ = the mobility of the charge carriers.

The impurity concentration typically lies between 10^{16} and 10^{20} atoms/cm^3. The resistivity of p-type silicon with a concentration of 10^{20} atoms/cm^3 is 500 $\mu\Omega$m, which is much higher than that of copper. The strain gauge material shows a change in resistivity with strain. This change of resistivity is called the piezo resistive effect. The mobility of the charge carriers changes due to the applied strain, thus causing a large change in resistivity.

Advantages

- A high gauge factor ==> useful for the measurement of micro strain.
- A small gauge length ==> very small strain gauge sensors.

Disadvantages

- High temperature sensitivity.
- Nonlinearity.
- Mounting difficulties.
- Performance deteriorates in the presence of moisture.

3.2.8 The temperature compensation circuit

The strain gauge is a resistive device and therefore its value depends on the ambient temperature. The ambient temperature effect can be nullified by a dummy gauge which is at the same temperature as the active gauge but is not subjected to any stress. The compensation circuit is shown in figure 3.2. The bridge's unbalanced voltage is now independent of ambient temperature changes.

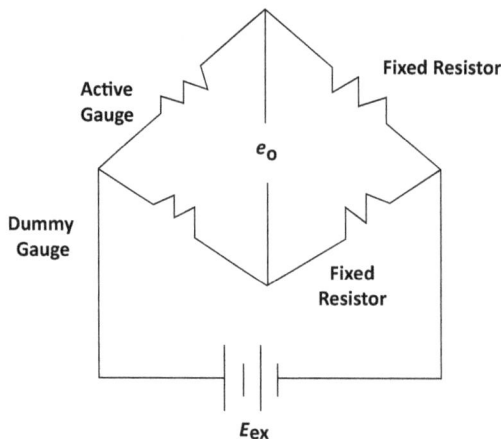

Figure 3.2. A strain gauge temperature compensation circuit.

3.3 The load cell

3.3.1 Introduction

Transducers that measure force or torque usually consist of an elastic member that converts the quantity to be measured into a deflection or strain. A strain gauge or a set of strain gauges can be used to measure force or torque. The shape and size of the elastic member, the material used for its fabrication, and the strain gauge determine the static characteristics of the transducer, such as range, linearity, and sensitivity. Transducers that measure force and torque are called load cells and torque cells, respectively. The most commonly used elastic members are shaped as columns, beams, rings, and numerous other shapes for special applications. The basic sensors used in all cases are strain gauges.

3.3.2 The use of the Wheatstone bridge in load cells

A Wheatstone bridge is commonly used to convert a change in resistance to an output voltage, as shown in figure 3.3.

With an initially balanced bridge, an output voltage e_o develops when resistances R_1, R_2, R_3, and R_4 are varied by amounts ΔR_1, ΔR_2, ΔR_3, and ΔR_4, respectively. Using the new resistance values, the change in output voltage can be expressed as

$$e_o = \frac{(R_1 + \Delta R_1)(R_3 + \Delta R_3) - (R_2 + \Delta R_2)(R_4 + \Delta R_4)}{(R_1 + \Delta R_1 + R_2 + \Delta R_2)(R_3 + \Delta R_3 + R_4 + \Delta R_4)} E_{ex}.$$

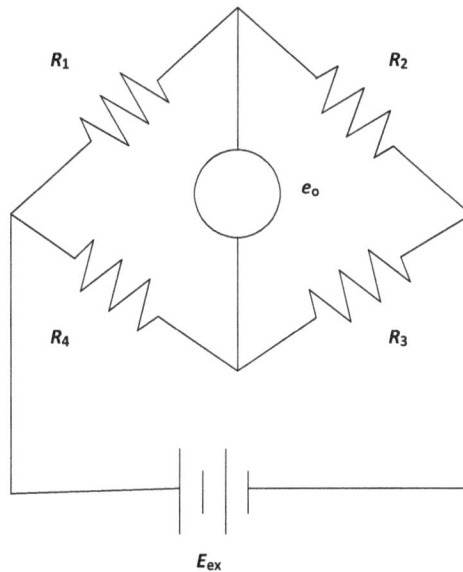

Figure 3.3. A Wheatstone bridge.

Substituting for an initially balanced bridge yields

$$R_1 R_3 = R_2 R_4$$

$$e_o = \frac{r}{(1 + r)^2}\left(\frac{\Delta R_1}{R_1} - \frac{\Delta R_2}{R_2} + \frac{\Delta R_3}{R_3} - \frac{\Delta R_4}{R_4}\right)E_{ex}, \qquad (3.7)$$

where

$$r = \frac{R_2}{R_1}.$$

3.3.3 The column load cell

A simple column load cell is shown in figure 3.4. The load can be either tensile or compressive. Transverse gauge 4 is located on the left side of the column and axial gauge 3 is located on its rear face. Strain gauges 1 and 3 are subjected to axial strains and gauges 2 and 4 are subjected to transverse strains. Here, all four strain gauges are active gauges. We can measure the load P with one strain gauge; however, the use of four strain gauges produces a larger bridge output and makes the load cell insensitive to ambient temperature variations.

The axial strain $\epsilon_{ax} = \frac{P}{AE}$.

The transverse strain $\epsilon_{tr} = -\frac{\nu P}{AE}$.

P = the tensile force (in N).

A = the cross-sectional area of the link in m^2.

E = the modulus of elasticity of the link material in N m^{-2}.

ν = Poisson's ratio of the link material.

Moreover,

$$\frac{\Delta R_1}{R_1} = \frac{\Delta R_3}{R_3} = \lambda \, \epsilon_{ax} = \frac{\lambda P}{AE}$$

$$\frac{\Delta R_2}{R_2} = \frac{\Delta R_4}{R_4} = \lambda \, \epsilon_{tr} = -\frac{\nu \lambda P}{AE},$$

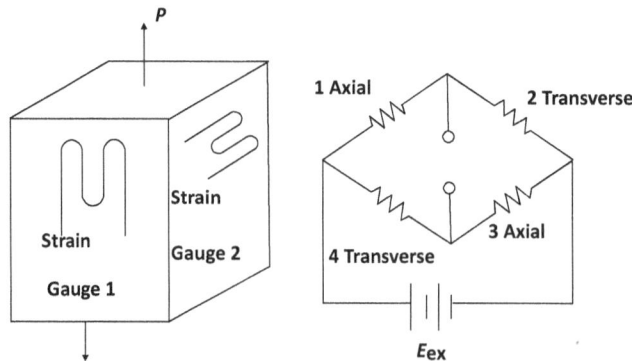

Figure 3.4. The column load cell.

where $\lambda \equiv$ the gauge factor of the strain gauges used on the load cell.

Therefore, the output voltage e_o can be expressed in terms of the load P, assuming $R_1 = R_2$ in equation (3.7), as follows:

$$e_o = \frac{\lambda P(1 + \nu)E_{ex}}{2AE}$$

$$\therefore P = \frac{2AE}{\lambda(1 + \nu)E_{ex}} e_o = Ce_o.$$

Thus the load P is linearly proportional to the output voltage e_o and the calibration constant C is equal to $\frac{2AE}{\lambda(1+\nu)E_{ex}}$.

The sensitivity of the load cell–Wheatstone bridge combination is given by

$$S = \frac{e_o}{P} = \frac{1}{C} = \frac{\lambda(1 + \nu)E_{ex}}{2AE}. \tag{3.8}$$

This arrangement gives a sensitivity that is $2(1 + \nu)$ times the sensitivity achieved using a single active gauge in the bridge. It also provides temperature compensation, since all four gauges are at the same temperature. Equation (3.8) indicates that the sensitivity of the column load cell depends upon its cross-sectional area A, the modulus of elasticity and Poisson's ratio (E and ν) of the column material, the gauge factor of the strain gauge (λ), and the excitation voltage (E_{ex}) of the bridge.

The range of the link load cell is given by

$$P_{max} = S_{fs} \cdot A, \tag{3.9}$$

where S_{fs} is the fatigue strength of the material of the load cell. Comparing equations (3.8) and (3.9), we can infer that the high sensitivity is associated with low range of the load cell, while low sensitivity is associated with a high range. The voltage ratio at maximum load $(\frac{e_o}{E_{ex}})_{max}$ for the column load cell–Wheatstone bridge combination is given by

$$\left(\frac{e_o}{E_{ex}}\right)_{max} = \frac{\lambda S_{fs}(1 + \nu)}{2E}.$$

Load cells are usually fabricated from steel with the following specifications:
$E = 2.0684 \times 10^{11}$ N m^{-2},
$\nu = 0.3$,
$S_{fs} = 5.516 \times 10^8$ N m^{-2},
$\lambda = 2$ (for Advance),
Therefore, $\left(\frac{E_o}{E_{ex}}\right)_{max} = 3.47$ mv V^{-1}.

3.3.4 Some important points

The deflection of the column load cell under full load is of the order of 0.025 mm to 0.38 mm because it has high stiffness. Therefore, the natural frequency of the load cell is unimportant and is determined by the force-carrying elements, such as a

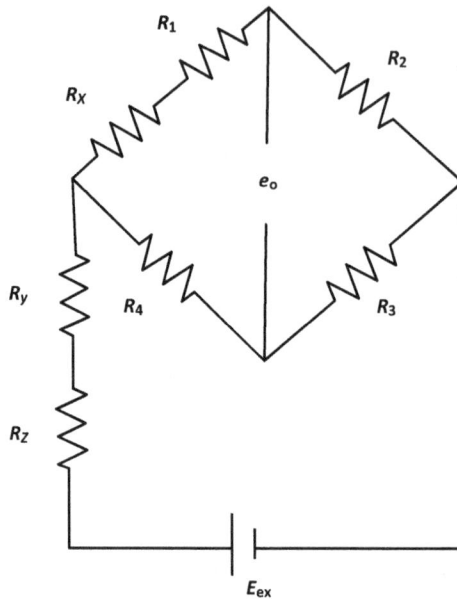

Figure 3.5. A modified Wheatstone bridge used for additional temperature compensation.

weighbridge. To achieve high accuracy, additional temperature compensation is required as shown in the following circuit of figure 3.5.

R_x and R_y are temperature-sensitive resistors.

R_x compensates for slightly different temperature coefficients of resistance.

R_y compensates for the temperature dependence of the modulus of elasticity of the load cell material.

R_z is a non-temperature-sensitive resistor that provides the desired sensitivity.

3.4 The cantilever beam load cell

A cantilever beam with strain gauges is shown is figure 3.6. The beam is fixed at one end and the load is applied at the other end.

All strain gauges are oriented in the axial direction. Gauges 2 and 4 are located on the bottom surface of the beam. The resistances of gauges 1, 2, 3, and 4 are R_1, R_2, R_3, and R_4, respectively. Moreover, in the unstrained condition, $R_1 = R_2 = R_3 = R_4$. The Wheatstone bridge is shown in figure 3.7. A load P is now applied such that it acts vertically downward at the free end of the cantilever beam. Strain gauges 1 and 3 become elongated and gauges 2 and 4 are compressed.

The load P produces a moment of $P{\cdot}x$, where x is the distance from the midpoint of the axial gauges to the point of application of load P. The strain developed at the gauges

$$\epsilon_1 = -\epsilon_2 = \epsilon_3 = -\epsilon_4 = \frac{6P.\,x}{Ewh^2},$$

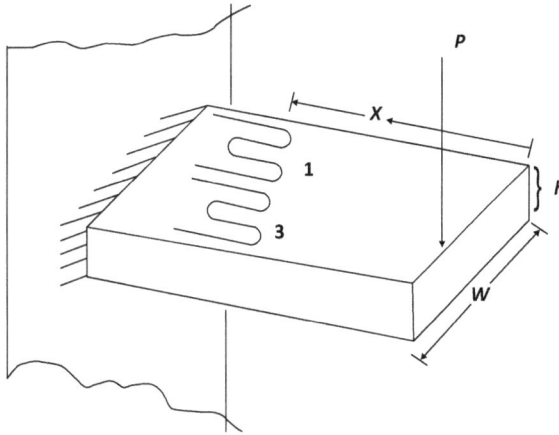

Figure 3.6. The cantilever beam load cell.

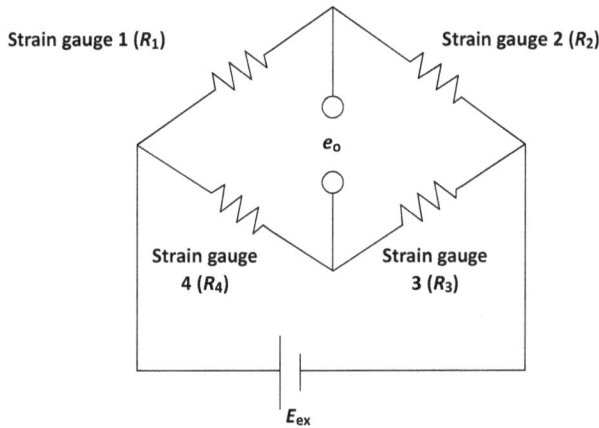

Figure 3.7. The Wheatstone bridge.

where w = the width of the cross-section of the beam and h = the height of the cross-section of the beam. Therefore, the changes of resistance in the gauges are given by

$$\frac{\Delta R_1}{R_1} = -\frac{\Delta R_2}{R_2} = \frac{\Delta R_3}{R_3} = -\frac{\Delta R_4}{R_4} = \frac{6\lambda Px}{Ewh^2}. \tag{3.10}$$

Substituting equation (3.10) into the expression for an unbalanced Wheatstone bridge, i.e. equation (3.7), the output voltage can be written as

$$e_o = \frac{6\lambda PxE_{ex}}{Ewh^2}$$

$$\therefore P = \frac{Ewh^2}{6\lambda xE_{ex}} \cdot e_o$$

$$P \propto e_{\mathrm{o}}.$$

The sensitivity, $S = \frac{e_{\mathrm{o}}}{P} = \frac{6\lambda x E_{\mathrm{ex}}}{Ewh^2}$.

The sensitivity of the load cell depends on:

- the shape of the beam,
- Young's modulus of elasticity of the beam material,
- the gauge factor, and
- the point of application of the load.

The range of the cantilever beam that can be used to measure load is

$$P_{\max} = \frac{S_{\mathrm{fs}}wh^2}{6x}.$$

The voltage ratio at the maximum load is $\left(\frac{e_{\mathrm{o}}}{E_{\mathrm{ex}}}\right)_{\max}$, which is given by

$$\left(\frac{e_{\mathrm{o}}}{E_{\mathrm{ex}}}\right)_{\max} = \frac{\lambda S_{\mathrm{fs}}}{E}.$$

Intelligent load cells

An intelligent load cell can be formed by adding a microcontroller to a standard cell. The resulting device can calculate and display the total cost from the measured weight, using stored cost-per-unit-weight information. The cost per weight can be stored for a large number of substances, making the instrument very flexible in its application.

3.4.1 Appendix

3.4.1.1 The balanced Wheatstone bridge

A Wheatstone bridge is shown in figure 3A.1.

For the bridge to be balanced, nodes b and d must be at the same potential (i.e. $e_{\mathrm{o}} = 0$)

$$\therefore \ I_1 R_1 = I_4 R_4 \tag{A.1}$$

$$I_2 R_2 = I_3 R_3. \tag{A.2}$$

Dividing equation (A.2) by (A.1), we get

$$\frac{I_1 R_1}{I_2 R_2} = \frac{I_4 R_4}{I_3 R_3}. \tag{A.3}$$

At balance, there is no current in the detector, therefore $I_1 = I_2$ and $I_4 = I_3$. Equation (A.3) reduces to

$$\frac{R_1}{R_2} = \frac{R_4}{R_3}.$$

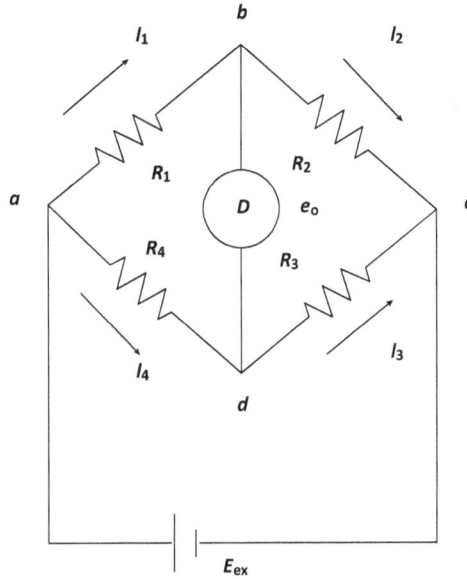

Figure 3A.1. A Wheatstone bridge.

If R_1 is the unknown resistance R_x, then

$$R_x = \frac{R_4}{R_3} \cdot R_2.$$

If R_x is the strain gauge, ambient temperature compensation can be performed by replacing R_2 with a dummy gauge. While the strain gauge (R_x) is subjected to stress, the dummy gauge is not; however, both have at the same temperature at all times.

3.4.1.2 The unbalanced Wheatstone bridge
When an Wheatstone bridge is unbalanced, then nodes b and d are not at the same potential. The potential drop across R_1 and R_4 can be calculated as follows:

$$e_{ab} = \frac{R_1}{R_1 + R_2} E_{ex}$$

and

$$e_{ad} = \frac{R_4}{R_3 + R_4} E_{ex}.$$

The bridge output, $e_o = e_{bd} = e_{ab} - e_{ad}$

$$= \left(\frac{R_1}{R_1 + R_2} - \frac{R_4}{R_3 + R_4} \right) E_{ex}$$

$$= \frac{R_1 R_3 - R_2 R_4}{(R_1 + R_2)(R_3 + R_4)} E_{ex}.$$

If $R_1 R_3 = R_2 R_4$, then $e_o = 0$ (the bridge is balanced).

Suppose all the resistances, i.e. R_1, R_2, R_3, and R_4, are varied by ΔR_1, ΔR_2, ΔR_3, and ΔR_4, respectively; the bridge output can now be expressed as

$$e_o = \frac{(R_1 + \Delta R_1)(R_3 + \Delta R_3) - (R_2 + \Delta R_2)(R_4 + \Delta R_4)}{(R_1 + \Delta R_1 + R_2 + \Delta R_2)(R_3 + \Delta R_3 + R_4 + \Delta R_4)} E_{ex}$$

$$= \frac{R_1 R_3\left(1 + \frac{\Delta R_1}{R_1}\right)\left(1 + \frac{\Delta R_3}{R_3}\right) - R_2 R_4\left(1 + \frac{\Delta R_2}{R_2}\right)\left(1 + \frac{\Delta R_4}{R_4}\right)}{(R_1 + \Delta R_1 + R_2 + \Delta R_2)(R_3 + \Delta R_3 + R_4 + \Delta R_4)} E_{ex}$$

$$\cong \frac{R_1 R_3\left(1 + \frac{\Delta R_1}{R_1} + \frac{\Delta R_3}{R_3}\right) - R_2 R_4\left(1 + \frac{\Delta R_2}{R_2} + \frac{\Delta R_4}{R_4}\right)}{(R_1 + \Delta R_1 + R_2 + \Delta R_2)(R_3 + \Delta R_3 + R_4 + \Delta R_4)} E_{ex}$$

Since $R_1 R_3 = R_2 R_4$ (when the bridge is initially balanced),

$$\simeq \frac{R_1 R_3\left(\frac{\Delta R_1}{R_1} - \frac{\Delta R_2}{R_2} + \frac{\Delta R_3}{R_3} - \frac{\Delta R_4}{R_4}\right)}{(R_1 + R_2)(R_3 + R_4)} E_{ex}$$

$$= \frac{R_1 R_3\left(\frac{\Delta R_1}{R_1} - \frac{\Delta R_2}{R_2} + \frac{\Delta R_3}{R_3} - \frac{\Delta R_4}{R_4}\right)}{R_1 R_3\left(1 + \frac{R_2}{R_1}\right)\left(1 + \frac{R_4}{R_3}\right)} E_{ex}.$$

Now $\frac{R_2}{R_1} = r$ (say); therefore, $e_o = \frac{\frac{\Delta R_1}{R_1} - \frac{\Delta R_2}{R_2} + \frac{\Delta R_3}{R_3} - \frac{\Delta R_4}{R_4}}{(1 + r)\left(1 + \frac{1}{r}\right)} E_{ex}$

$$= \frac{r}{(1 + r)^2}\left(\frac{\Delta R_1}{R_1} - \frac{\Delta R_2}{R_2} + \frac{\Delta R_3}{R_3} - \frac{\Delta R_4}{R_4}\right) E_{ex}.$$

3.5 Torque measurement

3.5.1 Introduction

The measurement of applied torque is of fundamental importance to rotating shafts. Such measurement ensures that the design of the rotating element is adequate to prevent failure under shear stress. Torque measurement is necessary to determine the power transmitted by a rotating shaft.

There are basically three methods of torque measurement, namely:

1. Measuring the reaction force in cradled shaft bearings.
2. The 'Prony brake' method.
3. Measuring the strain produced in a rotating shaft due to applied torque.

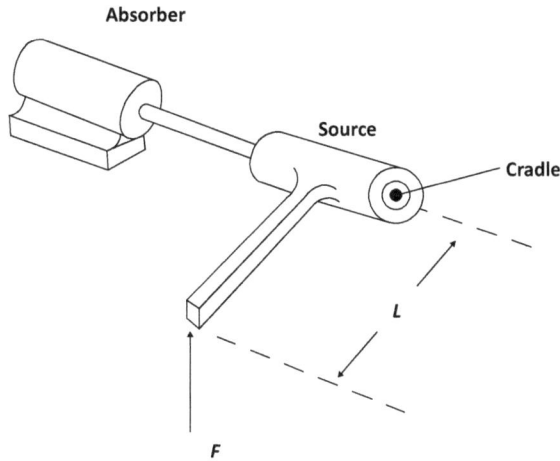

Figure 3.8. The measurement of reaction force using a cradled shaft bearing.

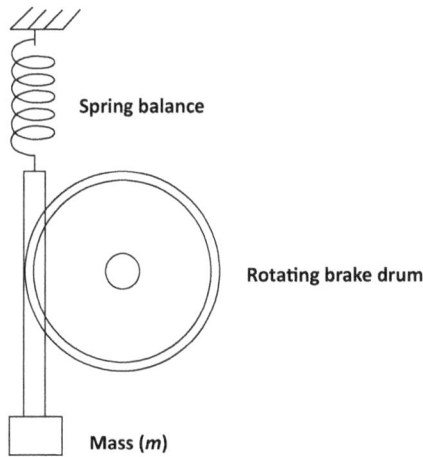

Figure 3.9. The Prony brake.

3.5.2 Reaction forces in shaft bearings

The transmission of torque through a shaft involves both a power source and a power absorber. The torque is measured by cradling either the power source or the power absorber end of the shaft bearing and then measuring the reaction force (F) and the arm length (L). The torque is given by $F \cdot L$. Such a measurement scheme is shown in figure 3.8.

3.5.3 The Prony brake

The Prony brake is used to measure the torque on a rotating shaft. A rope is wound round the shaft. One end of the rope is attached to a spring balance and the other end carries a load of mass m. A Prony brake is shown in figure 3.9.

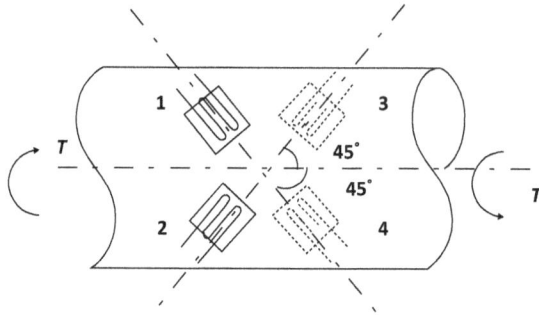

Figure 3.10. A torque cell with four strain gauges installed on it.

If the measured force in the spring balance is F_{sb}, then the effective force, F_{eff}, exerted by the rope on the shaft is given by

$$F_{eff} = mg - F_{sb}.$$

If the radius of the shaft is R_{sh} and that of the rope is R_r, then the effective radius

$$R_{eff} = R_{sh} + R_r.$$

The torque in the shaft is given by

$$T = F_{eff} \cdot R_{eff}.$$

3.5.4 The measurement of torque by strain gauges

Torque measurement is most commonly achieved using strain gauges, as they do not disturb the measured system by introducing frictional torque in the same way as the last two methods described. They convert the torque into an electrical signal. The approach used is similar to that of the load cell; the torque cell consists of a mechanical element (usually a shaft with a circular cross-section) and sensors (usually strain gauges). A cylindrical shaft with four strain gauges mounted on it is shown in figure 3.10.

3.5.5 Installation of the strain gauges

The four strain gauges are mounted on two perpendicular 45° helixes. Strain gauges 1 and 3 are mounted on the right-hand helix and are subjected to elongation. Strain gauges 2 and 4 are mounted on the left-hand helix and are subjected to compression.

3.5.6 The bridge output

The shearing stress τ on the circular shaft is given by

$$\tau = \frac{TD}{2J} = \frac{16T}{\pi D^3}, \tag{3.11}$$

where T is the applied torque,

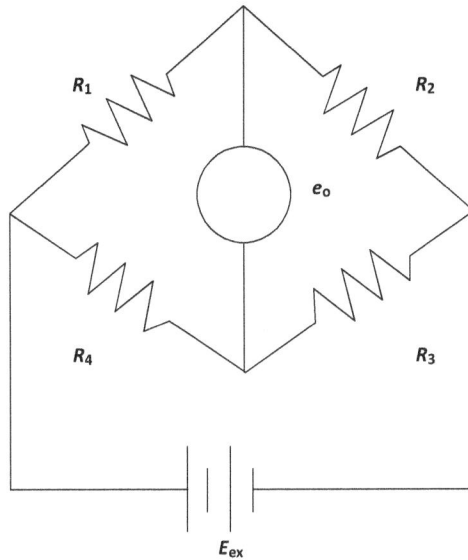

Figure 3.11. Strain gauges connected to form a Wheatstone bridge.

J denotes the polar moments of inertia of the circular cross-section, and D is the diameter of the shaft.

If the shaft is subjected to pure torsion, the normal stress is zero; in this case, for a circular shaft,

$$\sigma_1 = -\sigma_2 = \tau_{xz}\frac{16T}{\pi D^3}. \tag{3.12}$$

Applying Hooke's law, the principal strains $\in_1$ and $\in_2$ are obtained as follows:

$$\in_1 = \frac{1}{E}(\sigma_1 - \nu\sigma_2) = \frac{16T}{\pi D^3}\left(\frac{1+\nu}{E}\right)$$

and

$$\in_2 = \frac{1}{E}(\sigma_2 - \nu\sigma_1) = \frac{16T}{\pi D^3}\left(\frac{1+\nu}{E}\right).$$

However, $\Delta R/R = \lambda\in$.

Observing the installation of four strain gauges, we write

$$\frac{\Delta R_1}{R_1} = -\frac{\Delta R_2}{R_2} = \frac{\Delta R_3}{R_3} = -\frac{\Delta R_4}{R_4} = \frac{16T}{\pi D^3}\left(\frac{1+\nu}{E}\right)\lambda.$$

The strain gauges are now connected to form a Wheatstone bridge, as shown in figure 3.11.

Therefore, the unbalanced voltage of the Wheatstone bridge can be expressed as

$$e_o = \frac{16T}{\pi D^3}\left(\frac{1+\nu}{E}\right)\lambda E_{ex}$$

3-16

$$\therefore T = \frac{\pi D^3 E}{16(1 + \nu)\lambda E_{\text{ex}}} e_{\text{o}} \qquad (3.13)$$

The sensitivity of the torque cell

$$S = \frac{e_{\text{o}}}{T} = \frac{16(1 + \nu)\lambda E_{\text{ex}}}{\pi D^3 E}$$

The range of the torque cell can be expressed as

$$\therefore T_{\text{max}} = \frac{\pi D^3 S_\tau}{16},$$

where S_τ is the yield strength of the material.

The voltage ratio at the maximum torque $\left(\frac{e_{\text{o}}}{E_{\text{ex}}}\right)_{\text{max}}$ can be expressed as follows:

$$\left(\frac{e_{\text{o}}}{E_{\text{ex}}}\right)_{\text{max}} = \frac{S_\tau \lambda (1 + \nu)}{E}.$$

If $S_\tau = 4.14 \times 10^8$ N m^{-2} and $E = 2.068 \times 10^{11}$ N m^{-2}, $\nu = 0.3$, and $\lambda = 2$, then

$$\left(\frac{e_{\text{o}}}{E_{\text{ex}}}\right)_{\text{max}} = 5.2 \text{ mv V}^{-1}$$

Therefore, it is found that the output from the bridge is proportional to strain produced in the shaft and hence the torque applied. It is important to install the strain gauges precisely, otherwise the expression of torque given by equation (3.11) is no longer valid.

3.5.7 Data transmission from a torque cell

In a torque cell, the shaft along with the four strain gauges rotate at high speed. However, electrical connections must be made for bridge excitation and the output. This can be achieved using either slip rings and brushes or telemetry.

3.5.7.1 The slip ring method
The slip ring assembly contains insulated rings mounted on the shaft and brushes mounted in a case. The case remains stationary since high-speed bearings are placed between the shaft and the brushes. The slip ring method is shown in figure 3.12.
  Drawback of the slip ring method
  The contact resistances between the rings and the brushes vary. Thus, noise is developed in the bridge excitation as well as in the bridge output. Special rings made of copper–nickel alloy and brushes made of a silver–graphite mixture are used to reduce the noise.

3.5.7.2 The telemetry system
The telemetry system utilizes a split collar that fits over the shaft. A power supply to the bridge, a modulator, a voltage controlled oscillator (VCO) and an antenna

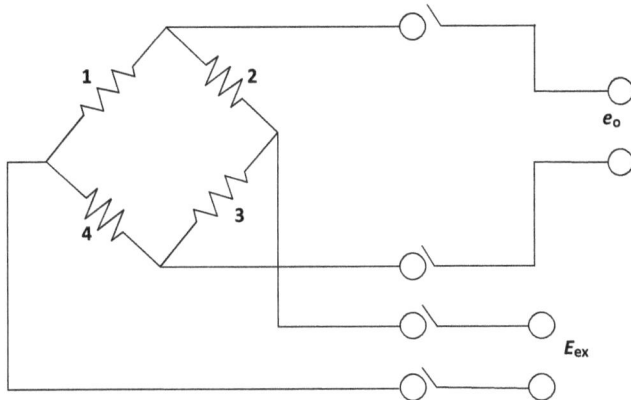

Figure 3.12. The slip ring method.

are placed in the split collar. The unbalanced voltage of the Wheatstone bridge is passed through a pulse width modulator (PWM) that modulates a constant-amplitude square wave signal at 5 kHz. The square wave varies the free-running frequency (10 MHz) of the VCO. The VCO signal is transmitted through the rotating antenna. The signal is received by a stationary loop antenna that encircles the split collar. The power supply to the bridge, the PWM, and the VCO is fed through an inductive coupling by a 100–150 kHz signal transmitted by the stationary loop antenna.

Problem 3.1. A steel bar of rectangular cross-section (2 cm × 1 cm) is subjected to a tensile force of 20 kN. A strain gauge is placed on the steel bar as shown in figure 3.13. Find the change of resistance of the strain gauge if it has a gauge factor of two and a resistance of 120 Ω in absence of an axial load. The Young's modulus of elasticity of steel is equal to 2×10^8 kN m^{-2}.

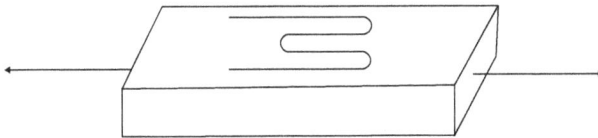

Figure 3.13.

Problem 3.2. A strain gauge whose resistance is 120 Ω and whose gauge factor is 2.0 is under zero strain. A 200 kΩ fixed resistance is connected in parallel with it. How much strain does the combination of the resistance and the strain gauge represent?

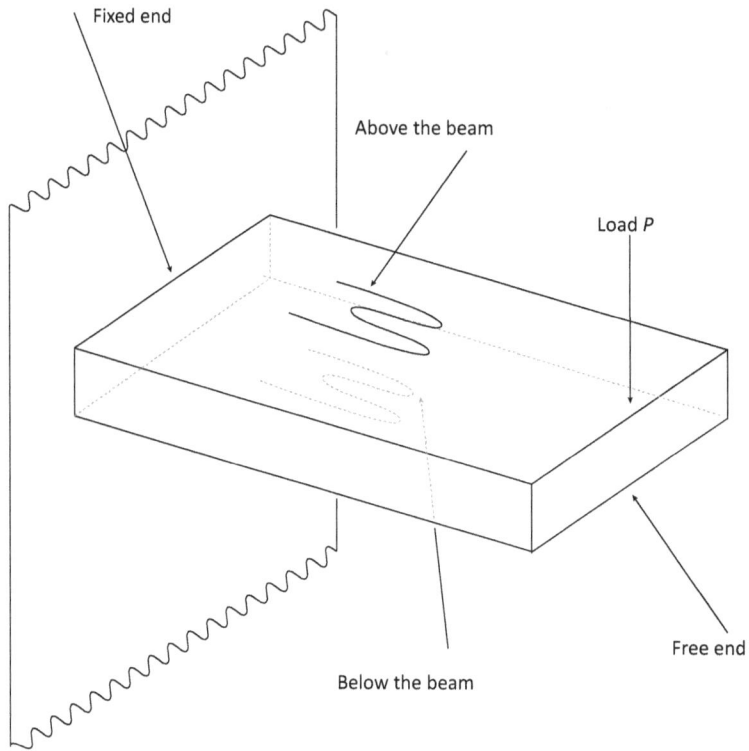

Figure 3.14.

Problem 3.3. Two strain gauges with resistance values of 120 Ω and gauge factor of $\lambda = 2$ are installed on a cantilever beam as shown in figure 3.14. The beam is fixed at one end and a load P is applied at the free end. One gauge is mounted above the beam and other below the beam. Both gauges are aligned in the axial direction The gauges are connected to a Wheatstone bridge that uses two more fixed resistances of 120 Ω each and the bridge excitation is 2.0 Volt DC.

   (a) Show the bridge arrangement used for ambient temperature change compensation.

   (b) The unbalanced voltage of the bridge is fed to a voltmeter that has a resolution of 2 µV. Find the minimum microstrain it can detect.

IOP Publishing

Fundamentals of Industrial Instrumentation (Second Edition)

Alok Barua

Chapter 4

Temperature sensors

Learning objectives:

- The thermistor and its application as a temperature measuring device.
- Other application of thermistors.
- Thermocouples and their connecting wires (leads).
- The range and sensitivity of the thermocouple.
- Cold junction compensation.
- Semiconductor temperature sensors.
- Details of the platinum resistance temperature detector (RTD).
- The signal conditioning circuits of the RTD.
- The construction of the RTD.

4.1 Introduction

Temperature is one of the most commonly used and measured process parameters. The sensation of heat or cold is a matter of daily experience. By the mere sense of touch, we can say whether a substance is hotter or colder than ourselves. Much of our lives is affected by the diurnal and seasonal variations in ambient temperature. The sense of touch is merely qualitative, hence we must find a way to quantify temperature. Although temperature is one of the most familiar engineering variables, it is not easy to define. Temperature must be defined in terms of the behaviour of a material as the temperature changes. Material behaviours are used to measure temperature. Some examples include the change in the length of a material, the change in the electrical resistance of a wire or semiconductor, the change of pressure in a gas at constant volume, the change in the voltage of two connected dissimilar metal wires, and the change in color of a lamp filament. For the quantitative measurement of temperature, a temperature sensor is needed. In this chapter, we discuss three basic temperature sensors, namely the thermistor, the thermocouple, and the RTD.

doi:10.1088/978-0-7503-3755-7ch4

4.2 The thermistor

The thermistor is a temperature-sensitive device. Unlike a metal, it shows a decrease in its resistance value for an increase in temperature. This means that it has a negative temperature coefficient of resistance. Thermistors are made of the oxides of nickel, cobalt, or manganese and the sulfides of iron, aluminum, or copper. The resistance–temperature relationship can be expressed as

$$R = R_0 e^{\beta\left(\frac{1}{T}-\frac{1}{T_0}\right)},$$

where
R = the resistance of the thermistor at T K,
R_0 = the resistance of the thermistor at T_0 K, and
β = a material constant that ranges from 3000 K to 5000 K.

The sensitivity S of a thermistor is given by

$$S = \frac{\Delta R/R}{\Delta T} = -\frac{\beta}{T^2}.$$

If $\beta = 4000$ K and $T = 298$ K, then

$$S = -0.045 \text{ K}^{-1}.$$

This is much higher than the sensitivity of a platinum RTD.

4.2.1 The manufacturing process of the thermistor

Two or more semiconductor powders are mixed with a binder to form a slurry. Small drops of the slurry are formed over the connecting leads, dried, and put into a sintering furnace. During sintering, the metallic oxides shrink onto the connecting wires and form an electrical connection. The beads are then sealed by coating them in glass. The glass coating improves stability by eliminating water absorption.

Size
- 0.125–1.5 mm

Shape
- Disks, wafers, flakes, or rods

General features

The resistance of a thermistor may vary from few ohms to several kilo ohms. Instrumentation engineers should pay attention to the minimum resistance. Other semiconductor temperature sensors include carbon resistors, silicon, and germanium. Silicon with a varying amount of boron impurities can have either a positive or negative temperature coefficient of resistance. Germanium doped with arsenic or gallium is used for cryogenic temperatures, at which it shows a large decrease in resistance for an increase in temperature. The applications of the thermistor in electronic circuits include limiting the large initial charging current of power supply filtering capacitors.

4.2.2 The thermistor in a Wheatstone bridge

A Wheatstone bridge is shown is figure 4.1. One arm of the bridge is replaced by a thermistor. If the Wheatstone bridge is initially balanced, then $R_T R_3 = R_2 R_4$. If R_2, R_3, R_4 are fixed-value resistors, then the output voltage e_o due to a change in resistance (ΔR_T) caused by a change in temperature can be derived as follows:

The Wheatstone bridge is redrawn in figure 4.2. All nodes are cabled in the circuit.

$$e_{ab} = \frac{R_T}{R_T + R_2} E_{ex}$$

$$e_{ad} = \frac{R_4}{R_3 + R_4} E_{ex}$$

The output voltage, $e_o = e_{bd} = e_{ab} - e_{ad} = \left(\frac{R_T}{R_T + R_2} - \frac{R_4}{R_3 + R_4}\right) E_{ex}$.

When a change in temperature occurs, the new resistance value is $R_T + \Delta R_T$.

$$\therefore e_o = \left(\frac{R_T + \Delta R_T}{R_T + \Delta R_T + R_2} - \frac{R_4}{R_3 + R_4}\right) E_{ex}$$

$$= \left(\frac{\cancel{R_T R_3} + \Delta R_T R_3 + \cancel{R_T R_4} + R_4 \Delta R_T - \cancel{R_4 R_T} - R_4 \Delta R_T - \cancel{R_2 R_4}}{(R_T + \Delta R_T + R_2)(R_3 + R_4)}\right) E_{ex}$$

And $\because R_T R_3 = R_2 R_4$

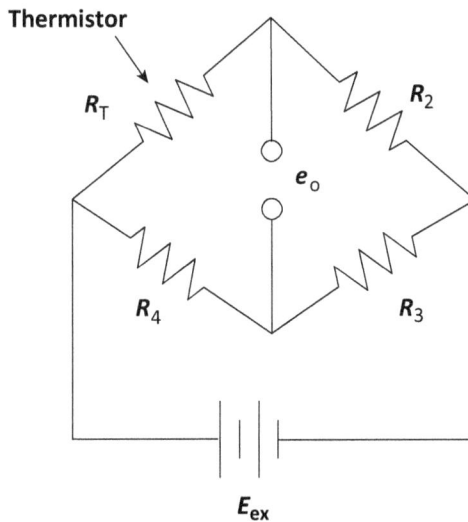

Thermistor

R_T R_2

e_o

R_4 R_3

E_{ex}

Figure 4.1. A thermistor in a Wheatstone bridge.

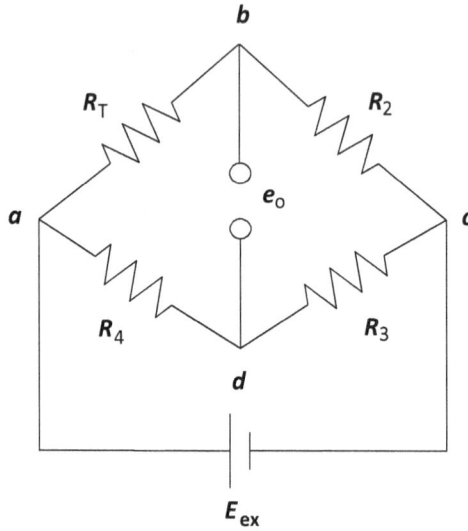

Figure 4.2. A redrawn Wheatstone bridge.

$$\frac{e_o}{E_{ex}} = \frac{\Delta R_T R_3}{(R_T + \Delta R_T + R_2)(R_3 + R_4)}.$$ (4.1)

If $R_2 = R_3$ and $R_T = R_4$, then equation (4.1) becomes

$$\frac{e_o}{E_{ex}} = \frac{\Delta R_T / R_T}{(1 + \Delta R_T / R_T + R_2 / R_T)(1 + R_T / R_2)}$$

$$= \frac{\Delta R_T / R_T}{2 + R_T / R_2 + R_2 / R_T + \Delta R_T / R_T + \Delta R_T / R_2}.$$ (4.2)

For thermistors, the terms $\Delta R_T / R_T$ and $\Delta R_T / R_2$ in the denominator are not small with respect to the other terms, therefore they cannot be neglected to simplify the solution of the equation for $\Delta R_T / R_T$. For the special case of an equal bridge ($R_T = R_2 = R_3 = R_4$), we can rearrange equation (4.2) to get

$$\frac{e_o}{E_{ex}} = \frac{\Delta R_T / R_T}{2 + R_T / R_2 + \frac{R_2}{R_T} + \frac{\Delta R_T}{R_T} + \frac{\Delta R_T}{R_2}}$$

$$= \frac{\Delta R_T / R_T}{2 + 1 + 1 + 2\frac{\Delta R_T}{R_T}}$$

$$= \frac{\Delta R_T / R_T}{4 + 2\frac{\Delta R_T}{R_T}}$$

$$\therefore 4\frac{e_o}{E_{ex}} + 2\frac{\Delta R_T}{R_T} \cdot \frac{e_o}{E_{ex}} = \frac{\Delta R_T}{R_T}$$

$$\therefore \frac{\Delta R_T}{R_T}\left(1 - 2\frac{e_o}{E_{ex}}\right) = 4\frac{e_o}{E_{ex}}$$

$$\frac{\Delta R_T}{R_T} = \frac{4e_o/E_{ex}}{1 - 2e_o/E_{ex}}. \tag{4.3}$$

The thermistor resistance R_T' at any temperature T is then given by the simple expression

$$R_T' = R_T + \Delta R_T$$
$$R_T' = R_T(1 + \Delta R_T/R_T). \tag{4.4}$$

Substituting equation (4.3) into (4.4), we get

$$R_T' = R_T\left(\frac{1 + 2e_o/E_{ex}}{1 - 2e_o/E_{ex}}\right). \tag{4.5}$$

The value of R_T' obtained from equation (4.5) is converted to a temperature using tables that list T as a function of R_T' for the specific thermistor used.

4.2.3 The thermistor in a potentiometer circuit

The thermistor can also be used in a potentiometer circuit as follows (figure 4.3):

$$\text{if } R_T = R_1, \text{ then}$$

$$R_T' = R_T\frac{(1 - 2e_o/E_{ex})}{(1 + 2e_o/E_{ex})}.$$

Some other circuits that are used to improve the linearity of operation are as follows (figure 4.4):

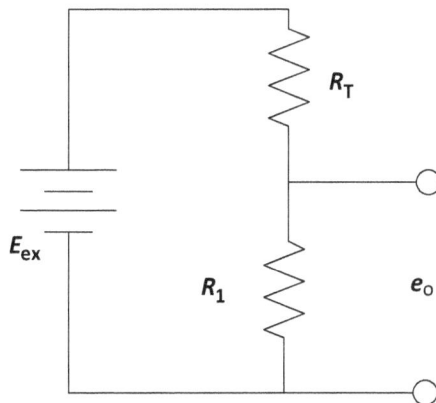

Figure 4.3. A thermistor in a potentiometer circuit.

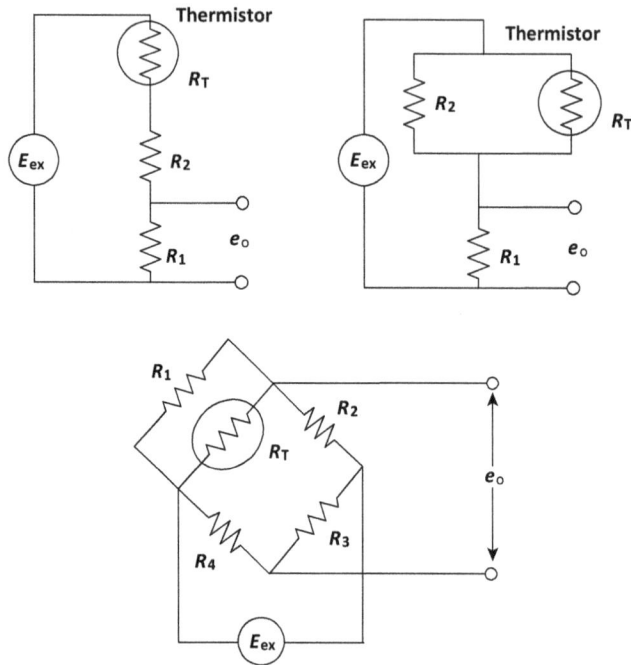

Figure 4.4. Circuits used to improve thermistor linearity.

4.2.4 The resistance of the connecting wire

When thermistors are used to measure temperature, errors resulting from the effect of the connecting wire are usually small enough to be neglected, even for relatively long wires. The sensitivity of a thermistor is high; therefore, the change in resistance ΔR_T resulting from a temperature change is much greater than the small change in resistance of the wires due to the temperature variation. In addition, the resistance of the thermistor is large compared to the wire resistance ($R_T/R_L \cong 1000$), where R_L is the resistance of the wire; consequently, any reduction in the sensitivity of the sensor due to a resistance change of the connecting wire is negligible.

4.2.5 Self-heating errors

Errors may occur due to the self heating of the thermistor. The recommended practice is to limit the current flow through the thermistor to a value such that the temperature rise due to the $I^2 R_T$ power dissipation is smaller than the precision to which the temperature is to be measured. Let us assume that $R_T = 5000\ \Omega$ is capable of dissipating 1 mW per °C above the ambient temperature. Thus, if the temperature is to be determined with an accuracy of 0.5 °C, the power to be dissipated should be limited to less than 0.5 mW. This limitation establishes a maximum current value of

$$I = \sqrt{P/R_T} = 316\ \mu A.$$

An adequate response can be obtained even at these low currents because the sensitivity is high.

4.3 The thermocouple

The thermocouple is a temperature sensor which relies on the physical principle that if any two different metals (A and B) are connected together, an emf that is the function of the temperature is developed at the junction of the two metals. A thermocouple is shown in figure 4.5.

$$e = a_1 t + a_2 t^2 + a_3 t^3 + \cdots.$$

Therefore the temperature–emf relationship is clearly nonlinear.

The values of the constants a_1, a_2, etc depend on the metals A and B.

- There are three emfs present in a thermocouple circuit.
- When two dissimilar metals are joined together, an emf exists between the two points that is the function of the junction temperature.
  ⇒ This is known as the Seebeck effect.
- If two metals are connected to an external circuit in such a way that a current is drawn, the emf may be altered slightly.
  ⇒ This is known as the Peltier effect.
- If a temperature gradient exists along either or both of the metals, the emf may undergo an additional modification.
  ⇒ This is known as the Thompson effect.

Two important rules are used to analyze thermocouple circuits.

4.3.1 The law of intermediate metals

If a third metal is connected in the circuit as shown in figure 4.6, the net emf of the circuit is not altered as long as the new connections are at the same temperature.

4.3.2 The law of intermediate temperature

If a thermocouple produces an emf e_1 when its junctions are at the temperatures t_1 and t_2 and e_2 when they are at the temperatures t_2 and t_3, then it will produce an emf of $(e_1 + e_2)$ if the junctions are at the temperatures t_1 and t_3 (figure 4.7).

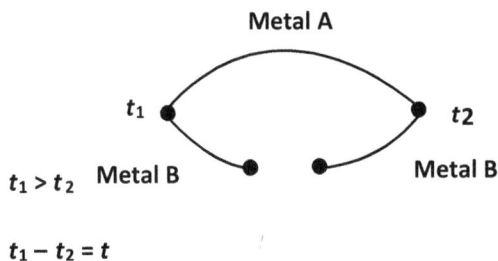

Metal A

t_1

t_2

$t_1 > t_2$ **Metal B**   **Metal B**

$t_1 - t_2 = t$

Figure 4.5. A basic thermocouple circuit.

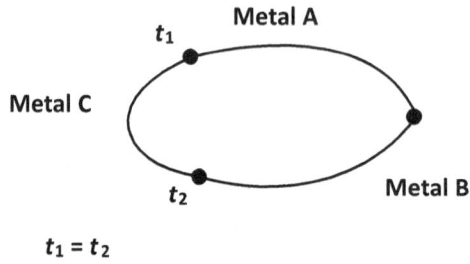

$t_1 = t_2$

Figure 4.6. The law of intermediate metals.

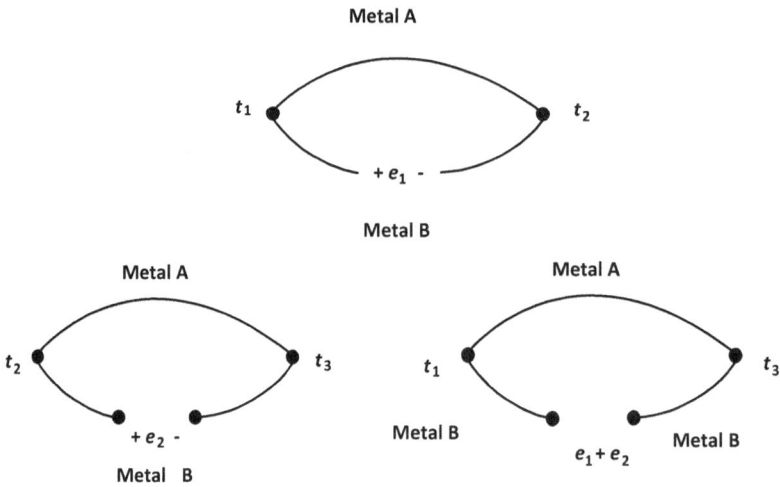

Figure 4.7. The law of intermediate temperature.

4.3.3 Cold junction compensation

If the ambient temperature variation of the cold junction is likely to cause significant error in the output of a thermocouple pair, there are two alternatives:

(1) Maintain the cold junction at a constant temperature using some technique such as an ice bath or a thermostatically controlled oven.

(2) Subtract a voltage that is equal to the voltage developed across the cold junction at a given temperature in the expected ambient temperature range.

Specifications of thermocouples are shown in table 4.1.

4.3.4 The junction semiconductor sensor

Junction diodes are suitable for temperature measurement. The junction potential of silicon transistors and diodes changes at about 2.2 mV per °C over a wide range of temperatures. This property can be used as the basis of an inexpensive sensor that has a fast response. The Analog Devices AD590 is a two-terminal temperature-sensitive

Table 4.1. Specifications of Thermocouple

Type	K	T	J	S	B	R	E
Positive	Chromel	Copper	Iron	Platinum/10% Rhodium	Platinum/30% Rhodium	Platinum/13% Rhodium	Chromel
Negative	Alumel	Constantan	Constantan	Platinum	Platinum/6% Rhodium	Platinum	Constantan
Can be represented by	Eighth degree polynomial	Eighth degree polynomial	Seventh degree polynomial	Second/third degree polynomia.	Eighth degree polynomial	Second/third degree polynomial	Ninth degree polynomial
Application	−200 to 1300 °C. The main application however is from 700 to 1200 °C in reducing atmosphere	−200 to 350 °C. Beyond this temperature oxidation of copperwill occur	−150 to 1000 °C. It is usable in oxidizing atmospheres to about 760 °C and reducing atmosphere to 1000 °C	0 to 1538 °C. Main features are its chemical inertness and stability at high temperatures in oxidizing atmospheres. Reducing atmosphere cause rapid deteriorationat high temperatures	38 to 1800 °C	0 to 1593 °C	0 to 982 °C
Voltage swingover range (mv)	56.0	26.0 For −184 °C to 400 °C	50.0 For −184 °C to 760 °C	16.0	13.6	18.7	75; Highest sensitivity
Lead wires (+)	Iron	Copper	Iron	Copper	Copper	Copper	Iron
(−)	Copper Nickel alloy	Constantan	Constantan	Copper–Nickel alloy	Copper–Nickel alloy	Copper–Nickel alloy	Constantan

Note: Constantan: % Copper with 45% nickel:

Chromel: 90% Nickel with 10% chromium;

Alumel: 94% Nickel with 3% manganese, 2% aluminium and 1% silicon;

Typical sensitivity of the thermocouple lies between 10 µv to 60 µv/°C.

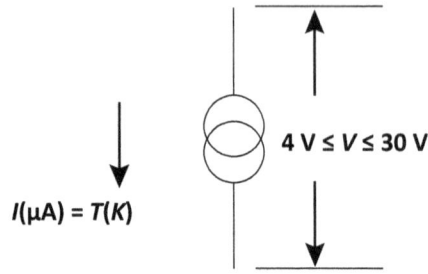

$I(\mu A) = T(K)$

$4\text{ V} \leq V \leq 30\text{ V}$

$-55\,^{\circ}\mathrm{C} \leq T \leq +150\,^{\circ}\mathrm{C}$

$218\text{ K} \leq T \leq +423\text{ K}$

$218\,\mu A \leq T \leq 423\,\mu A$

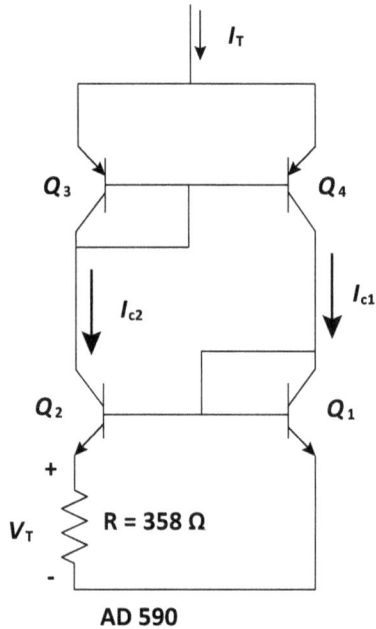

AD 590

Figure 4.8. The junction semiconductor sensor.

current source (figure 4.8). This device can also be used for cold junction compensation in a thermocouple circuit.

The junction semiconductor is a current source which passes a current numerically equal (μA) to the absolute temperature (K) when excited by a voltage from $+4$ V to $+30$ V at temperatures from $-55\,^{\circ}\mathrm{C}$ to $150\,^{\circ}\mathrm{C}$.

If transistors Q_3 and Q_4 are identical, the current I_T is divided into two equal parts I_{C1} and I_{C2}. Q_2 consists of eight transistors in parallel. So the current in Q_1 is eight times the current in each transistor of Q_2. Again, the difference between the V_{BE} values of two identical transistors with different collector currents is proportional to the absolute temperature.

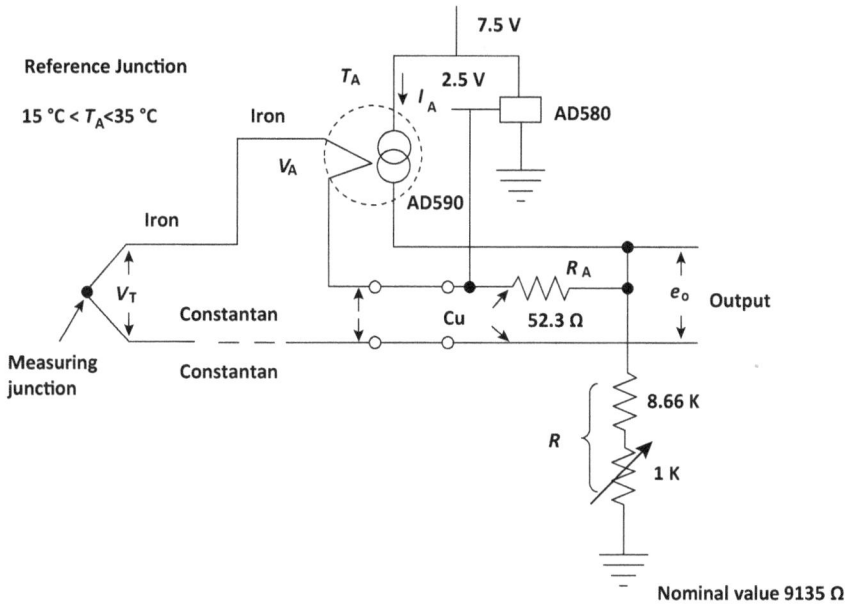

Figure 4.9. Cold junction compensation using the AD590.

$$V_T = V_{BE_1} - V_{BE_2} = \frac{kT}{q} \ln \frac{I_1}{I_2} = \frac{k}{q}(\ln 8)T$$
$$= 179 \times 10^{-6} T \text{ volts,}$$

where V_{BE_1} and V_{BE_2} are the base–emitter drops of transistor Q_1 and Q_2, respectively, I_1 is the current passing through transistor Q_1, and I_2 is the current passing through each of the eight transistors that make up Q_2.

The voltage V_T across the resistor R is thus proportional to the absolute temperature. Therefore, the current passing through R, I_{C2}, or I_2 must also be proportional to the absolute temperature. Since $I_T = 2I_{C2}$, the total current through the device, I_T must be proportional to the absolute temperature. If $R = 358\ \Omega$, $I_T\ T^{-1} = 1\ \mu A\ K^{-1}$.

Figure 4.9 shows a simple application in which the variation of the cold junction voltage of a type J thermocouple (iron–constantan) is compensated by a voltage developed in series by the temperature-sensitive output current of an AD590 (a semiconductor temperature sensor).

$$e_o = V_T - V_A + \frac{52.3 I_A + 2.5}{1 + \frac{52.3}{R}} - 2.5$$
$$\cong V_T.$$

The circuit is calibrated by adjusting R for the correct output voltage while maintaining the measuring junction at a known reference temperature and the circuit near 25 °C. If resistors with a low temperature coefficient are used, the compensation accuracy is within ±0.5 °C for temperatures between +15 °C and +35 °C. Other T/C values may be accommodated using the standard resistance values shown in table 4.2.

Table 4.2. Nominal values of standard resistance R_A.

Type	Nominal values of R_A for different thermocouples
J	52.3 Ω
K	41.2 Ω
E	61.4 Ω
T	40.2 Ω
S, R	5.76 Ω

4.3.5 Desirable properties of a thermocouple

The desirable properties of a thermocouple for industrial use are:

(1) A large thermal emf
(2) Precision of calibration
(3) Resistance to corrosion and oxidation

The interchangeability of thermocouples is a principal reason for their wide use and application. The best accuracy is obtained with platinum/platinum-rhodium thermocouples, which have an accuracy of 0.5% of the standard emf temperature calibration curve. Thermocouples are most commonly made in the form of wires that are insulated and welded together at the measuring junction. The measuring junction is formed in two different ways, namely a twisted weld or a butt weld. The twisted weld is used for wires of larger size and the butt weld is used for wires of smaller size. In order to prevent the formation of a second junction, the wires of a thermocouple are insulated from each other by being threaded through porcelain insulators which retain their shape at temperatures of up to 1500 °C.

4.3.6 Multiple-junction thermocouple circuits or thermopiles

'Thermopile' is a term used to describe a multiple-junction thermocouple circuit that is designed to amplify the output voltage of the thermocouple. Since thermocouple voltages are typically in the millivolt range, increasing the voltage output may be a key element in reducing the uncertainty in the temperature measurement. In a thermopile, the measuring junctions are usually located at the same physical location to measure one temperature.

Figure 4.10 shows a thermopile used to provide an amplified output signal. In this case, the output voltage would be N times the voltage produced by a single thermocouple, where N is the number of junctions in the circuit. The average output voltage corresponds to the average temperature level sensed by the N junctions. In transient measurements, a thermopile may have a more limited frequency range than a single thermocouple due to its increased thermal capacity. Thermopiles are particularly useful for reducing the uncertainty in measuring small temperature differences between the measuring and reference junctions.

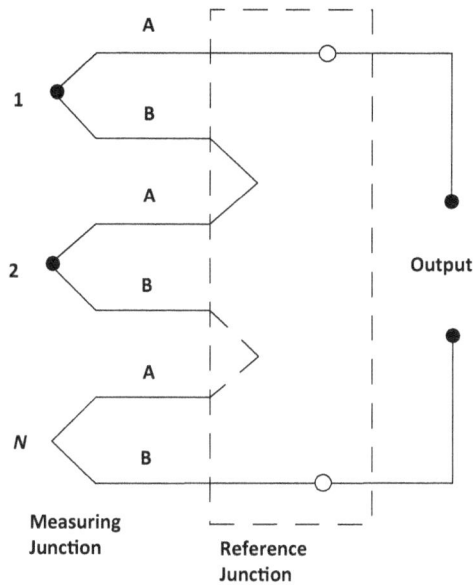

Figure 4.10. A thermopile.

4.3.7 The thermocouple sheath

Thermocouple junctions are prone to contamination by gases, liquids, and other metals. Metallic contamination alters the thermoelectric behaviour of the device such that its characteristic varies from that published in the standard tables. The effect of the sheath is to increase the time constant of the thermocouple and hence make the response more sluggish. Table 4.3 shows some common sheath metals and their temperature limits.

4.3.8 Grounded thermocouple circuits

Thermocouples are influenced by external effects. The main examples of external effects not always considered are the effects of electric and magnetic fields, crosstalk effects, and effects connected with common-mode voltage rejection. A brief review of these effects follows:

1. Voltage sources are capacitively coupled to the thermocouple extension wires through parasitic capacitance. This causes an alternating noise signal to be superimposed on the desired signal. The noise is minimized by shielding the thermocouple extension wire and grounding the shield.
2. The magnetic fields associated with current-carrying conductors produce noise current and hence noise voltage in the thermocouple circuit. Such noise is minimized by twisting the thermocouple extension wires.
3. Adjacent pairs of cables tend to pick up noise when pulsating DC signals are transmitted. Crosstalk noise is minimized by shielding the individual pairs of thermocouple extension wires.
4. Electrical connections made between a thermocouple and a grounded instrument may introduce common-mode noise if a difference in ground

Table 4.3. Some of the common sheath metals.

Material	Maximum operating temperature (°C)
Mild steel	900
Fused silica	1000
Recrystallized alumina	1850
Magnesia	2400
Thoria	2600

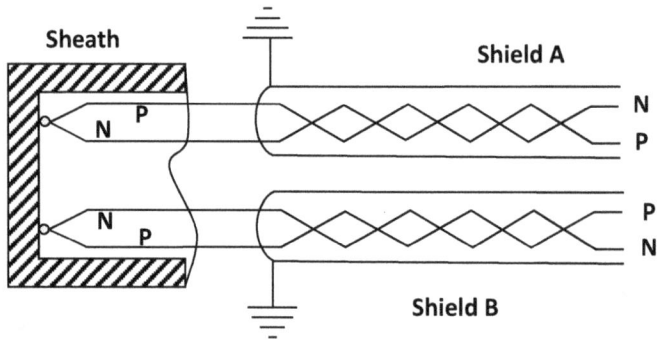

Figure 4.11. A grounded circuit used when the measuring junctions are grounded to the sheath.

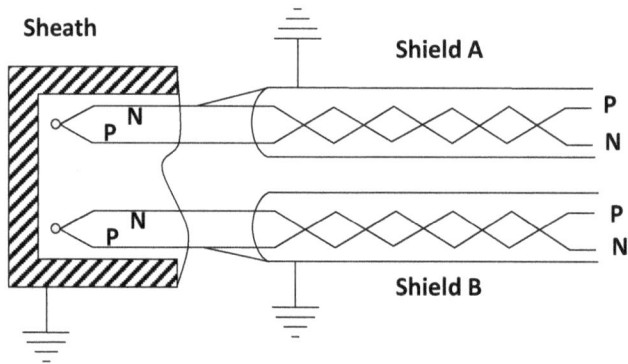

Figure 4.12. A grounded circuit used with ungrounded measuring junctions.

potential exists between the sensor and the instrument. Common-mode noise is minimized by grounding the thermocouple and its shielding at a single point as close as practical to the measuring junction.

Two different arrangements of thermocouple extension wires and shield–ground combinations that are acceptable from the noise reduction viewpoint are shown in figures 4.11 and 4.12.

4.4 The resistance thermometer

The resistance thermometer, which is alternatively known as the RTD, is one of the most accurate temperature sensors. The variation in the resistance of a metal versus temperature can be represented by the following relationship:

$$R_t = R_o\left(1 + \alpha t + \beta t^2 + \gamma t^3 + \cdots\right) \tag{4.6}$$

where R_o = the resistance at 0 °C, R_t = the resistance at t °C, and α, β, γ, etc. are constants. The purity of the platinum can be checked by measuring $\frac{R_{100}}{R_o}$. It should be higher than 1.390.

We can approximate R_t as follows:

$$R_t \approx R_0(1 + \alpha t + \beta t^2). \tag{4.7}$$

For pure platinum,

$$\alpha = 3.94 \times 10^{-3} \, /°C$$
$$\beta = -5.8 \times 10^{-7} \, /(°C)^2$$

Equation (4.7) is nonlinear. However, the nonlinearity of a platinum RTD at 100 °C is 0.76% of the full-scale deflection. To calculate the actual temperature, we do not have to solve the quadratic equation.

The resistance–temperature relation is rewritten as

$$R_t = R_o\left(1 + kt_{\mathrm{pt}}\right), \tag{4.8}$$

where k is the mean temperature coefficient of resistance between 0 °C and 100 °C and we define the platinum temperature t_{pt} (which is nearly equal to the true temperature t) using the relation

$$t_{\mathrm{pt}} = \frac{R_t - R_o}{R_{100} - R_o} \times 100, \tag{4.9}$$

where R_t, R_o, and R_{100} respectively denote the resistances at t °C, 0 °C, and 100 °C. The quantity $(R_{100} - R_o)$ is called the fundamental interval (FI) of the thermometer. The difference between the true temperature t and the platinum temperature t_{pt} is accurately given by the parabolic formula

$$t - t_{\mathrm{pt}} = \delta\left\{\left(\frac{t}{100}\right)^2 - \frac{t}{100}\right\}, \tag{4.10}$$

where δ is a constant for that particular specimen of wire.

To derive (4.10), we proceed as follows:

$$t - t_{\mathrm{pt}} = t - \frac{\alpha t + \beta t^2}{(100)\alpha + (100)^2\beta} \times 100 \quad \text{from (4.7) and (4.9)}$$

$$= -\frac{\beta(100)^2}{\alpha + 100\beta}\left[\left(\frac{t}{100}\right)^2 - \frac{t}{100}\right]$$

Table 4.4. Ranges of some common RTDs.

RTD	Temperature range
Platinum	-100 °C to 650 °C $\Rightarrow$ good linearity and chemical inertness.
Nickel	-180 °C to 430 °C $\left.\right\}$ Nickel and copper are susceptible to corrosion and oxidation.
Copper	-200 °C to 260 °C
Tungsten	-270 °C to 1100 °C

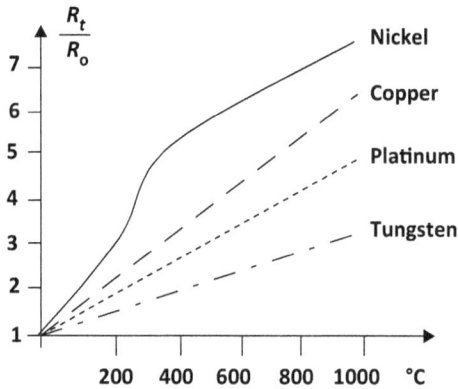

Figure 4.13. The typical resistance–temperature characteristics of some selected metals.

Thus δ in equation (4.10) is

$$\delta = -\frac{\beta(100)^2}{\alpha + 100\beta}.$$

The value of δ for the specimens employed lies between 1.488 and 1.498. The higher the purity, the larger the value of α and the smaller the value of δ. δ is determined by finding the platinum temperature t_{pt} for the boiling point of sulfur, whose true temperature is known (444.6 °C), and then substituting it into equation (4.10).

The effective range of RTDs depends on type of wire used. The ranges of some common RTDs are shown in table 4.4.

Resistance–temperature characteristics for different metals are shown in figure 4.13.

4.4.1 RTD circuits

An RTD can be connected to form one arm of a Wheatstone bridge as shown in figure 4.14.

At balance, $R_t = \left(\frac{R_2}{R_1}\right)R_3$.

R_3 is an adjustable potentiometer. All resistors and potentiometers are made of Manganin in order to avoid any effect due to ambient temperature changes.

The potentiometer R_3 might be calibrated for temperature. Three improvements must be made to the simple Wheatstone bridge:

- The contact resistance of potentiometer R_3 is part of the bridge circuit. The contact resistance may be a fraction of an ohm but is quite variable, depending on the condition of the potentiometer.
- The RTD is connected into the bridge by leads of some length, which can be 5 m long or even more. Variations in the temperature of the lead wire cause variations in the resistance, which is part of the bridge circuit.
- The current passing through the RTD causes a heating effect which varies depending on the magnitude of the current. The heat thus generated raises the temperature of the RTD sensor.

4.4.2 The self-heating effect

There is an I^2R heat loss in the RTD, because it is a resistor. This heating of a resistor due to the flow of current in it is called self heating. The effect of self heating is to introduce an error into the reading. Therefore, the current passing through the RTD must be low to avoid self-heating error. RTD manufacturers provide the dissipation constant of the sensor. The dissipation constant is defined as the power required to raise the RTD temperature by 1 °C.

The self-heating temperature rise can be found from the following expression:

$$\Delta T = \frac{P}{P_d}$$

ΔT = the temperature increase because of self heating in °C
P = the power dissipated in the RTD (in watts).
P_d = the dissipation constant of the RTD in W /°C.

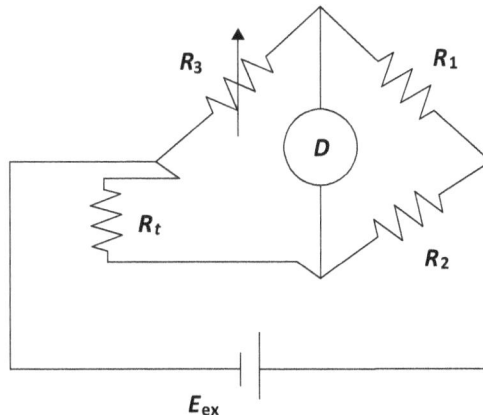

Figure 4.14. A simple Wheatstone bridge circuit with an RTD in one arm.

4.4.3 The three-wire method of temperature measurement

In this method, the potentiometer R_3 is placed in the detector circuit where it may lie in both arms of the bridge. The bridge is balanced by adjusting R_3.

$$I_4 R_4 = I_2[R_2 + (1 - f)R_3] \tag{4.11}$$

$$I_4 R_t = I_2(R_1 + f R_3), \tag{4.12}$$

where f is the fraction of the potentiometer R_3 lying in the R_1 arm of the bridge, and I_4 and I_2 are the currents through the corresponding arms of the bridge at balance. Dividing equation (4.12) by equation (4.11) gives

$$R_t = R_4 \frac{\frac{R_1}{R_3} + f}{\frac{R_2}{R_3} + 1 - f}.$$

Balance is obtained by adjusting the potentiometer R_3 which is calibrated in temperature. The effect of contact resistance is avoided because the wiper of R_3 is now not in the bridge but in the detector circuit. The only effect of contact resistance is to cause a negligible change in the sensitivity of the detector, while the accuracy of the bridge remains high. The three-wire method is used to compensate for variable connection wire resistance. The detector circuit connection is now made at the RTD and forms the third connection wire.

4.4.3.1 The specifications of the connecting wires

Each connecting wire is made of copper wire of the same diameter and length so that each has equal resistance. Three connecting wires are run in the same cable so that they have common temperature conditions. If the resistance of each connecting wire is R_L, then one connecting wire is in the R_t arm of the bridge and other is in the R_4 arm of the bridge. The resistance of the detector connection has a negligible effect. Including the connecting wire resistances, the bridge balance equations are

$$I_4(R_4 + R_L) = I_2[R_2 + R_3(1 - f)] \tag{4.13}$$

$$I_4(R_t + R_L) = I_2(R_1 + f R_3) \tag{4.14}$$

Dividing equation (4.14) by equation (4.13) gives

$$\frac{R_t + R_L}{R_4 + R_L} = \frac{\frac{R_1}{R_3} + f}{\frac{R_2}{R_3} + 1 - f}$$

If the right-hand side of the above equation is made unity, then

$$R_t = R_4$$

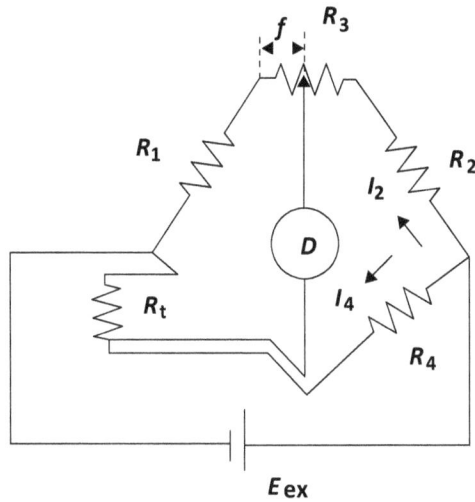

Figure 4.15. The RTD is connected to the bridge by three wires.

Therefore, the balance of the bridge is independent of the connecting wire resistance only at one particular setting of the potentiometer R_3. At one temperature of the scale (i.e. 50%), the compensation is exact, and the resistances must be equal in magnitude. At other temperatures, there is a slight and almost negligible connecting wire error when there are changes of ambient temperature along the connecting wires (figure 4.15).

4.4.4 The Mueller bridge or four-wire resistance method

The effect of the connecting wires cannot be totally eliminated by the three-wire method, because its compensation is not exact at all positions of the potentiometer. However, the effect of the connecting wire resistance can be totally eliminated through the use of the four-wire resistance circuit method. This method involves switching the resistance thermometer bulb wire from one arm of the bridge to the other arm.

A Mueller bridge is shown in figure 4.16.

The potentiometer R_{P2} is used to make R_3 equal to R_4.

When the switch is in position 'a,'
   point 1 is connected to 2′,
   point 2 is connected to 1′,
   point 4 is connected to 5,

and the bridge balance equation is

$$R_{P1a} + R_2 = R_t + R_1. \tag{4.15}$$

When the switch is in position 'b,'
   point 1 is connected to 1′,

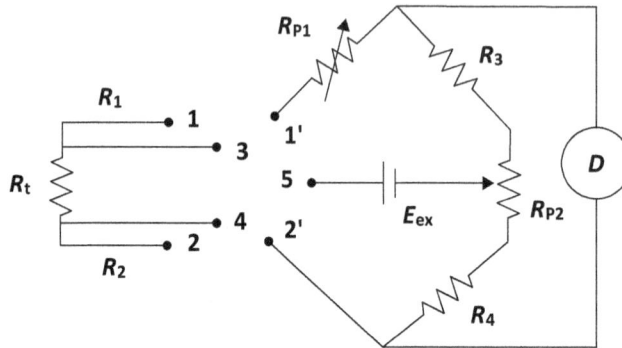

Figure 4.16. A Mueller bridge.

point 2 is connected to 2′,
point 3 is connected to 5,

and the bridge balance equation is

$$R_{P1b} + R_1 = R_t + R_2. \tag{4.16}$$

By adding equations (4.15) and (4.16), we get

$$R_t = \frac{R_{P1a} + R_{P1b}}{2}.$$

Therefore, the measurement of the RTD resistance is independent of the connecting wires' resistances.

4.4.5 Why is platinum universally used to make RTDs?

- Being a noble metal, it is used for precision resistance thermometry.
- Platinum is stable (it resists corrosion, it has chemical inertness, and it is not readily oxidized).
- It is easily workable (i.e. it can be drawn into fine wires).
- It has a high melting point.
- It can be obtained at a high degree of purity.

However, these desirable and necessary features do not come without effort. The resistance–temperature relationship of platinum alters if small amounts of impurities are present in the platinum.

4.4.6 The construction details of the platinum RTD bulb

The platinum RTD bulb consists of a coil of fine wire wound on a notched mica-cross frame. The wire is arranged so that good thermal conductivity is provided and a high rate of heat transfer is obtained. The windings should be made in such a manner that physical strain is negligible as the wire expands and contracts due to temperature changes.

The construction details of the platinum RTD bulb is shown in figure 4.17.

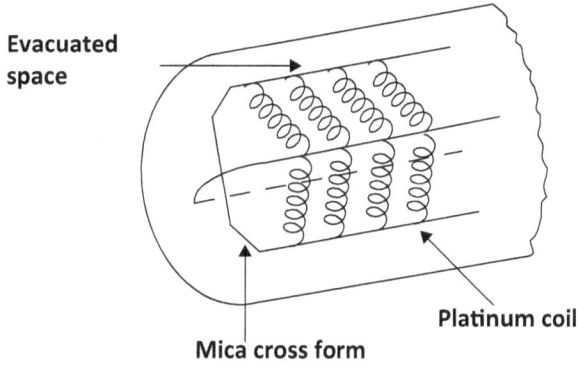

Figure 4.17. Construction details of a platinum RTD.

Problem 4.1. Based on figure 4.18, answer the following questions:
(a) Determine the sensitivity of the bridge for small x (<0.1).
(b) Also determine the linearity of the output voltage with respect to x when x is large.
(c) What happens to the linearity and sensitivity when the following circuit of figure 4.19 is used instead of using the Wheatstone bridge signal conditioning circuit?

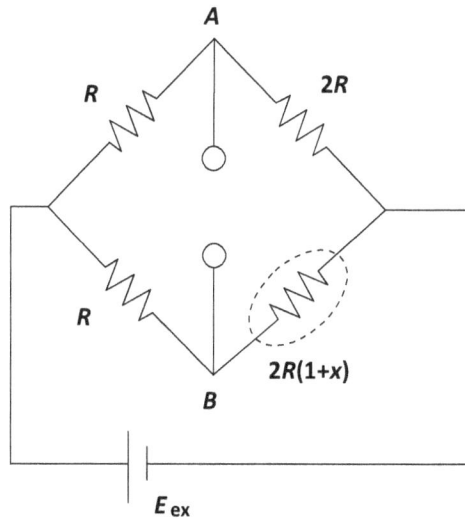

Figure 4.18.

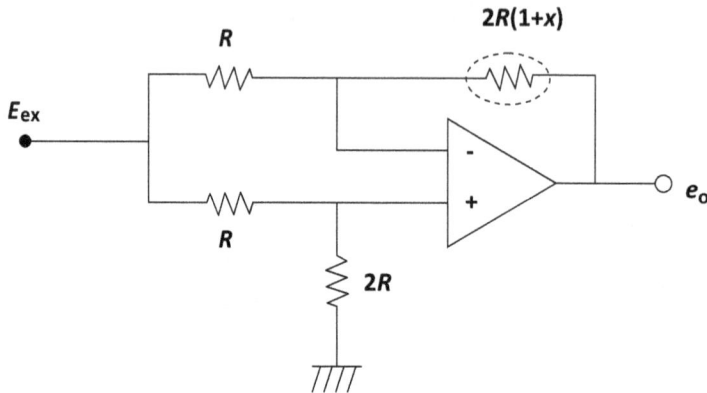

Figure 4.19.

Problem 4.2. An RTD is required to measure temperature correctly in the temperature range of 75 °C to 125 °C with a resolution of 0.5 °C. Design a signal conditioning circuit for the above.

The following data are given:

- α of the RTD = 0.002 /°C,
- the resistance of the RTD = 350 Ω at 25 °C, and
- P_D = 30 mW /°C.

Problem 4.3. Find the temperature versus output voltage relation for a thermistor using the following linearization techniques:
  (a) Using a resistance in parallel to a thermistor.
  (b) Using a thermistor in a Wheatstone bridge.

Problem 4.4. A K-type thermocouple is used to measure a certain temperature range. The signal conditioning circuit is kept at a distance from the place of measurement. Design a signal conditioning circuit, taking into account the noise as well as the cold junction compensation. Use a thermistor for cold junction compensation

Problem 4.5. A simple potentiometer circuit, as shown in figure 4.20, is used to measure the emf of an iron–constantan thermocouple. A fixed voltage of 1.215 V is applied across points A and B. A current of 3 mA flows through the resistors. The range of the temperature variation is from 150 °C to 650 °C. Find the values of R_1, R_2, and R_G for an ambient temperature of 25 °C.

The following data are provided:

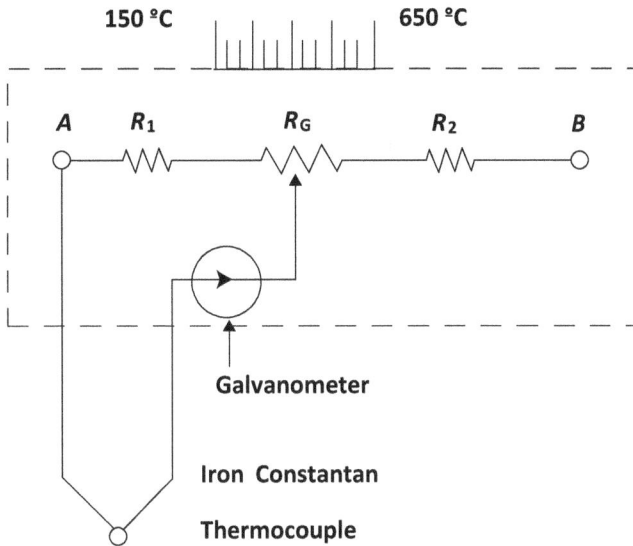

150 °C   **650 °C**

A R_1 R_G R_2 B

Galvanometer

Iron Constantan

Thermocouple

Figure 4.20.

the emf at 25 °C = 1.022 mV,
the emf at 150 °C = 14.682 mV, and
the emf at 650 °C = 53.512 mV.

Problem 4.6. An RTD has a resistance of 600 Ω at 25 °C and a temperature coefficient of 0.005 /°C. The RTD is used in a Wheatstone bridge circuit with $R_1 = R_2 = 600$ Ω figure 3.3. The variable resistance R_3 nulls the bridge. If the bridge excitation is 12 V and the RTD is in a bath at 0 °C, find the values of R_3 required to null the bridge when:
(a) The self-heating effect of the RTD is not considered.
(b) The self-heating effect of the RTD is considered and it is known that the dissipation constant for the RTD is 20 mW /°C.

Problem 4.7. An iron–constantan thermocouple is to be used to measure temperatures between 0 °C and 200 °C. With the reference junction at 0 °C, the emf outputs at temperatures of 100 °C and 200 °C are 5.268 and 10.777 mV, respectively. Find the nonlinearity at 100 °C as a percentage of the full-scale reading.

Problem 4.8. The resistance R_T of a resistive transducer is modeled as $R_T = R (1 -Kx)$, where K is the constant of transformation and x is the input quantity being sensed. The resistor is connected to an op-amp as shown in figure 4.21. The value of R is 100 Ω, K is 0.004, and x is 75. Find the output voltage V_o. (Assume the op-amp as an ideal device.)

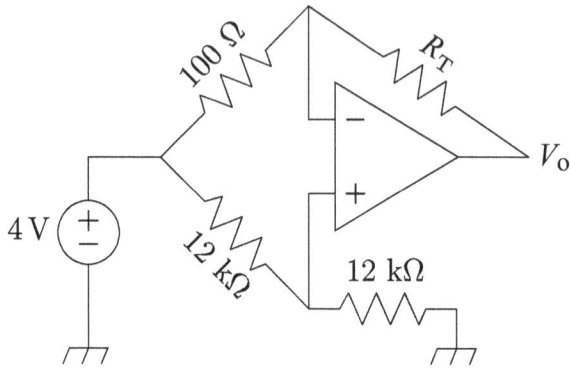

Figure 4.21.

Problem 4.9. A two-wire platinum RTD with resistance R_0 of 10 Ω at 0 °C is shown in figure 4.22. This is an experimental setup used to measure temperatures between 0 °C and 100 °C. All other arms of the bridge are resistances of fixed value $R = 10$ Ω. To avoid the self-heating effect, the power dissipation of the RTD should be less than 1 mW. Calculate the maximum bridge excitation voltage E_{ex} that can be applied in the circuit.

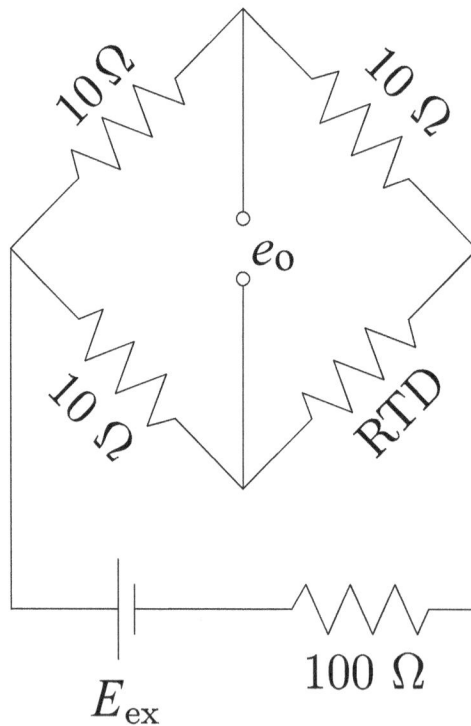

Figure 4.22.

Problem 4.10. An RTD is represented using a linear model over the range of 0 °C to 100 °C by the equation

$$R_T = R_0 (1 + 0.003T)$$

where T is the temperature in °C. $R_0 = 100 \ \Omega$ with a tolerance of $\pm 3 \ \Omega$. The actual model of the RTD is

$$R_T = R_0 (1 + 0.003T + 6 \times 10^{-7}T^2).$$

What is the worst-case error magnitude in the measurement system?

Chapter 5

Displacement measurement

Learning objectives:

- The basic principle of capacitance sensors.
- Linearization of the sensor.
- Differential measurement.

5.1 Introduction

This chapter discusses some basic displacement measurement techniques that use the potentiometer, the linear variable differential transformer (LVDT), and the capacitance transducer. Although these sensors are directly used to measure displacement, the latter two can be used for many other process parameter measurement viz. pressures, flows, etc. The sensors and physical principles used to measure displacement are highly dependent upon the particular application; therefore, the two most common methods will be discussed. The measurement of displacement involves the determination of the relative motion of two points, one of which is usually fixed.

5.2 The potentiometer

The potentiometer or variable electrical resistance transducer discussed in section 2.2 can be used as a displacement sensor. The sensor is composed of a sliding contact and a winding. The winding is made of many turns of wire wrapped around a nonconducting material. The sensor must be excited by a voltage source, and the output signal from the device is proportional to the fraction of the total distance the contact point has moved along the substrate. The potentiometer can be also used for rotational displacement measurement when it is configured in a rotary form. The output from the sliding contact is actually discrete, and its resolution is limited by the number of turns per unit distance. Loading error occurs unless the output voltage is measured by a voltmeter with a reasonably high input impedance.

5.3 The linear variable differential transformer (LVDT)

The LVDT as shown in figure 5.1 produces an electrical output that is proportional to the displacement of a movable core. It has one primary coil and two secondary coils which are in series opposition. If the primary is excited by an alternating current source, the movement of the core causes the mutual inductance in between the coils to change, giving a variable output. If the two secondary windings are identical and the core is at the geometrically central position, the sensor has zero voltage at the output since the two secondary windings are series opposition. If the core is moved to either side of the null position, the output voltage progressively increases. The output voltages on either side of the null position are 180° out of phase with respect to each other. Using a phase-sensitive demodulation circuit, it is possible to distinguish between positive and negative displacements of the core. However, due to harmonic distortion in the supply voltage and the fact that the two secondary windings are never identical, the output voltage obtained with the coil at the central position is never zero but has a minimum value called the null voltage. As the core only moves in the air gap between the windings, there is no friction or wear during operation. Therefore, the resolution of this instrument and its life expectancy are very high.

5.3.1 The theory of the linear variable differential transformer

The LVDT is a sensor used for displacement measurements.

The flux formed by the primary coil is linked to the two secondary coils, inducing an AC voltage in each. An iron core which moves through the bobbin without contact provides a path for the magnetic flux linkage between the coils. The position of the iron core controls the mutual inductance between the primary and the two secondary coils. The excitation voltage (e_{ex}) and the two secondary voltages (e_1 and e_2) are plotted in figure 5.2. The two secondary windings are connected in series opposition, as shown in figure 5.3. The object whose translational displacement is to be measured is physically connected to the central iron core of the transformer so that all motions of the body are transferred to the core. Because of the series opposition of the secondary windings, the voltage e_o is theoretically zero for the central position of the core. The voltage e_o is nonzero for movement to either side of the null position. The secondary voltages for the different positions of the core are

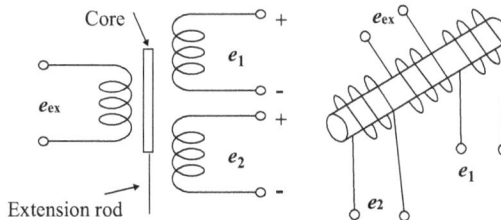

Figure 5.1. A schematic of the LVDT.

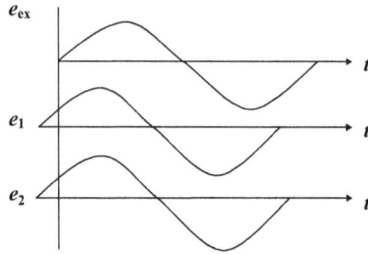

Figure 5.2. The primary and secondary voltages of an LVDT with core at null position.

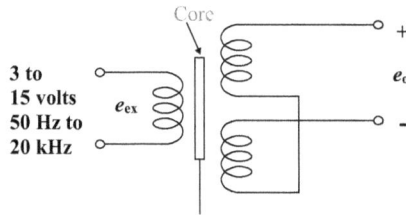

Figure 5.3. The secondary windings are in opposition.

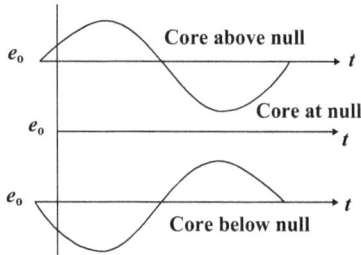

Figure 5.4. The secondary voltages for different positions of the core.

plotted in figure 5.4 and the absolute values of the output voltage, $|e_o|$ are plotted in figure 5.5 for different positions of the core.

An equivalent circuit for an LVDT is shown in figure 5.6. The value of the primary resistance, R_p, should be as low as possible without loading the excitation source. Applying the KVL to the primary side yields

$$i_p R_p + L_p \frac{di_p}{dt} - e_{ex} = 0$$

$$\therefore i_p = \frac{e_{ex}}{R_p + L_p D}.$$

The induced secondary voltages are

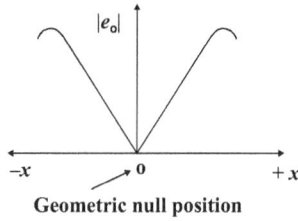

Figure 5.5. The absolute value of the output voltage for different positions of the core.

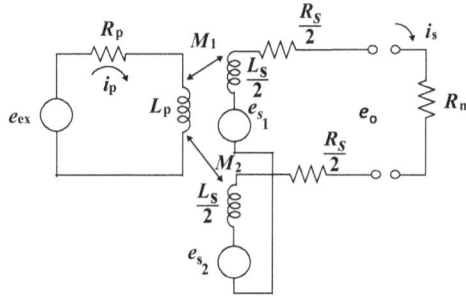

Figure 5.6. An equivalent circuit for an LVDT.

$$e_{s_1} = M_1 \frac{di_p}{dt}$$

and

$$e_{s_2} = M_2 \frac{di_p}{dt}.$$

e_{s_1} and e_{s_2} try to induce a voltage in the primary. However, this is nullified, since the secondary coils are in series opposition.

The net secondary voltage

$$e_s = e_{s_1} - e_{s_2} = (M_1 - M_2) \frac{di_p}{dt}. \tag{5.1}$$

The net mutual inductance $(M_1 - M_2)$ varies linearly with the core motion. For a particular core position,

$$e_o = e_s = (M_1 - M_2) \frac{D}{L_p D + R_p} e_{ex}$$

$$\therefore \frac{e_o}{e_{ex}} = \frac{[(M1 - M2)/R_P]D}{\tau_p D + 1} \text{ where } \tau_p = \frac{L_p}{R_p}$$

$$\therefore \frac{e_o}{e_{ex}}(j\omega) = \frac{(M_1 - M_2)/R_P}{\sqrt{(\omega \tau_p)^2 + 1}} |\Phi, \tag{5.2}$$

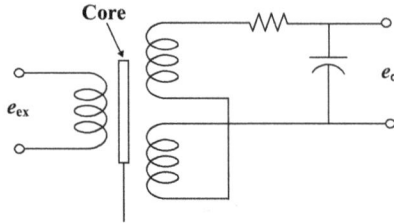

Figure 5.7. A lag network for compensation of the leading phase angle.

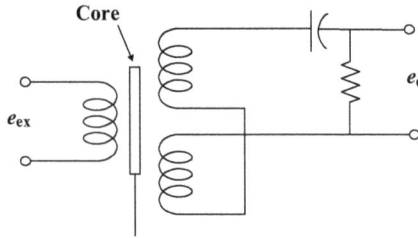

Figure 5.8. A lead network for compensation of the lagging phase angle.

$$\text{where } \Phi = \pi/2 - \tan^{-1} \omega \tau_p.$$

If the input impedance of the voltmeter is finite, R_m, a current i_s flows; then

$$i_p R_p + L_p D i_p - (M_1 - M_2) D i_s - e_{ex} = 0$$

$$(M_1 - M_2) D i_p + (R_s + R_m) i_s + L_s D i_s = 0$$

$$\frac{e_o}{e_{ex}}(D) = \frac{R_m(M_1 - M_2)D}{\left[(M_1 - M_2)^2 + L_p L_s\right]D^2 + \left[L_p(R_s + R_m) + L_s R_p\right]D + (R_s + R_m)R_p}. \quad (5.3)$$

The frequency of the excitation voltage applied to the primary winding can range from 50 Hz to 20 kHz. The power requirement is usually less than 1 W. The sensitivities of different LVDTs vary from 0.02 to 0.2 V mm^{-1} of displacement per volt of excitation applied to the primary coil. There is no contact between the core and the coils; therefore, friction is eliminated, thereby giving infinite resolution and no hysteresis. The absence of contact also ensures that the life of the device will be very long and that there will be no significant deterioration in performance over this period. The small core mass and freedom from friction give the sensor some capability for dynamic measurements. As can be observed from equation (5.3), there is a phase difference between the excitation voltage (e_{ex}) and the output voltage (e_o). The phase difference can be compensated by simple RC lead or lag networks. Such lag and lead networks are shown in figures 5.7 and 5.8, respectively.

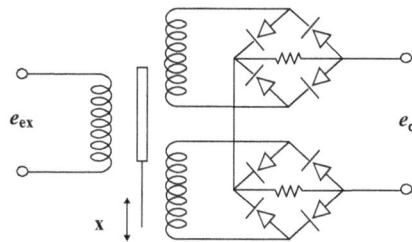

Figure 5.9. A circuit used for phase-sensitive demodulation.

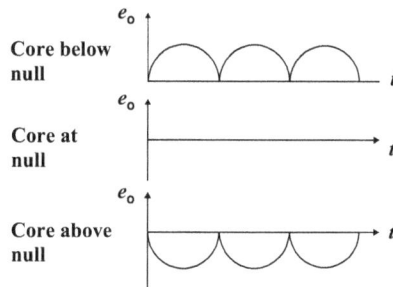

Figure 5.10. Demodulator outputs for different positions of the core.

5.3.2 Phase-sensitive demodulation

The output of the secondary coils which are series opposition is a sine wave whose amplitude is proportional to the core motion. The output voltage can be calibrated in displacement units. However, it gives the same output voltage amplitude for displacements of the core on both sides of the null. Therefore, it is not possible to state to which side of the null the reading applies without some other independent check. The phase-sensitive demodulator is a circuit that can resolve this problem. A phase-sensitive demodulator circuit using diodes is shown in figure 5.9. Both the secondary coil outputs are full-wave rectified and the current passes through the resistors only in one direction. The output voltage (e_o) is now DC and a sign change occurs whenever the core crosses the null position.

The demodulator outputs for three different positions of the core are shown in figure 5.10.

5.3.3 Dynamic displacement measurement

If the LVDT is to be used to measure dynamic displacement, the carrier frequency should be ten times greater than highest frequency component in the dynamic signal. The excitation voltage ranges from 3 to 15 V.

For dynamic measurement, the frequency of the excitation voltage should be much greater than the frequency of the core movement. If a frequency ratio of 10:1

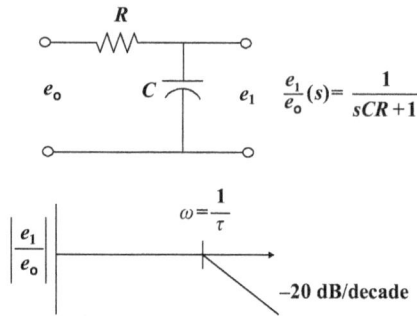

Figure 5.11. A low-pass filter and its frequency response.

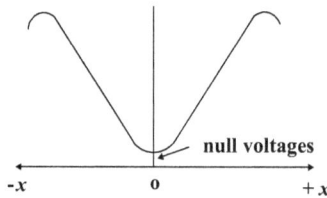

Figure 5.12. The null voltage.

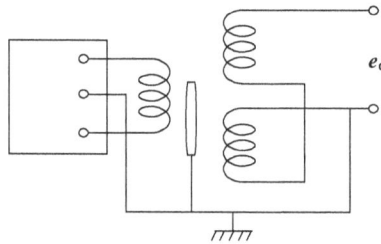

Figure 5.13. A null voltage reduction circuit used when a center-tapped primary excitation voltage source is available.

or more is feasible, a simple RC filter may be adequate. A low-pass filter and its frequency response are shown in figure 5.11

5.3.4 Null voltage and its reduction

The harmonies in the excitation voltage and stray capacitance between the primary coil and the secondary coils lead to a nonzero voltage at the geometric null position of the core. The mismatch between the two secondary windings also plays a role in generating this null voltage, as shown in figure 5.12. The circuits used to reduce the null voltage are shown in figures 5.13 and 5.14.

Figure 5.13 shows a method used to reduce the null voltage when a center-tapped excitation voltage is available. This will reduce the stray capacitance effect.

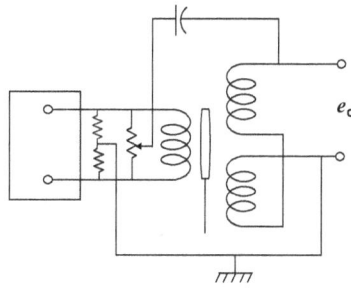

Figure 5.14. A null voltage reduction circuit used when a center-tapped primary voltage source is unavailable.

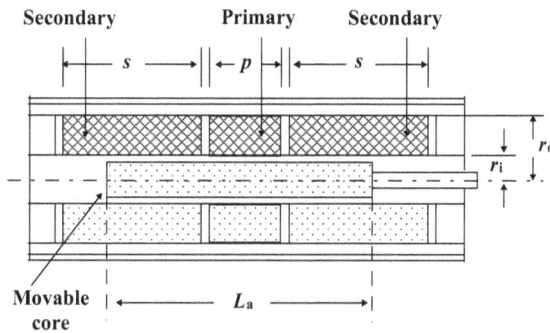

Figure 5.15. An LVDT.

Figure 5.14 shows a method used for null voltage reduction when a centre-tapped excitation voltage is unavailable. Here, the potentiometer is adjusted until the minimum null reading is obtained.

5.3.5 The LVDT design equation

An LVDT and its dimensions are shown in figure 5.15.

The net induced emf $e_o(j\omega)$ of the secondary coils is given by

$$e_o(j\omega) = j\omega I_p \left[\frac{4\pi N_p N_s \mu_o \, px}{3 \sin(r_o / r_i)} \left(1 - \frac{x^2}{2p^2} \right) \right], \tag{5.4}$$

where

ω = the frequency of the excitation signal (in rad),
I_p = the primary current,
N_p, N_s = the numbers of turns in the primary and secondary windings,
r_o, r_i = the outer and inner radii of the LVDT assembly,
x = the displacement of the core from the null position, and
μ_o = the permeability of free space ($4\pi \times 10^{-7}$ H m^{-1}).

The nonlinearity term $x^2/2p^2$ in equation (5.4) is dependent on the length of the primary winding p. For a desired range of x_{max} and nonlinearity error ε, the length of the primary winding is given by

$$p = x_{max}/\sqrt{2\varepsilon}.$$

The length of the secondary winding is

$$s = p + x_{max}.$$

The length of the core and the length of the secondary are increased by a little to account for the small space between the primary and each secondary winding. The ratio of r_i/L_a is about 0.05 and the ratio of r_o/r_i varies between two and eight. The number of secondary turns should be as large as possible to achieve greater sensitivity. Since they are likely to be connected to a high-input impedance amplifier and meter, the secondary windings can be made from finer wire.

5.4 Capacitive transducers

Like variable resistance and mutual inductance, variable capacitance also can be used in displacement measuring transducers in various ways. The capacitance of a variable capacitor can be changed by translational or rotational motion. However, capacitance transducers are also used to measure other process parameters such as levels and pressures.

5.4.1 Variation in capacitance

A capacitor consists of two conducting metal plates separated by an insulator. When a voltage is applied to the metal plates, equal and opposite electric charges appear on the plates. The ratio of that charge to the voltage is the capacitance. The capacitance of a parallel plate capacitor is proportional to the area A of the plates and inversely proportional to their separation d. Neglecting fringing, it can be expressed as

$$C = \epsilon_o \, \epsilon_r \, \frac{A}{d} F, \qquad (5.5)$$

where ϵ_o is the permittivity of free space (a vacuum), which has a value of 8.854×10^{-12} F m^{-1}, and ϵ_r is the dielectric constant of the material in the gap (for air, $\epsilon_r = 1$). A is given in m^2 and d is given in meters. Thus, capacitance is a function of shape, size, and permittivity, and an alteration in A, d, or ϵ_r causes a change in capacitance. Figures 5.16(a) and (b) show how the principle of varying the effective area and the separation between the plates can be used to vary the capacitance. If we can measure the change in the capacitance value, we can use this to make a displacement transducer. However, the principle of capacitance change can also be incorporated into level and differential pressure sensors. Figures 5.17(a) and (b) show a schematic of permittivity variation. In figure 5.17(a), the plates as well as the dielectric material are semicircular in shape. If the dielectric moves on a common axis, a variation in the capacitance value takes place between the two semicircular plates. Such a capacitor is used in LC tuned circuits to vary the tuning frequency.

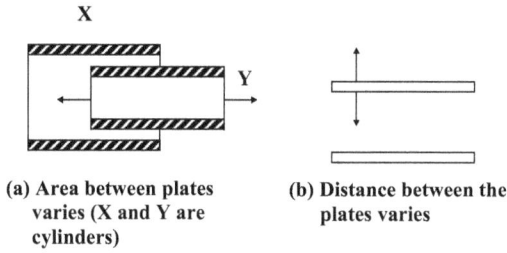

(a) Area between plates
varies (X and Y are
cylinders)

(b) Distance between the
plates varies

Figure 5.16. The use of capacitance based on geometrical variation in displacement sensors.

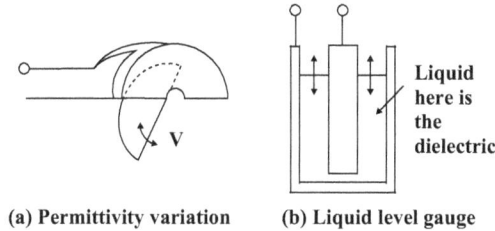

(a) Permittivity variation

(b) Liquid level gauge

Figure 5.17. Capacitance with dielectric variation.

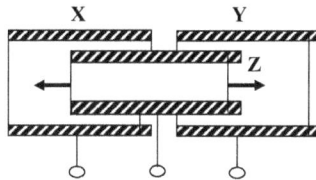

Figure 5.18. X, Y, and Z are cylinders. The areas between the plates vary (geometrical variation).

Figure 5.17(b) shows a schematic for level measurement based on the principle of capacitance variation. A solid rod is placed in a hollow cylinder which contains a liquid. Two capacitors are formed in parallel, one with an air dielectric and the other with a liquid dielectric. As the liquid level changes, the resultant capacitance also changes. For a nonconductive liquid, this arrangement is satisfactory since the liquid resistance is sufficiently high. For conductive liquids, the inner probe must be insulated to prevent the liquid from short circuiting the capacitance.

To obtain a differential output, the arrangement shown in figure 5.18 can be used. Here, the cylinders X and Y are fixed, and the inner concentric cylinder Z can move. If Z moves to the left-hand side, the effective area and hence the capacitance between X and Z increase while those between Y and Z decrease and vice versa. The differential arrangement provides a higher bridge output than that of the single arrangement shown in figure 5.16(a).

Another differential arrangement is shown in figure 5.19, in which the separation between the plates varies. Here, plates P and Q are fixed while plate M can move up

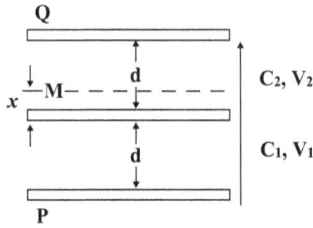

Figure 5.19. The distance between the plates varies (geometrical variation).

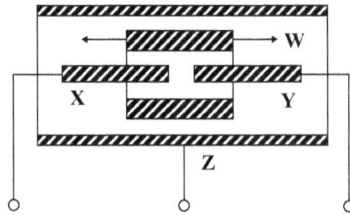

Figure 5.20. A variable capacitor with differential output.

or down. If plate M moves upward, the capacitance between plates M and Q increases and that between P and M decreases.

Differential variation of the capacitance is also possible by varying the permittivity. As shown in figure 5.20, the dielectric W can change position, which leads to a differential variation in capacitance. In all differential arrangements, the capacitance change is either $C \pm \Delta C$ or $C \mp \Delta C$.

One or more of the plates is not connected to ground and therefore electrostatic screening is required to avoid noise, usually at the mains supply frequency. The screened cable connection to the capacitive sensor can be a source of error because it may change its capacitance when there is movement between the cable conductors and the cable dielectric. Variation of the area A or the separation d requires a physical connection to the moving part, while variation of the permittivity ϵ_r does not. The variations in A or ϵ_r have linear operating ranges of 1–10 mm, although the capacitance transducer is mostly used for small displacements in which d varies. The most common form of variable capacitor used in displacement transducers is the parallel plate capacitor with a variable air gap. The problem of nonlinearity between the distance between the plates d and the capacitance C is illustrated in figure 5.21.

Assuming that the op-amp is an ideal device,

$$\frac{e_o}{e_{ex}} = -\frac{C_f}{C_x},$$

where C_f is the fixed capacitor and C_x is the variable capacitor or capacitance sensor

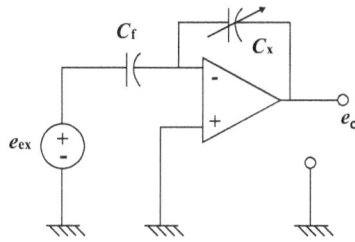

Figure 5.21. The linearization of a capacitance sensor

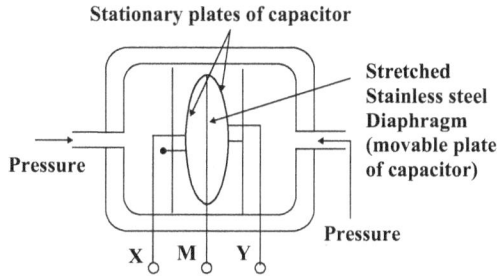

Figure 5.22. A differential-capacitor pressure pickup.

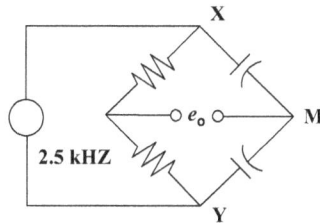

Figure 5.23. The bridge circuit of a differential pressure pickup.

or

$$e_{\mathrm{o}} = -\frac{C_{\mathrm{f}}d}{\in_{\mathrm{o}} \in_{\mathrm{r}} A} \tag{5.6}$$

It can clearly be seen that the output voltage is now directly proportional to the plate separation d; thus, linearity is achieved for both large and small motions. In commercial instruments, e_{ex} is a 50 kHz sine wave of fixed amplitude. The output e_{o} is also a 50 kHz sine wave, which is rectified and applied to a DC voltmeter calibrated directly in distance units. In a differential-capacitance pressure transducer (figure 5.22), plates X and Y are circular while plate M is a thin diaphragm across which the pressure differential to be measured is applied. The equivalent bridge circuit is shown in figure 5.23. When equal pressures are applied to both pressure ports, the diaphragm is in a neutral position, the bridge is balanced, and the output voltage e_o produced by the bridge is zero.

If one pressure is greater than the other, the diaphragm deflects in proportion, giving an unbalanced output at e_o in proportion to the differential pressure. For the opposite pressure difference, e_o shows a 180° phase change. Therefore, phase-sensitive demodulation is necessary. This method allows static deflection to be measured. Such differential capacitor arrangements also exhibit considerably greater linearity than that of single capacitor types.

5.4.2 Capacitance pickups (capacitive microphones)

The variation of capacitance by applied pressure gives us the capacitor microphone or pickup. In this type of pressure transducer, the elastic element is typically a diaphragm that serves as one plate of the capacitor. When pressure is applied, the diaphragm moves with respect to a fixed plate, thus changing the thickness of the dielectric between the plates. The variation in capacitance can be measured by simple bridge circuit or by allowing the change of current to flow through a resistor as shown in figure 5.24. Here, C is the value of the capacitance for a plate separation of d, and ΔC is the change of capacitance for a change in plate separation of Δd.

We know that

$$\frac{\Delta C}{\Delta d} = -\frac{C}{d},$$

thus

$$\frac{\Delta C}{C} = -\frac{\Delta d}{d}.$$

When the capacitor plates are stationary with separation d_o, no current flows and $e_o = 0$.

If there is then a relative displacement d_i from the position d_o, a voltage e_o is produced; their values are related by

$$\frac{e_o}{d}(D) = \frac{K\tau D}{\tau D + 1},$$

where $K = \frac{e_{ex}}{d_o} \text{V m}^{-1}$.

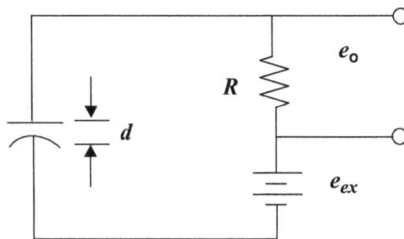

Figure 5.24. A variable capacitor with a fixed plate area.

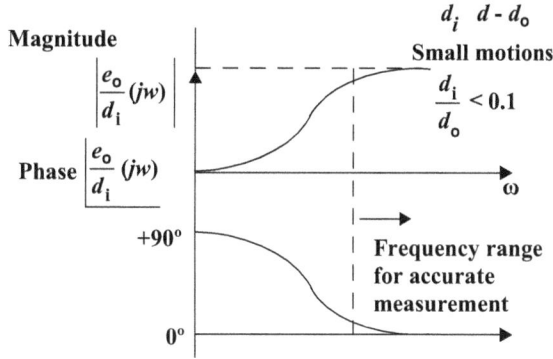

Figure 5.25. A gain and phase plot for a capacitor microphone.

The circuit of figure 5.24 does not allow the measurement of static displacements, since e_o is zero in the steady state for any value of d_i. For sufficiently rapid variations in d_i, however, the signal e_o faithfully measures the motion. The gain and phase plots of capacitor microphone are shown in figure 5.25.

As we know

$$\frac{e_o}{d_i}(j\omega) = \frac{Kj\omega\tau}{j\omega\tau + 1},$$

for $\omega\tau \gg 1$,

$$\frac{e_o}{d_i}(j\omega) \approx K.$$

A microphone does not usually need to measure sound pressures at less than 20 Hz and so the above arrangement is perfectly satisfactory. To make $\omega\tau \gg 1$ for low frequencies, a large τ is required. For a given capacitance and d_o, the value of τ can be increased only by increasing R. Typically, R is 1 MΩ or more. Thus, to prevent loading of the capacitance transducer circuit, a high input impedance amplifier is required.

Capacitive sensors in general are now widely used to measure the levels of liquids and solids in powdered or granular form.

- Such sensors are suitable for use in extreme conditions:

    1. Liquid metals (high temperature).
    2. Liquid gases (low temperature).
    3. Corrosive liquids (acids).
    4. High-pressure processes.

Problem 5.1. In figure 5.26, let x_i be a periodic motion with a significant frequency content at up to 500 Hz. The excitation frequency is 5000 Hz. The output signal

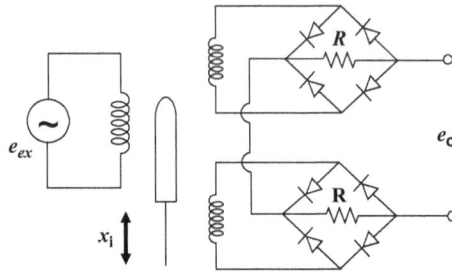

Figure 5.26.

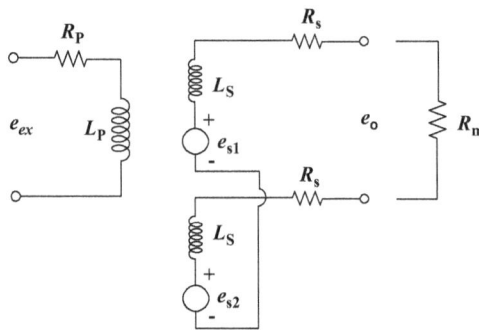

Figure 5.27.

obtained is then passed through a low-pass filter and then to an oscilloscope with an input impedance of 10^6 Ω. It is desired that ripple due to higher frequencies should not be more than 5% of the unfiltered value.

(a) Find the frequency range after the modulation process.
(b) Design a low-pass filter for the above application. Also calculate the value of R in figure 5.26.
(c) Calculate the time lag introduced by the low-pass filter.

Problem 5.2. In figure 5.27, a voltmeter with a finite input impedance R_m is connected. Find the expression for frequency at which the phase shift is zero.

Problem 5.3. Design a linear variable differential transformer (LVDT) based on the following details:
- the supply frequency = 10 kHz,
- the length of the core (L_a) = 20 cm,
- the maximum distance from the core to the null position (x_{max}) = 6 cm,
- the error introduced due to nonlinearity (ε) = 10%,

- the maximum emf induced in the secondary coils (e_o) = 5 V, and
- the current in the primary required to obtain the maximum emf in the secondary = 20 mA.

Assume that the number of turns in the secondary (N_S) is four times the number of turns in the primary (N_P).

IOP Publishing

Fundamentals of Industrial Instrumentation (Second Edition)

Alok Barua

Chapter 6

Pressure sensors

Learning objectives:
- The working principle of the C-type Bourdon tube.
- Metals and alloys used for pressure sensors.
- Detailed analysis of the diaphragm gauge.
- The working principle of vacuum gauges.

6.1 Introduction

The measurement of pressure and vacuum or low pressure has always been important in process control instrumentation. Pressure is expressed as a force per unit area. It has same units as stress. We shall consider the force per unit area exerted by a fluid on the wall of the container or hollow closed body to be pressure. The origin of pressure involves a consideration of the forces acting between fluid molecules and solid boundaries. When molecules with certain amounts of kinetic energy collide with a solid boundary, they rebound in different directions. We know that a change in the linear momentum of the molecule produces an equal but opposite force on the wall of the container. Pressure is the net effect of these collisions which produce the pressure sensed at the boundary surface.

Absolute pressure is the total pressure exerted by the fluid. *Differential pressure* is the algebraic difference between two pressures. As a special case of differential pressure, gauge pressure is the algebraic difference between the total pressure exerted by a fluid and the pressure exerted by the atmosphere.

One of the oldest means of pressure measurement is the liquid column manometer. It is the simplest and most direct method of measurement of pressure; it is ideal for laboratory use but has limited application in industrial pressure measurements.

The most commonly used pressure transducers are the Bourdon, bellows, and diaphragm gauges. The diaphragm gauge is preferred because of its electrical output, although the bellows gauge can also be used in conjunction with a precision

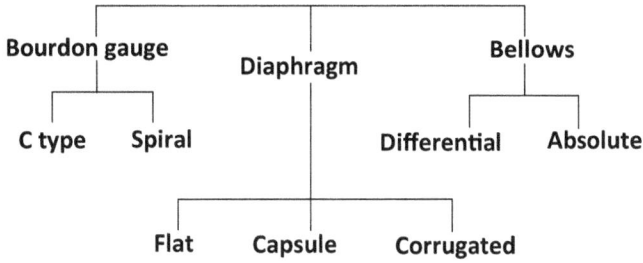

Figure 6.1. Pressure transducers.

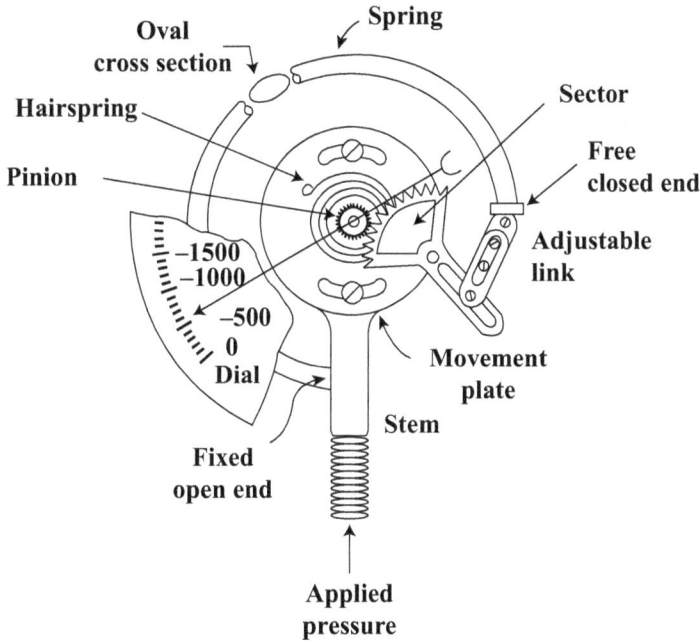

Figure 6.2. A common Bourdon tube transducer.

potentiometer to convert a pneumatic signal into an electrical output. The Bourdon gauge is one of the most rugged pressure measurement devices, though it is used only as a monitoring instrument. The entire gamut of pressure transducers is shown in figure 6.1.

6.2 The Bourdon gauge

Bourdon tube pressure gauges are used for the measurement of wide range of static pressures. They are also an inexpensive method of measurement. A C-type Bourdon tube transducer is shown in figure 6.2. It is an elliptical cross-sectional tube that has a C-shaped configuration. The tube has a fixed open end and a free closed end. When

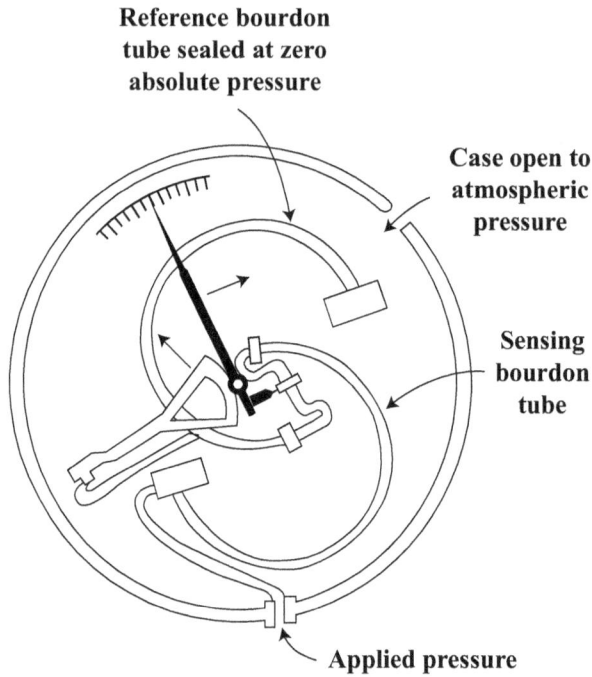

Figure 6.3. Bourdon tubes arranged for absolute pressure measurements.

pressure is applied inside the tube through the open end, the oval-shaped tube tends to become circular, which is accompanied by an increase in the radius of the arc. The free end is connected to a sector and spring-loaded linkage which converts the movement of the free end into angular rotation of the pointer. The linkage is adjustable for linearity and minimum hysteresis. The reference pressure is atmospheric, therefore the Bourdon gauge indicates gauge pressure.

6.2.1 Absolute pressure measurement

A two-C-tube Bourdon gauge used for the measurement of absolute pressure is shown in figure 6.3. The absolute pressure can be measured by biasing the sensing Bourdon tube against a reference Bourdon tube that is evacuated and sealed.

6.2.2 The twisted Bourdon tube

A twisted Bourdon tube is shown in figure 6.4. The large angular displacement encountered in the C-type Bourdon tube can be avoided by using a twisted tube which is derived from a flattened tube twisted about its own longitudinal axis.

Units of pressure measurement

$$1 \text{ pound per square inch (PSI)} = 6894.76 \text{ Nm}^{-2}$$
$$= 6.894 \text{ kPa}$$

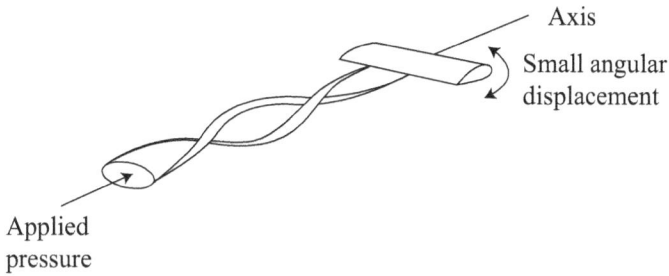

Figure 6.4. Twisted bourdon tube for use with passive electric elements.

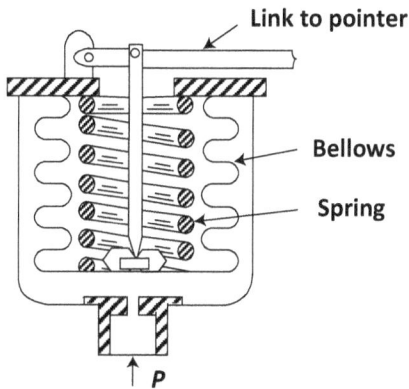

Figure 6.5. The bellows pressure element.

The ranges of pressure measurement of the different materials used to make Bourdon tubes are as follows:
- the bronze pressure spring—up to 4136856 N m^{-2}
- the beryllium–copper spring—up to 68947600 N m^{-2}
- the steel and alloy steel spring—pressure ranges greater than 6894×10^4 N m^{-2}

6.3 The bellows gauge

The bellows gauge is a metallic bellows that works with the pressure to be measured on one side and a spiral spring on the other. A bellows pressure gauge is shown in figure 6.5. The spring counterbalances the deflector. The pressure measurement range of the system is determined mainly by the material of the bellows, the effective area of the bellows, and the spring gradient. The bellows are commonly made of brass or phosphor bronze. Bellows gauges are commonly employed to measure gauge pressures ranging up to 689476 N m^{-2} (100 PSI). Through the use of suitable linkages, the pressure can be shown on the scale with better resolution. A bellows gauge with such a linkage and a pointer is shown in figure 6.6.

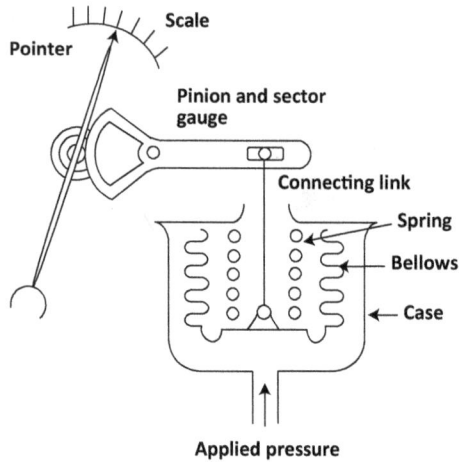

Figure 6.6. A common bellows gauge.

6.4 The diaphragm pressure transducer

When a differential pressure exists across a flat stretched metal diaphragm, it deflects. The deflection can be sensed either by an LVDT or by strain gauges. Diaphragms are used for the low and medium pressure ranges (1–20×10^4 kPa). Both tension and compression stresses are simultaneously present in diaphragm gauges. This allows the use of a four-active arm bridge in which all effects are additive, which provides both temperature compensation and a large output. In computing the overall sensitivity, the expressions for tangential and radial stresses cannot be applied to find the overall sensitivity. The stress distribution in a stretched diaphragm is shown in figure 6.7. Since the diaphragm surface is in a state of biaxial stress and both radial and tangential stresses contribute to the radial or tangential strain at any point, the general biaxial strain relations are as follows:

the radial strain (ϵ_r) $= \frac{S_r - \nu S_t}{E}$, and the tangential strain (ϵ_t) $= \frac{S_t - \nu S_r}{E}$.

$$S_r = \frac{3PR^2\nu}{8t^2}\left[\left(\frac{1}{\nu} + 1\right) - \left(\frac{3}{\nu} + 1\right)\left(\frac{r}{R}\right)^2\right] \tag{6.1}$$

$$S_t = \frac{3PR^2\nu}{8t^2}\left[\left(\frac{1}{\nu} + 1\right) - \left(\frac{1}{\nu} + 3\right)\left(\frac{r}{R}\right)^2\right], \tag{6.2}$$

where

- P = the pressure difference across the diaphragm,
- t = the thickness of the diaphragm,
- R = the radius of the diaphragm,
- r = the positional parameter,
- S_r = the radial stress,
- S_t = the tangential stress, and
- E = Young's modulus of elasticity,
- ν = Poisson's ratio of the material of the diaphragm.

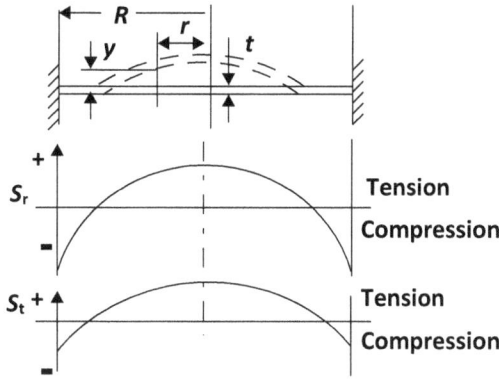

Figure 6.7. A diaphragm gauge and its stress distribution.

Once the strains have been calculated, the change of resistance ΔRs of an individual gauge are obtained from the gauge factors; e_o can then be determined from the bridge circuit sensitivity. The tangential strain ϵ_t reaches its maximum value at $r = 0$. The radial strain ϵ_r is positive in some regions but negative in others and reaches its maximum value at $r = R$.

Flat diaphragms show nonlinearity at large deflections, since a stretching action is added to the basic bending, causing a stiffening effect. This nonlinearity in the stresses closely follows the nonlinearity in the central point deflection (y_c) of the diaphragm, for which the following relation is available:

$$P = \frac{16Et^4}{3R^4(1-\nu^2)}\left[\frac{y_c}{t} + 0.488\left(\frac{y_c}{t}\right)^3\right]. \qquad (6.3)$$

By designing for a sufficiently small value of y_c/t, the desired nonlinearity may be achieved. However, note that small values of y_c/t also lead to small strains and small output voltages.

The deflection at any point is given by

$$y = \frac{3P(1-\nu^2)(R^2-r^2)^2}{16Et^3}. \qquad (6.4)$$

A special-purpose strain gauge rosette which has been designed to take advantage of this strain distribution is widely used in diaphragm pressure transducers. A four-element strain gauge rosette for the diaphragm pressure transducer is shown in figure 6.8. Two strain gauges are located at the center of the diaphragm, where the tangential strain (ϵ_t) is maximized. Similarly, two strain gauges are installed at the edge of the diaphragm, where the radial strain (ϵ_r) is maximized.

Strain gauges R_2 and R_4 are oriented to read the radial strain and placed as close to the edge as possible, since the radial strain reaches its maximum negative value at this point. Strain gauges R_1 and R_3 are installed as close to the center as possible and

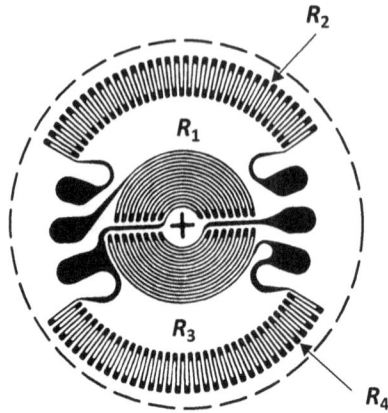

Figure 6.8. A four-element strain gauge rosette for a diaphragm pressure transducer.

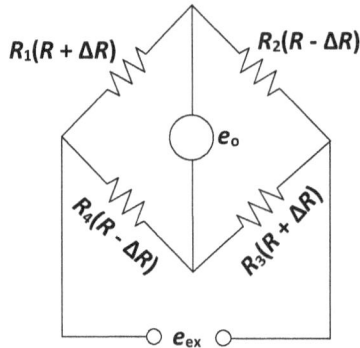

Figure 6.9. A strain gauge rosette in a Wheatstone bridge.

read the tangential strain, since it is maximized at this point. A strain gauge rosette in a Wheatstone bridge is shown in figure 6.9.

In the Wheatstone bridge, the tangential elements are in arms R_1 and R_3 and the radial elements are in arms R_2 and R_4. The output voltage e_o (with $\lambda = 2$) supplied by the manufacturer for the diaphragm pressure sensor with the strain gauge rosette is as follows:

$$e_o = \frac{0.82PR^2(1-\nu^2)}{Et^2}e_{ex} \tag{6.5}$$

Or,

$$P = \frac{1.22Et^2e_o}{R^2(1-\nu^2)e_{ex}} = Ce_o$$

$$\therefore \ C = 1.22\frac{Et^2}{R^2(1-\nu^2)e_{ex}}.$$

Diaphragm strain gauge rosettes installed on backing materials are commercially available in several sizes ranging from 0.45 to 3.2 cm. The sensitivity of the diaphragm pressure transducer and Wheatstone bridge combination is given by

$$S = \frac{e_o}{P} = \frac{1}{C} = 0.82\frac{R^2(1-\nu^2)e_{ex}}{Et^2}.$$
(6.6)

Here, diaphragm deflection rather than the yield strength determines the limit of

$$\left(\frac{R}{t}\right)_{max},$$

which is not the case for other transducers. It can be mathematically shown that the relationship between pressure and voltage is linear to within 0.3% if the deflection y_c at the center of the diaphragm is less than $t/4$. Neglecting the nonlinear term, the deflection at the center of the diaphragm (y_c) can be expressed in terms of pressure as follows:

$$y_c = \frac{3PR^4(1-\nu^2)}{16t^3E}.$$

With the restriction that $y_c \leqslant t/4$,

$$P_{max} \leqslant \frac{4}{3}\left(\frac{t}{R}\right)^4 \frac{E}{1-\nu^2}.$$

Pressure transducers that use a diaphragm sensor are well suited for either static or dynamic pressure measurement. The diaphragm sensor has a very high natural frequency with a small damping ratio because of its low mass and relative stiffness. However, the high-frequency limit of the diaphragm pressure sensor depends primarily upon the degree of damping provided by the fluid in contact with the diaphragm. The resonant frequency of the diaphragm should be three to five times higher than the highest frequency of the applied dynamic pressure. A diaphragm has an infinite number of natural frequencies; however, the lowest is the only one of interest here. For a clamped edge diaphragm vibrating in contact with a fluid that has a density of ρ_f, the lowest natural frequency is given by

$$f = \frac{10.21}{\pi R^2}\sqrt{\frac{Et^2}{12(1-\nu^2)\rho}}\ \text{Hz},$$
(6.7)

where
- E = the modulus of elasticity (in Pa),
- t = the thickness of the diaphragm (m),
- R = the radius of the diaphragm (m),
- ρ = the density of the diaphragm material (kg m^{-3}),
- ν = Poisson's ratio.

For a steel diaphragm, equation (6.7) can be simplified to

$$f = 4.912 \times 10^4 \frac{t}{\pi R^2}.$$

6.4.1 The semiconductor diaphragm gauge

Because of its small gauge length, the semiconductor strain gauge can be used to construct very small transducers. These transducers may, for example, be used for the measurement of pressure, and due to their small size they can measure strain at higher frequencies. The diffusion process employed for integrated circuit manufacture is utilized to produce diffused semiconductor gauges. In a monolithic pressure transducer, the diaphragm is made of silicon instead of metal and the strain gauges are formed by depositing impurities at specific locations. The use of semiconductor technology in pressure transducer construction has led to the development of a variety of very fast, very small, highly sensitive strain gauge diaphragm transducers. Silicon piezo resistance strain gauges can be diffused into a single silicon wafer which forms the diaphragm. Semiconductor strain gauges have a sensitivity that is 50 times greater than those of conventional metallic strain gauges. And because the piezo resistive gauges are integral to the diaphragm, they are relatively immune to the thermoelastic strains prevalent in conventional metallic strain gauge diaphragm construction. They solve the problem of bonding the strain gauges to the diaphragm, and they are now the most commonly used type of pressure transducer. These pressure sensors can be made very small and are often called as micro sensors. Furthermore, unlike a metallic gauge, a silicon diaphragm will not creep with age, thus minimizing calibration drift over time. However, gauge failure is catastrophic and silicon is not well suited to wet environments.

6.5 Low-pressure measurement

The technology of low-pressure measurement is a rather specialized area in which different techniques are used to obtain accurate readings. At the upper end of the low-pressure measurement range, Bourdon tubes, bellow gauges, and diaphragm gauges are used. In this chapter, we are primarily interested in measuring pressures below 1 torr. The McLeod gauge, the Pirani gauge, the thermocouple gauge, and the Knudsen and ionization gauges all find application in measuring this range of pressures.

Units of low-pressure measurement

$$\begin{aligned}
1 \text{ micrometer} &= 10^{-6} \text{ m of mercury } (\mu m, \text{ micron}) \\
&= 10^{-3} \text{ mm of mercury} \\
&= 0.133\,322 \text{ N m}^{-2}. \\
1 \text{ torr} \qquad &= 1 \text{ mm of mercury (mm Hg)} \\
&= 1333.22 \text{ microbar}. \\
1 \text{ bar} \qquad &= 10^5 \text{ N m}^{-2}.
\end{aligned}$$

6.5.1 The McLeod gauge

The McLeod gauge is considered to be a vacuum standard. The pressure can be computed from the dimensions of the gauge. A schematic of the McLeod gauge is shown in figure 6.10. The principle of the McLeod gauge is the compression of a sample of the low-pressure gas to a pressure sufficiently high to read with a simple manometer. The reading is independent of the gas composition.

The movable reservoir is lowered until the mercury column drops below the opening 0p. The bulb B and capillary C are then at the same pressure as the vacuum source p. The reservoir is subsequently raised until the mercury fills the bulb and rises in the capillary to a point at which the level in the reference capillary is located at the zero point. The area of the capillary is denoted by a, so the volume of the gas in the capillary is $V_c = ay$, where y is the length of the capillary occupied by the gas. Let the volume of the capillary, bulb, and tube down to the opening be V_B. If we assume isothermal compression of the gas in the capillary, we have

$$p_c = p\frac{V_B}{V_c}, \tag{6.8}$$

and the pressure indicated by the capillary is

$$p_c - p = y \tag{6.9}$$

Here, the pressures are expressed in terms of the height of the mercury column.

Combining equations (6.8) and (6.9) gives

$$p = \frac{ay^2}{V_B - ay} = \frac{yV_c}{V_B - ay}$$

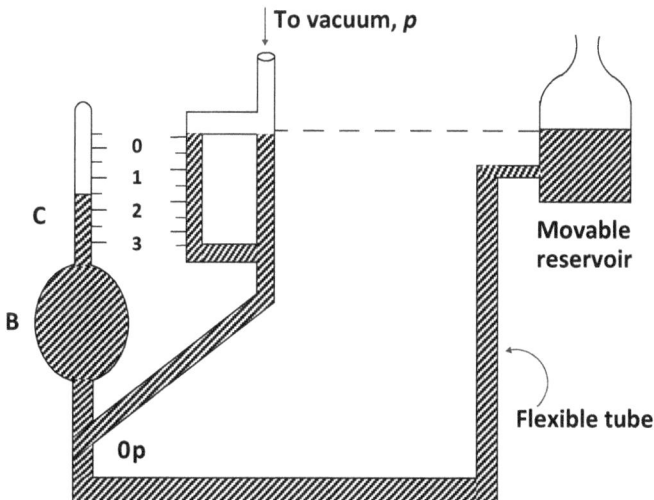

Figure 6.10. The McLeod gauge.

For most applications,

$$ay \ll V_B$$

$$\therefore p \cong \frac{ay^2}{V_B}$$

Commercial McLeod gauges have the capillary calibrated directly in micrometers. The McLeod gauge is sensitive to condensed vapors that may be present in the sample because they can condense upon compression and invalidate equation (6.8). For dry gases, the gauge is applicable in the range of 10^{-2} to 10^{+2} μm (0.0013–13.3 Pa).

6.5.2 The Pirani gauge

The Pirani gauge is a device that measures pressure using a change in the thermal conductance of a gas. A schematic of a Pirani gauge is shown in figure 6.11. It works on the principle that at low pressure, the effective thermal conductivity of gases decreases with pressure. An electrically heated filament is placed inside the vacuum space. The heat loss from the filament is dependent upon the thermal conductivity of the gas and the filament temperature. The lower the pressure, the lower the thermal conductivity and consequently the higher the filament temperature. The resistance of the filament is measured by a bridge circuit.

The resistance element takes the form of four coiled tungsten or platinum wires. The cold surface is the glass tube. The measurement circuit is shown in figure 6.12. Two identical tubes are generally connected in the bridge circuit to avoid changes in bridge output due to changes in ambient temperature. One of the tubes is evacuated to a very low pressure and then sealed off while the other has the gas admitted to it. The balance potentiometer is then adjusted to produce the null condition. To initially balance the bridge, the pressure in the measuring element is made very small and the balance potentiometer is set for zero output. Generally, the bridge is used as a deflection device rather than a null device. Any change in pressure causes a bridge imbalance.

The variation of the unbalanced bridge current for pressure variation inside the Pirani gauge is shown in figure 6.13.

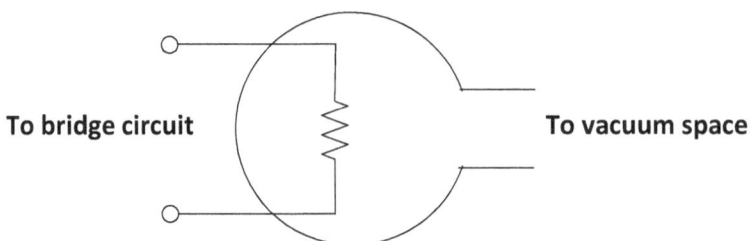

Figure 6.11. A schematic of the Pirani gauge.

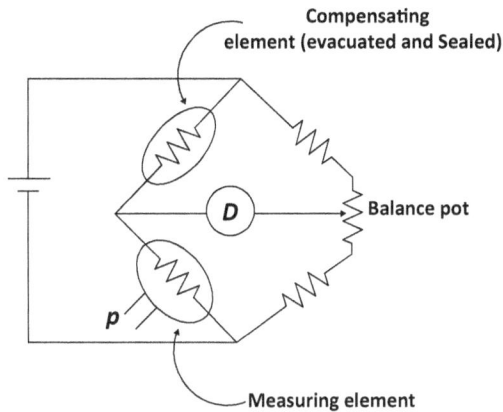

Figure 6.12. Bridge circuit used for the measurement of low pressures by Pirani gauge.

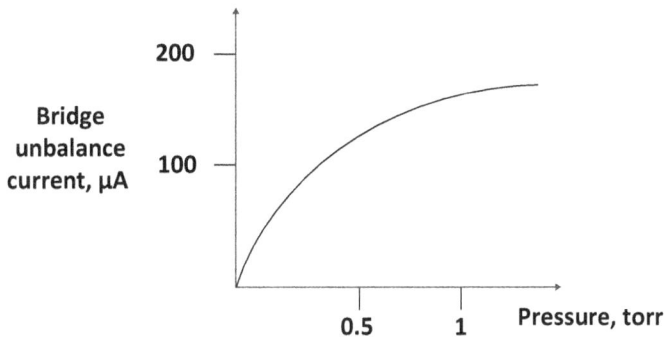

Figure 6.13. Unbalance current of the bridge.

6.5.2.1 Calibration of the Pirani gauge

The gauge must be calibrated against a standard: it is unsuitable for pressures of less than 1 μm. The upper limit is 1 torr (133 Pa), giving an overall range about 0.1–100 Pa. For higher pressures, the thermal conductance changes very little with pressure. It must be noted that heat loss from the filament is also a function of the conduction losses to the filament supports and radiation losses to the surroundings. The lower limit of low-pressure measurement by the Pirani gauge is the point at which these effects outweigh conduction into the gas. The Pirani gauge has a large time constant; the establishment of thermal equilibrium may need several minutes.

6.5.3 The thermocouple gauge

The hot surface of the thermocouple gauge is a thin metal strip, and the cold surface is a glass tube. A thermocouple gauge is shown in figure 6.14. Its temperature may be varied by varying the heating current. For a given heating current and gas, the temperature assumed by the hot surface depends on the pressure. This temperature is

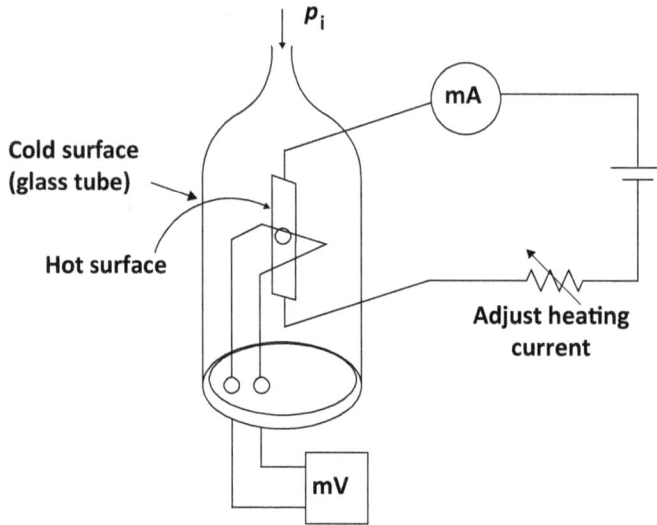

Figure 6.14. The thermocouple gauge

measured by a thermocouple welded to the hot surface. This type of gauge is not independent of ambient temperature changes. Its range is 10^{-4} torr to 1 torr.

6.5.4 The ionization gauge

A schematic of the ionization gauge is shown in figure 6.15. It is used to measure very low pressures in the range of 10^{-12}–10^{-10} torr. A gas at an unknown pressure is introduced into a glass vessel containing free electrons emitted by a heated cathode. The electrons that pass through the grid are collected by a plate or ion collector.

In the ionization gauge, a stream of electrons is emitted from the cathode. Some of these electrons strike gas molecules and knock out secondary electrons, leaving the molecules as positive ions. Under normal gauge operation, the secondary electrons form a negligible part of the total electron current. Therefore, the electron current i_e is same for all vacuum pressures. The number of positive ions formed is directly proportional to i_i. The rate of production of positive ions (the ion current i_i) for a given gas is a direct measure of the number of gas molecules per unit volume and thus the gas pressure. The positive ions are attracted to a negatively charged electrode, which collects them and carries the ion current.

Therefore, the pressure of the gas is proportional to the ratio of the ion current and electron current, and it is expressed as follows:

$$p = \frac{i_i}{S i_e},$$

where S is the sensitivity of the gauge. However, the electron current is almost constant for a given system and therefore the gas pressure is determined solely by measuring the ion current. A typical value of S for nitrogen is $S = 20$ torr^{-1} (2.67 kPa^{-1}).

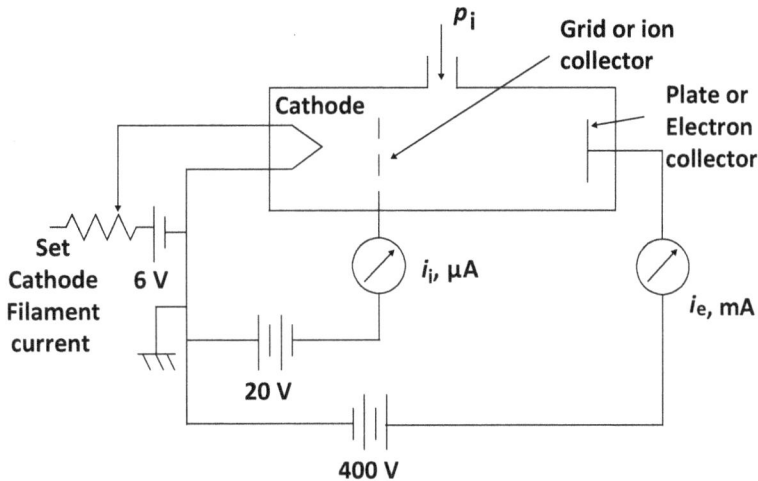

Figure 6.15. A schematic of an ionization gauge.

The exact value of S must be determined by calibrating each particular gauge. The value of S is a function of the tube geometry and the type of gas. The current output is usually linear.

Disadvantages:
- The filament may burn out if exposed to air while hot.
- Some gases may be decomposed by the hot filament.
- The measured gas can be contaminated by gases forced out of the hot filament.

6.5.5 The Knudsen gauge

The Knudsen gauge is relatively insensitive to the gas composition and thus offers promise for development into a standard for pressures too low for the McLeod gauge. A schematic of a Knudsen gauge is shown in figure 6.16.

The fixed plates are heated to an absolute temperature T_f, a temperature that must be measured. The temperature T_v of the spring-restrained movable vane must also be known. The space between the fixed and movable plates must be smaller than the mean free path of the gas whose pressure is being measured. The kinetic theory of gases states that gas molecules rebound from the heated plates with greater momentum than from the cooler movable vane, thus exerting a net force on the movable vane whose deflection is measured by a light-and-mirror arrangement. Analysis shows that the force is directly proportional to pressure for given values of T_f and T_v and takes the form

$$p_i = \frac{KF}{\sqrt{\frac{T_f}{T_v} - 1}}$$

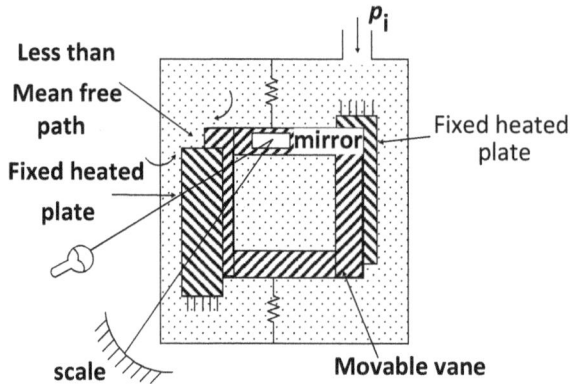

Figure 6.16. A schematic of a Knudsen gauge.

where F is force and K is a constant. The range of the Knudsen gauge is 10^{-8} to 10^{-2} torr.

Problem 6.1. In the figure 6.17, four strain gauges are placed over a diaphragm.

For the above arrangement, draw the signal conditioning circuit.
1. Find the sensitivity of the circuit in mV Pa^{-1}.
2. Calculate its natural frequency in vacuum.
3. Find the maximum allowable pressure for a nonlinearity of 2%.
4. Find the full-scale output.

The following data are given:
- $r_t = 0.015$ m, $r_r = 0.064$ m.
- The diameter of the diaphragm $(D) = 0.15$ m.
- The thickness of the diaphragm $(t) = 1.28$ mm.
- Poisson's ratio $(\nu) = 0.26$.
- The excitation voltage (E_{ex}) of the Wheatstone bridge $= 5$ V.
- The gauge factor $= 2$.
- The gauge resistance $= 120$ Ω
- The density of the diaphragm material $(\rho) = 6 \times 10^3$ kg m^{-3}.
- The modulus of elasticity of the diaphragm material $(E) = 2.1 \times 10^{11}$ Pa.

Problem 6.2. A McLeod gauge has a bulb of volume 110 cm^3. Its capillary diameter is 1.2 mm. Initially, the reading was found to be 3 cm. Later, it was found that the observed reading was wrong. Find the error in the measured pressure if the true reading is 2.5 cm.

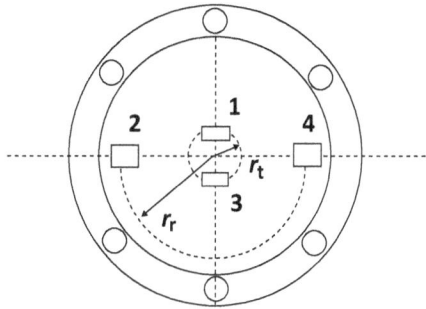

Figure 6.17.

Problem 6.3. A pressure sensor has the following specifications:
- its sensitivity at the design temperature = 10 V MPa^{-1},
- the zero drift = 0.01 V/°C, and
- the sensitivity drift = 0.01 V MPa/°C.

What will the true value of the pressure be when the sensor is used at an ambient temperature of 20 °C above the design temperature and the device output is 7.4 V?

IOP Publishing

Fundamentals of Industrial Instrumentation (Second Edition)

Alok Barua

Chapter 7

Flowmeter

Learning objectives:
- The operational principle of a differential pressure flowmeter.
- The orifice meter.
- The Venturi meter.
- Discharge coefficient.
- Pressure tapping.
- Pipe bending near the differential pressure flowmeter.
- Pressure recovery.
- The working principle of the Pitot tube.
- The advantages and disadvantages of the elbow meter.
- The rotameter as a linear sensor.
- The operational principle of the open channel meter (weir).
- The magnetic pickup.
- Turbine flowmeters with direct electrical output.
- The operational principle of the electromagnetic flowmeter.
- Types of pipe used.
- Magnetic field excitation for electromagnetic flowmeter.
- The advantages and disadvantages of electromagnetic flowmeter.
- The operational principle of the ultrasonic flowmeter.
- The working principle of the hot wire anemometer (both constant temperature and constant current types).

7.1 Introduction

A flow rate can be stated in terms of a volume per unit time, known as the volume flow rate, or as a mass per unit time, known as the mass flow rate. The total quantity of liquid is given by the time integral of the flow rate. Measurements of the volumetric flow rate and mass flow rate are necessary for the purpose of determining the proportions of materials introduced into a process and the amounts of materials produced by the process. For example, in a heat exchange process, the rate at which energy can be can be removed or added to a moving fluid is directly proportional to the mass flow rate of the fluid. The transducers used to measure flows are diverse in nature. Flow metering is necessary for applications ranging from the measurement of the blood flow rate in a human artery to measurement of the total volume of water supplied to a municipality on a single day.

This chapter discusses some of the most commonly used flow transducers. The flow rate, accuracy, cost, pressure loss, and compatibility with the fluid are important design considerations for flow metering transducers. All transducers have both advantages and disadvantages that necessitate some compromise in the selection of the best sensor for a particular application. For example, a high degree of precision is necessary in the measurement of flow for a gas pipeline, since a small error in measurement could cost huge amount of money over a period of time. However, the measurement of the water flow in a canal used for irrigation purposes does not need much precision. The operational principles of all the flow measuring devices are discussed in this chapter. To conclude the chapter, a set of problems on flow measurement is given. The solutions to the problems are also presented at the end of this book. Although the vortex shedding flowmeter is not discussed in the text, it is briefly introduced in the relevant problem. The flow rates of fluids in closed channels can successfully be measured by the following flow transducers:

1. Differential pressure flowmeters;
   - The Venturi meter,
   - The orifice meter,
   - The Pitot tube,
   - The elbow meter.
2. Variable area flowmeters.
3. Turbine flowmeters.
4. Electromagnetic flowmeters.
5. Ultrasonic flowmeters.
6. Laser Doppler flowmeters.

Flows of liquid and gas in open channels can be measured by Weirs and Pitot tube, respectively.

'The hot wire anemometer' can be used for both open and closed channel.

7.2 Differential pressure flowmeters

They are most widely used flowmeters for liquids and gases. A restriction is placed in the pipe and the differential pressure developed across the restriction is measured. The differential pressure output is calibrated in terms of volume flow rate. A differential pressure flowmeter is shown in figure 7.1.

Let us consider two cross-sectional areas, one and two. The total energy is conserved at all points in a liquid. Therefore, the total energy per unit mass is same at both sections:

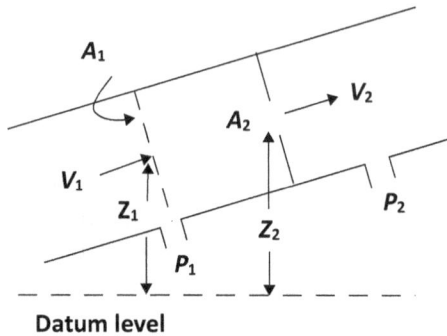

Figure 7.1. A differential pressure flowmeter

$$\frac{P_1}{\rho_1} + \frac{v_1^2}{2} + Z_1 g = \frac{P_2}{\rho_2} + \frac{v_2^2}{2} + Z_2 g. \tag{7.1}$$

P = pressure (in units of N m^{-2})
v = average velocity (m s^{-1})
ρ = fluid density (kg m^{-3})
A = cross-sectional area (m^2)
Z = elevation above datum (m)

The following assumption have been made in calculating the volume flow rate:

1. The flow is frictionless. This means there is no loss of energy due to friction either in the fluid itself or between the fluid and the pipe walls.
2. There are no heat losses or gains due to heat transfer between the fluid and its surrounding medium.
3. There is conservation of total energy (pressure + kinetic + potential energy) at all points in the liquid.
4. The fluid is incompressible, i.e. $\rho_1 = \rho_2 = \rho$,
   where ρ_1 and ρ_2 are the densities of the fluids in the upstream and downstream flows.
5. The pipe is horizontal, i.e. $Z_1 = Z_2$. Therefore equation (7.1) reduces to

$$\frac{v_2^2 - v_1^2}{2} = \frac{P_1 - P_2}{\rho}. \tag{7.2}$$

6. There is conservation of volume flow rate; that is, $Q_1 = Q_2 = Q$; also, $Q_1 = A_1 v_1$ and $Q_2 = A_2 v_2$,

where Q_1 and Q_2 are the volume flow rates in the pipe and at the restriction respectively. A_1 and A_2 are the cross-sectional areas of the pipe and the restriction, respectively

- Since $A_2 < A_1$, if follows that $v_2 > v_1$ and $P_2 < P_1$.

Therefore, the theoretical value of the volume flow rate in a differential pressure flowmeter Venturi and orifice) is

$$Q_{th} = \frac{A_2}{\sqrt{1-\left(\frac{A_2}{A_1}\right)^2}} \sqrt{\frac{2(P_1-P_2)}{\rho}}. \tag{7.3}$$

The theoretical value of the volume flow rate always differs from the actual flow rate for two main reasons:

1. Frictionless flow never occurs in a pipe. However, for turbulent flows in a smooth pipe, the friction losses are small. Laminar and turbulent flows are characterized by the Reynolds number. The Reynolds number is given by:

$$R_e = \frac{VDP}{\eta}.$$

When D is the pipe diameter and η is the viscosity of the fluid flowing in the pipe, V is the velocity of the fluid in the pipe and P is the differential pressure across the section of the pipe.

2. The cross-sectional area of the pipe is $\frac{\pi D^2}{4}$ and the cross-sectional area of the meter is $\frac{\pi d^2}{4}$, where D and d are the respective diameters.

If the fluid fills the pipe, then $A_1 = \frac{\pi D^2}{4}$.

However, the area of minimum cross-section is given by $A_2 = \frac{0.99\pi d^2}{4}$ for a Venturi. The restriction of a Venturi meter is gradual, whereas the orifice plate is a sudden restriction which causes the fluid's cross-sectional area to have a minimum value at the vena contracta. Therefore, the theoretical expression for the volume flow rate is corrected as follows:

$$Q_{act} = \frac{C}{\sqrt{1-\beta^4}} \cdot A_2 \sqrt{\frac{2(P_1-P_2)}{\rho}} \tag{7.4}$$

where C = discharge coefficient,
β = flowmeter pipe diameter ratio, $\frac{d}{D}$, and
A_2 = flowmeter cross-sectional area.

The value of the discharge coefficient depends on type of flowmeter, i.e. orifice or Venturi, the Reynolds number R_e and the diametric ratio β. Therefore, for a given flowmeter, $C = f(R_e, \beta)$. The values of C have been found experimentally for several types of flowmeter over a wide range of fluid velocities. For a given fluid and a known volumetric rate of fluid flow, C can be found from equation (7.4).

General features of the differential pressure flowmeter

1. It has no moving parts; it is robust, reliable, easy to maintain, and widely established.

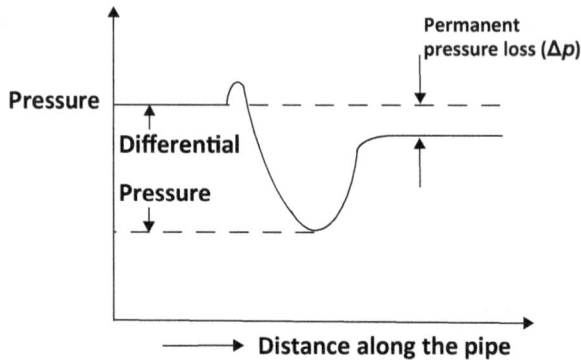

Figure 7.2. Pressure variation versus distance along the pipe for a pipe obstructed by an orifice plate.

2. There is always a permanent pressure loss, and thus extra pumping energy is necessary to compensate for it. The differential pressure drop along the length of an orifice meter is shown in figure 7.2.

3. Both Venturi meters and orifice meters are nonlinear. The volumetric flow rate is proportional to the square root of the pressure differential. This limits the useful range of a meter to between 25% and 100% of the full-scale output reading. For lower flows, the differential pressure measurement is less than 6% of the full-scale output reading and is not accurate enough for measurement. Its useful range is shown in figure 7.3. The measurement error is shown in figure 7.4.

4. It can be used for turbulent flows, i.e. $R_e > 10^4$.

As shown in figure 7.5, a typical flowmeter system consists of a differential pressure sensor, a differential pressure transmitter, a data acquisition system (DAS), and a computer. The transmitter produces a current output signal (4–20 mA) and the DAS consists of an amplifier, an I-to-V converter, and an analog-to-digital converter (ADC).

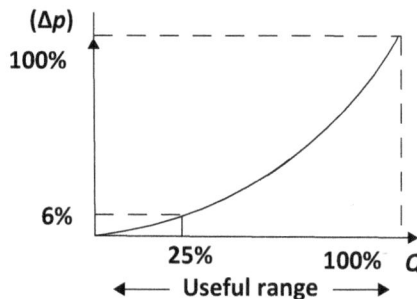

Figure 7.3. The useful range of a differential pressure flowmeter.

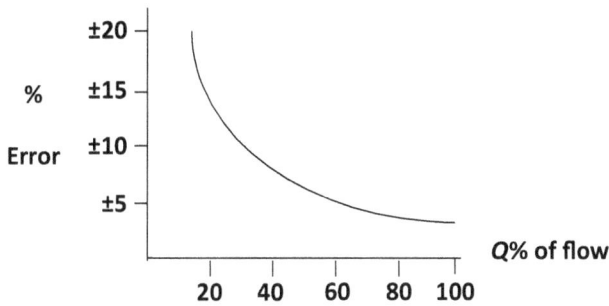

Figure 7.4. The measurement error.

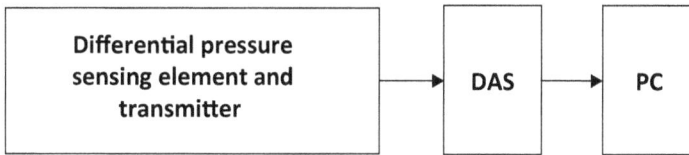

Figure 7.5. Differential pressure measurement system.

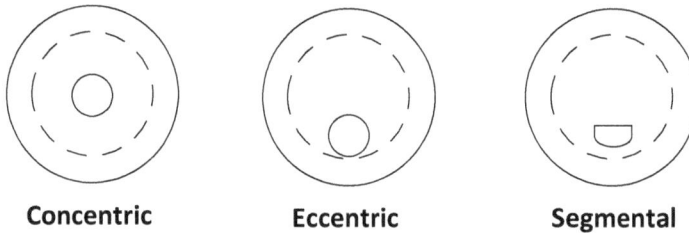

Figure 7.6. Types of orifice plate.

7.3 The orifice meter

The thin plate square-edge orifice is the most widely used differential pressure flowmeter in process industry, mainly because of its simplicity and low cost.

Moreover it has become well established over the years, and data are available for its behaviour. Typically, three types of orifice plates are available, as shown in figure 7.6.

The concentric orifice is the most widely used plate. The eccentric and segmental orifices are employed to measure the flows of fluids containing solids. In both cases, the bottom of the hole is located in such a way that the bottom of the hole is at same level as the internal bottom of the pipe installation. These two orifice plates need separate calibration, because their discharge coefficients differ from that of the concentric orifice type. The concentric orifice plate is installed in the pipe with its hole concentric to the pipe. It is a flat metal circular plate made of steel, stainless steel, or phosphor bronze. Its thickness is only sufficient to withstand the buckling forces caused by the differential pressure that exists across the plate. The circular hole is made with a 90°, square, and sharp edge on the upstream side. Any change from a sharp edge will modify the discharge coefficient of the orifice meter. It is advisable to replace the orifice during routine maintenance of the plant for better measurement accuracy.

7.3.1 Orifice pressure taps

The pressure taps are necessary to measure the pressure differential across the orifice plate. There are two taps, one on the upstream side and the other on the downstream side, called the upstream and downstream pressure taps, respectively. Wear and abrasion of the sharp edge of the orifice plate modifies the value of the discharge coefficient and changes the calibration of the meter. Therefore, it is advisable to replace the orifice at regular intervals to maintain the accuracy of the system. The taps are not placed very close to the plate, so that orifice plate can easily be replaced. Though there are various taps, only the most frequently used taps are discussed here. Three different pressure taps for orifice are shown in figures 7.7–7.9.

(i) **Flange taps**

These are constructed so that the taps used to measure the differential pressure are integral parts of the orifice plate assembly. The pressure taps are usually located 2.5 cm away from the sides of the orifice plate. The advantages of flange taps are that entire orifice assembly is easily replaced and the pressure taps are accurately located.

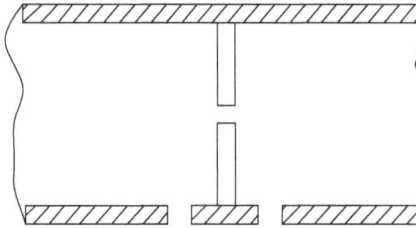

Figure 7.7. An orifice plate with flange taps.

(ii) **D and D/2 taps**

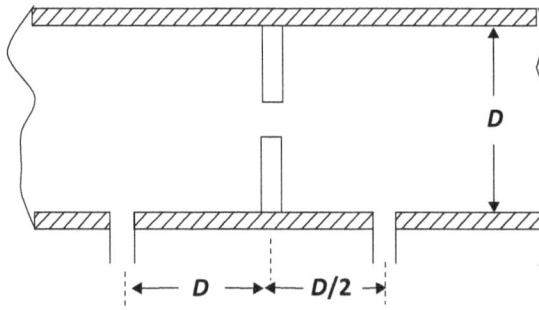

Figure 7.8. An orifice plate with D and D/2 taps.

(iii) **Vena contracta taps**

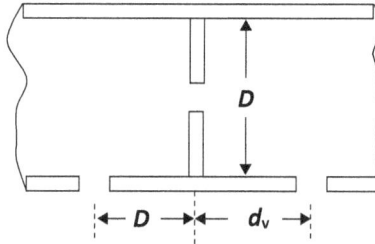

Figure 7.9. An orifice plate with vena contracta taps.

These are arranged in such a way that the downstream pressure tap is located at a variable distance from the orifice, depending on the pipe and orifice size. The upstream tap is located at a distance equivalent to one pipe diameter, and the downstream tap is located at the vena contracta. In a vena contracta tap, the pressure differential is maximised for a given flow rate.

7.3.2 Restrictions on pipe fittings adjacent to an orifice meter

The discharge coefficient is experimentally determined for straight pipes. Flow disturbances in the pipeline adjacent to the orifice alter the value of the discharge coefficient. Therefore elbows, pipe bends, tees, or valves are not allowed near the orifice. There should be no fittings closer than five pipe diameters from the orifice on the downstream side. There should not be any fittings closer than twenty pipe diameters on the upstream side. If the minimum distance is infeasible, specially on the upstream side, flow straightners can be installed. A flow straightner consists of a bundle of smaller tubes welded inside the pipe.

7.4 Flow nozzles, Dall tubes, and Venturi meters

The flow nozzle, Venturi meter and Dall tube are shown in figures 7.10, 7.11, and 7.12 respectively.

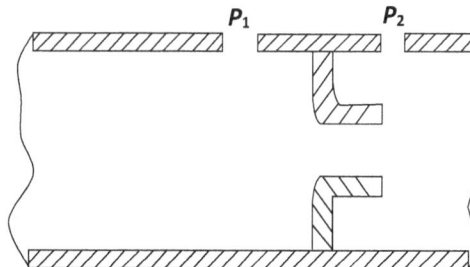

Figure 7.10. The flow nozzle.

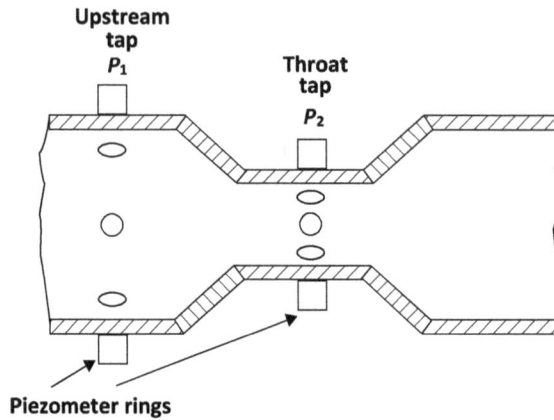

Figure 7.11. The Venturi tube.

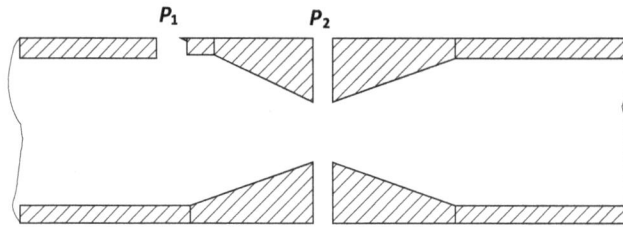

Figure 7.12. The Dall tube.

7.4.1 The flow nozzle

The flow nozzle, Venturi tube, and Dall flow tube are based on the same principle as the orifice. The Dall tube is a modified Venturi tube that has a low permanent pressure loss. The flow nozzle is more expensive than the orifice meter but cheaper than the Venturi meter. It is also a variation of Venturi in which the exit section is omitted, so that it is similar to an orifice with a well-rounded upstream edge. The upstream tap is located at about one pipe diameter from the entrance to the nozzle. The downstream tap is located on the pipe opposite the straight portion of the nozzle. The flow nozzle is used for high-velocity steam flows, and it is dimensionally more stable than an orifice at high temperatures and velocities. The permanent pressure loss in the flow nozzle is same as that of the orifice.

Solid particles or bubbles of a gas in flowing fluid do not stick to the flow restriction, and so, in this case, it is superior to the orifice plate. It has a longer working life because it does not get worn away in the same way as an orifice plate. So it has better measurement accuracy than that of the orifice plate.

7.4.2 The Venturi meter

The Venturi meter is an expensive instrument but offers very good accuracy (±1%).

The construction of the Venturi tube and its advantages
- It is made of cast iron or steel.
- Large Venturi tubes are usually made of concrete.
- Sometimes the throat is made of bronze.
- The upstream section has an angle of 20°.
- The downstream section has an angle of 7°.
- The pressure taps are made of piezometer rings so as to average the measurement around the periphery.
- The diametric ratio ($\frac{d}{D}$) for the Venturi typically lies between 0.25 and 0.50.
- It has almost no maintenance requirement and its working life is very long.
- It is widely used in high flow situations such as municipal water systems, where large savings in pumping costs are possible due to its low permanent pressure loss.
- The smooth internal shape of the Venturi tube means it is unaffected by solid particles or gaseous bubbles in flowing fluid, and it can measure the flow of liquids such as slurries.
- Its ranges are extremely large. It is possible to measure water flow rates as high as $1.5 \times 10^6 \, \text{m}^3 \, \text{h}^{-1}$.

7.4.3 The Dall tube

A Dall tube consists of two conical reducers inserted into the fluid-carrying pipe. It has a shape very similar to a Venturi without a throat. The construction of the Dall tube is much simpler than that of the Venturi, which needs complex machinery. It is much shorter in length, which makes its insertion into the flow line easier. Its permanent pressure loss is only half of the throat loss of a Venturi. So far as maintenance and operational life are concerned, the Venturi and the Dall tube are similar.

7.5 The Pitot tube

The Pitot tube measures the velocity at a point in a fluid. It is an open channel meter. It is suitable for investigating around an aerofoil in a wind tunnel or measuring the velocity profile in a pipe prior to the installation of a permanent flowmeter.

7.5.1 The operational principle operation of the Pitot tube

If a solid body is held stationary in a pipe in which a fluid is flowing with a velocity, as the fluid in the tube approaches the body, the fluid particles are decelerated until, at a point directly in front of the solid body, the velocity of the fluid is zero. Accompanying the deceleration is an increase in pressure. This is due to the process of converting the 'velocity' head into an additional static head. The velocity of the fluid can be found by measuring the differential pressure. At the impact hole or stagnation point, the fluid is brought to rest; this point therefore has no kinetic energy. A schematic of the Pitot tube is shown in figure 7.13.

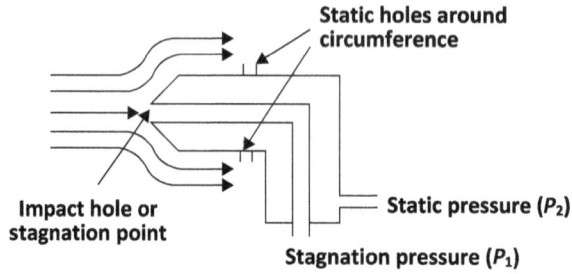

Figure 7.13. The Pitot tube.

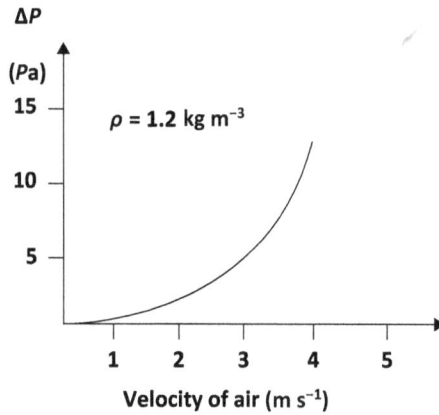

Figure 7.14. The differential pressure versus the air velocity measured by a Pitot tube.

Both the kinetic energy and pressure energy are present at the static holes, because fluid is moving at those positions. Assuming energy conservation, no frictional loss, and no heat loss, we can write

$$\frac{P_1}{\rho} + o + gz_1 = \frac{P_2}{\rho} + \frac{v^2}{2} + gz_2, \tag{7.5}$$

where z_1, z_2 are the elevations of the holes above a datum line and $g = 9.81$ m s^{-2}. If $z_1 = z_2$ then

$$v = \sqrt{\frac{2(P_1 - P_2)}{\rho}}. \tag{7.6}$$

From equation (7.6), we can write

$$\Delta P = \frac{1}{2}\rho v^2,$$

where ΔP = differential pressure = $P_1 - P_2$

$$\therefore v \propto \sqrt{\Delta P}.$$

The differential pressure versus the air velocity is shown in figure 7.14.

For air at 20 °C and pressure $P_2 = 10^5$ Pa with $\rho = 1.2$ kg m^{-3}, we find that $\Delta P = 0.6\, v^2$.

Thus, at $v = 10$ m s^{-1}, we have $\Delta P = 60$ Pa and $\frac{\Delta P}{P_2} = 6 \times 10^{-4}$, while at $v = 100$ m s^{-1} and $\Delta P = 6 \times 10^3$ Pa, $\frac{\Delta P}{P_2} = 6 \times 10^{-2}$. Thus the low value of $\Delta P/P_2$ ratio means that for $v < 100$ m s^{-1}, the difference in density between the air at the stagnation point and static holes is negligible. The error introduced by considering incompressible fluid is less than 1%.

7.5.2 The differential pressure transmitter

Due to its low differential pressure, a Pitot tube must be accompanied by a special differential pressure (D/P) transmitter. The transmitter is based on an LVDT sensor or differential capacitor arrangement which senses the central point of deflection of a diaphragm capsule. It produces an output of 4 to 20 mA for differential pressure. A scheme for velocity measurement employing pitot tube is shown in figure 7.15.

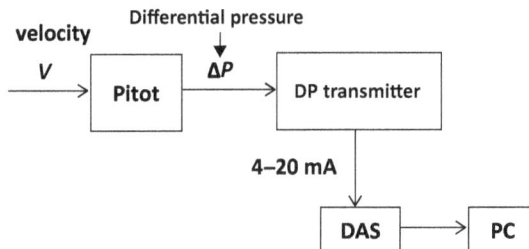

Figure 7.15. A Pitot tube measurement system

7.6 The elbow meter

The orifice meter, flow nozzle, Dall tube, and Venturi meter cause permanent pressure loss in the system. The elbow meter does not introduce any additional losses into the system, since it simply replaces an existing elbow or pipe bend that is used to change the direction of flow. An elbow meter is shown in figure 7.16.

The velocity, pressure, and elevation above the datum level for pressure taps on the inside and outside surfaces of a 90° elbow are related by the following expression:

$$C_k \frac{v^2}{2g} = \frac{P_o}{\rho g} + Z_o - \frac{P_i}{\rho g} - Z_i, \tag{7.7}$$

where C_k is a coefficient that depends upon the size and shape of the elbow. The normal value of C_k ranges from 1.3 to 3.2.

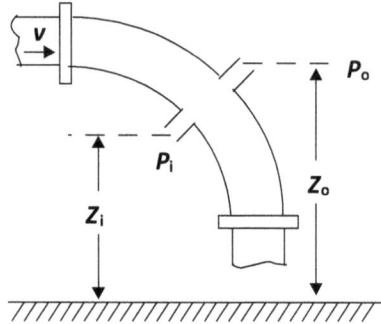

Figure 7.16. An elbow meter.

The variables are expressed in the following units:

$P \Rightarrow N\,m^{-2}$

$\rho \Rightarrow kg\,m^{-3}$

$g \Rightarrow m\,s^{-2} \therefore \dfrac{P}{\rho g} \Rightarrow m$

The volume flow rate is expressed as follows:

$$Q = Av = \frac{A}{\sqrt{C_k}}\sqrt{2g\left(\frac{P_o}{\rho g} + Z_o - \frac{P_i}{\rho g} - Z_i\right)}$$
$$= C.\,A\sqrt{2g\left(\frac{P_o}{\rho g} + Z_o - \frac{P_i}{\rho g} - Z_i\right)} \tag{7.8}$$

where A is the cross-sectional area of the pipe in m^2.

The value of C ranges from 0.56 to 0.88. The primary advantage of the elbow meter is the reduction in the extra pumping cost. The primary disadvantage is that each meter must be calibrated on site. The low operating cost can usually justify the calibration cost. The elbow meter requires a minimum of 20–30 pipe diameters of unobstructed upstream flow (to reduce turbulence and swirl) for accurate measurement, otherwise flow straightners are to be installed to stabilize the flow, as used in the orifice meter.

7.7 The rotameter

The rotameter is widely used for flow rate indication. The meter consists of a float (typically called a 'bob' by engineers in the process industry) within a vertical/ transparent tube tapered to an increasing cross-sectional area at the outlet. A rotameter is shown in figure 7.17.

The fluid entering through the bottom passes over the float, which is free to move only in the vertical direction. The rotameter is always installed in the upright position. When the fluid is flowing through the meter, three forces act on the bob; these are weight of the bob acting downward, a buoyant force acting upward, and a drag force acting upward on the bob. The vertical tube of the rotameter is made of glass to make it a monitoring instrument. Rotameters are used in applications where accuracy is not the prime concern.

1. Drag force, F_d
2. Weight of the bob, W
3. Buoyancy force acting on the bob, F_b

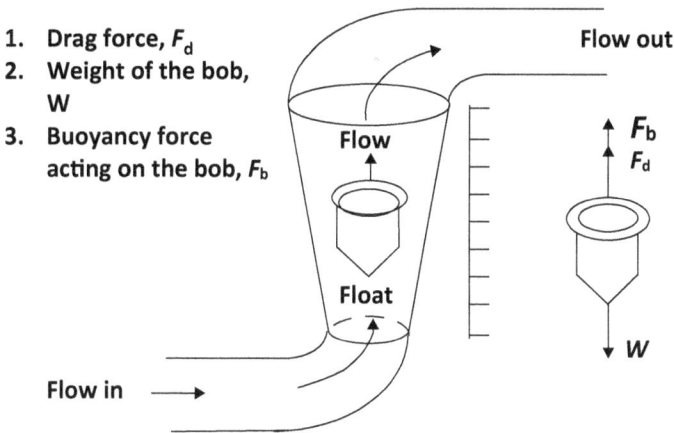

Figure 7.17. A rotameter

7.7.1 The operational principle of the rotameter

For a given flow rate, the float remains stationary when the weight of the float is balanced by the buoyancy and the drag force. The rotameter is an auto-balancing system. The annular area between the float and the vertical tube varies continuously with the vertical displacement of the float or bob. For a particular liquid, the weight of the float and the buoyant force are constant, therefore the drag force is maintained at a constant level. Since the cross-sectional area of the float is constant, the pressure drop across it should be constant.

When the float is in a particular position for a flow rate, the differential pressure varies with the square of the flow rate. Therefore, to keep the differential pressure constant for some other flow rate, the annular area in between the float and the vertical tube must change. A variable area is provided by the vertical tapered tube. The position of the float can be made essentially linear with the flow rate by making the cross-sectional area of the tube vary linearly with the vertical height.

Considering the incompressible flow, the volume flow rate is expressed as

$$Q = \frac{C_d(A_t - A_b)}{\sqrt{1 - \left(\frac{A_t - A_b}{A_t}\right)^2}} \sqrt{2g V_b \frac{(\rho_b - \rho_f)}{A_b \rho_f}}, \tag{7.9}$$

where Q = volume flow rate (m^3 s^{-1})

C_d = discharge co-efficient

A_t = cross-sectional area of the tube (m^2)

A_b = cross-sectional area of the float or bob (m^2)

V_b = volume of the float (m^3)

ρ_b = density of the float material (kg m^{-3})

ρ_f = density of the flowing fluid (kg m^{-3})

If we assume that there is no variation of the discharge coefficient (C_d) with float position and if

$$\left(\frac{A_t - A_b}{A_t}\right)^2 << 1,$$

then equation (7.9) can be simplified to $Q = K(A_t - A_b)$,

where $K \cong C_d \sqrt{2gV_b \frac{(\rho_b - \rho_f)}{A_b \rho_f}}$.

Now if the cross-sectional area of the vertical tube varies linearly with the float position, then $Q = K_1 + K_2 x$.

The rotameter usually has an accurate range of 10:1, which is better than that of the square root sensor.

7.7.2 The shape of the rotameter float

Floats with sharp edges are less sensitive to fluid viscosities that change with temperature.

7.8 The weir

The weir is an open channel meter. If flow rate of a liquid in an open duct or channel such as a river or canal is required, a weir can be used. Weirs are extensively used to get an estimate of the flow of water in a canal or duct for irrigation purposes.

7.8.1 The operational principle of the weir

The flow rate over a weir is a function of the weir geometry and of the weir head. (The weir 'head' is defined as the vertical distance between the weir crest and the liquid surface in the undisturbed region of the upstream flow).

A schematic of a weir is shown in figure 7.18. A rectangular weir is shown in figure 7.19. A flow of liquid over the sharp edge of a weir is shown in figure 7.20.

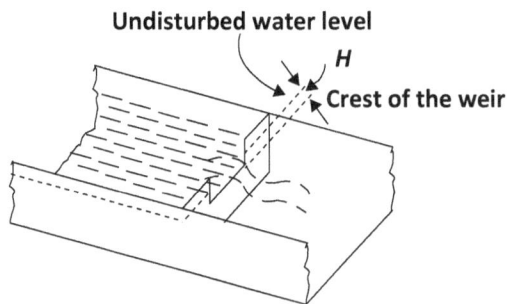

Figure 7.18. A schematic of a weir.

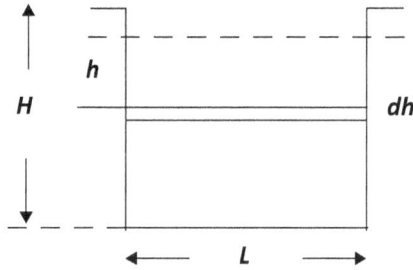

Figure 7.19. A rectangular weir.

Undisturbed region of upstream flow

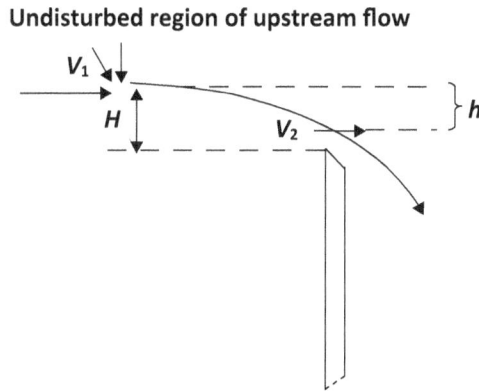

Figure 7.20. Flow over a sharp rectangular weir.

In figures 7.19 and 7.20, H is the head on the weir and h is the distance below the free surface of water where V_2 exists.

Applying Bernoulli's equation at the undisturbed region of the upstream flow and at the crest (sharp edge) of the weir, we get

$$H + \frac{V_1^2}{2g} = (H-h) + \frac{V_2^2}{2g}, \tag{7.10}$$

where V_1 is the upstream flow of the liquid and V_2 is the flow at the crest of the weir;

$$\therefore V_2 = \sqrt{2g\left(h + \frac{V_1^2}{2g}\right)}.$$

If V_1 is small compared to V_2, we get

$$V_2 \approx \sqrt{2gh}. \tag{7.11}$$

The ideal flow rate over the weir is obtained by integrating the quantity $V_2 \times dA$ over the area (A) of the flow plane just above the crest of the weir:

$$Q_{\text{theoretical}} = \int_A V_2 dA = \int_o^H \sqrt{2gh}\ Ldh$$
$$= \frac{2}{3}\sqrt{2g}\ LH^{3/2} \qquad (7.12)$$

However, the actual flow rate is less than the ideal flow rate due to vertical contraction from the top, friction loss, and the presence of non-horizontal liquid flow. Therefore, the actual flow rate is given by

$$Q_{\text{actual}} = C_D Q_{\text{theoretical}} = \frac{2}{3}\sqrt{2g}\ C_D\ LH^{3/2}, \qquad (7.13)$$

where C_D is the discharge coefficient of a rectangular weir, which lies between 0.62 and 0.75. The weir must be sharp for this coefficient to be valid over a long period of time before the flowmeter is recalibrated.

When the flow rate is low, a triangular (v-notch) weir is more suitable for flow measurements. This type of weir is more accurate than the rectangular weir. A v-notch weir is shown in figure 7.21.

The ideal flow rate for a v-notch weir is given by

$$Q_{\text{theoretical}} = \frac{8}{15}\sqrt{2g}\ \tan(\theta/2)H^{\frac{5}{2}}$$
$$= \frac{4L}{15H}\sqrt{2g}\ H^{\frac{5}{2}}, \qquad (7.14)$$

where L is the width of the weir.

However, the actual flow is given by

$$Q_{\text{actual}} = C_D\frac{4L}{15H}\sqrt{2g}\ H^{\frac{5}{2}}, \qquad (7.15)$$

where C_D is the discharge coefficient that lies between 0.58 and 0.60. Equations (7.13) and (7.15) indicate that the flow rate of the liquid depends on the head (H). The weir head is measured by a float-activated displacement transducer which is placed on the upstream side. It is usually used as an indicating instrument.

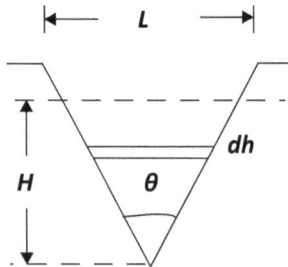

Figure 7.21. A v-notch weir.

7.9 The variable reluctance tachogenerator

The variable reluctance tachogenerator can be used for the measurement of linear and angular velocities. The magneto motive force (*mmf*) is the force that causes flux to be established; it is analogous to the electromotive force for electric circuits. The unit of the *mmf* is the ampere in SI units, but it refers to a coil of one turn.

The opposition to the establishment of magnetic flux is called reluctance. The reluctance is defined as

$$\Re = \frac{mmf}{\phi}.$$

Rearranging, $mmf = \Re \times \phi$, where *mmf* is in ampere turns and ϕ (the flux) is in weber.

7.9.1 The operational principle of the variable reluctance tachogenerator

If the time-varying flux ϕ is linked by a single turn of coil, then the back emf developed in the coil can be expressed as

$$e = -\frac{d\phi}{dt}.$$

A variable reluctance tachogenerator is shown in figure 7.22. It consists of a toothed wheel that is made of a ferromagnetic material and a coil wound as a permanent magnet.

The permanent magnet is extended by a soft iron pole piece. The teeth of the wheel move in close proximity to the pole piece. Therefore, the flux linked by the coil changes over time, and a voltage is developed across the coil. The total flux (ϕ_T) linked by a coil of *m* turns is given by

$$\phi_T = m\phi = m\frac{mmf}{\Re}$$

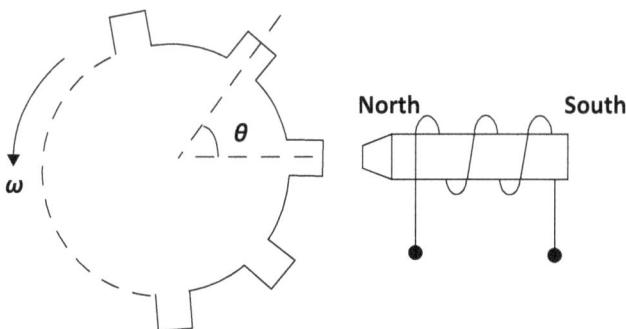

Figure 7.22. A variable reluctance tachogenerator.

Moreover, we know when the reluctance is at a minimum, the flux is at a maximum and vice versa. The variation of flux (ϕ_T) with the angular position (θ) of the wheel can be expressed as

$$\phi_T(\theta) = \alpha + \beta \cos(n\theta),$$

where α is the mean flux, β is the amplitude of the time-varying flux, and n is the number of teeth on the wheel. The induced emf is given by

$$e = -\frac{d\phi_T}{dt} = -\frac{d\phi_T}{d\theta}\frac{d\theta}{dt}$$

again, $\frac{d\phi_T}{d\theta} = -\beta n \sin(n\theta)$

and $\theta = \omega t, \ \therefore \frac{d\theta}{dt} = \omega,$

where ω is the rotational velocity of the wheel.
Therefore,

$$e = \beta n\omega \sin n\omega t.$$

7.10 The turbine flowmeter

The turbine flowmeter produces a direct electrical output. This flowmeter consists of a wheel with multiple blades that is installed inside a pipe in which clean fluid is flowing. A schematic of a turbine flowmeter is shown in figure 7.23.

The turbine meter must be installed along an axis parallel to the direction of fluid flow in the pipe. The flow of the fluid past the wheel causes it to rotate at a rate proportional to the volume flow rate. The wheel or blade of the meter is shown in figure 7.24.

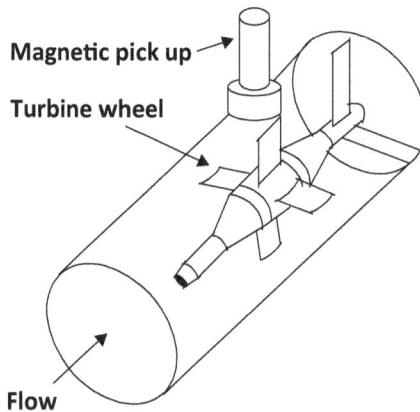

Figure 7.23. A schematic of a turbine flowmeter

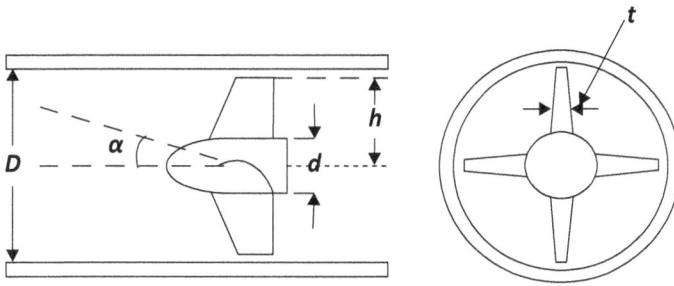

Figure 7.24. The blade assembly of a turbine flowmeter

7.10.1 The construction of the turbine flowmeter

The construction of a turbine flowmeter is similar to that of the variable reluctance tachogenerator. The blades or teeth are made of ferromagnetic material. The magnetic pickup consists of a coil wound on a permanent magnet. A voltage pulse is obtained at the output of the pickup whenever a turbine tooth/or blade passes the pickup coil. The flow rate can be determined by measuring the pulses using a pulse counter.

7.10.2 Theory

If the number of rotations of the blade per second is f and the volume flow rate is Q, then $f = kQ$, where k is the sensitivity of the flowmeter. Therefore, Q can be determined by measuring f. The total volume of liquid (V_T) that flows through the pipe over a given time T can be expressed as

$$V_T = \int_0^T Q dt.$$

The total number of rotations of the blade during the time T is expressed as

$$N_T = \int_0^T f dt = \int_0^T kQ dt = kV_T,$$

therefore, the total count is proportional to the total volume flow of liquid.

The voltage induced in the magnetic pickup is periodic in nature, and its frequency is proportional to the angular velocity of the blades.

Assuming the drag forces due to the bearings and viscous friction are negligible, the rotor's angular velocity (ω) is proportional to volume flow rate (Q), i.e.

$$\omega = k_1 Q,$$

where k_1 is a constant that depends on the geometry of the blade assembly.

The velocity of the fluid, u, is given by

$$u = \frac{Q}{A},$$

where A = cross-sectional area of the pipe minus the cross-sectional area of the hub minus the area of the blades:

$$= \frac{\pi D^2}{4} - \frac{\pi d^2}{4} - n\left(h - \frac{d}{2}\right)t,$$

where n is the number of blades and t is their average thickness; h and d are the dimensions of the blade system, as shown in figure 7.24.

Moreover, $\frac{\omega h}{u} = \tan\alpha$,

where ωh is the velocity of the blade tip perpendicular to the direction of flow of the fluid through the pipe and α is the inlet angle at the tip.

Again, $k_1 = \dfrac{\omega}{Q} = \dfrac{\tan\alpha}{Ah}$.

The output voltage produced by the magnetic pickup is given by

$$e = \beta n k_1 Q \sin(n k_1 Q)t, \tag{7.16}$$

where β is the amplitude of the angular variation of magnetic flux, n is the number of blades, and Q is the volume flow rate.

Equation (7.16) indicates that the output of the magnetic pickup is a sinusoidal signal of amplitude '$\beta n k_1 Q$' and frequency

$$'f = \left(\frac{n k_1 Q}{2\pi}\right)' = M_f Q.$$

Therefore, both the amplitude and frequency of the output signal vary with the flow rate. The flowmeter output signal is passed through an integrator and a Schmitt trigger. The output is a constant-amplitude square wave whose frequency is proportional to the flow rate.

Salient features
- Provided that the turbine wheel is mounted in low-friction bearings, the measurement accuracy can be as high as ±0.1%.
- It is less rugged and reliable than restriction-type differential pressure flowmeters.
- The blades and bearings can be damaged if solid particles are present in the liquid.
- It is more expensive and it also imposes a permanent pressure loss on the measurement system.
- The typical range is usually 10%–100% of the full-scale output reading. Below 10% of the full-scale reading, bearing friction makes the meter reading inaccurate.
- Turbine meters are particularly prone to large errors when there is any significant second phase in the fluid being measured.
- They have similar cost and market share to positive displacement meters. The former are smaller and lighter than the latter.
- They are preferred for low-viscosity, high flow measurements.

7.11 The electromagnetic flowmeter

The electromagnetic flowmeter is limited to measuring the volume flow rates of electrically conductive fluids (the conductivity of the fluid should be greater than $10\ \mu\text{mho.cm}^{-1}$). A schematic of an electromagnetic flowmeter is shown in figure 7.25. The instrument consists of a cylindrical tube made of stainless steel fitted with an insulating liner.

The typical lining materials are neoprene, polyterafluroethylene (PTFE) and polyurethane. A magnetic field that is perpendicular to the direction of the liquid flow is created in the tube by installing energized field coils on diametrically opposite sides of the pipe. The voltage induced in the fluid is measured by two electrodes inserted into the opposite sides of the tube. The ends of the electrodes are flush with the inner surface of the fluid-carrying pipe. The materials used to make the electrodes are stainless steel, platinum–iridium alloys, titanium, and tantalum. In accordance with Faraday's law of electromagnetic induction, the voltage e induced across the length L of the flowing fluid moving at a velocity of V in a magnetic field of flux density B is given by

$$e = BLV\ \text{Volt,}$$

where B (the flux density) is in Wb m^{-2} or Tesla.

L (the distance between the electrodes or the inner diameter of the pipe) is in m, and

V (the velocity of the fluid) is in m s^{-1}.

7.11.1 Advantages of the electromagnetic flowmeter

- There is no obstruction of the fluid flow. Therefore, no pressure loss is associated with the measurement.
- It is suitable for installations that can tolerate only a small pressure drop.
- The absence of any internal parts is very attractive when measuring the velocity of corrosive and dirty fluids.

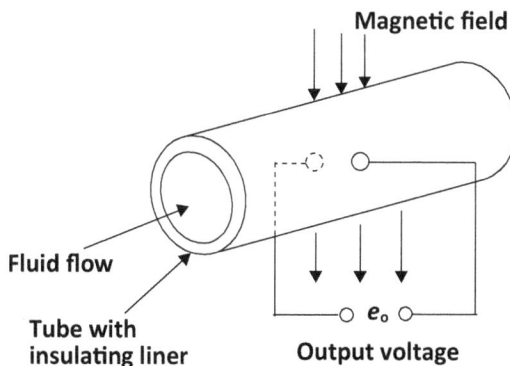

Figure 7.25. An electromagnetic flowmeter

- The operating principle is independent of fluid density and viscosity, and the output voltage is proportional to the average velocity of the fluid.
- There is no difficulty in measuring either laminar or turbulent flows.
- It has an accuracy of ±1% of the indicated flow.

The flux density (B) is related to the magnetic field intensity (H) by the relation

$$B = \mu H,$$

where μ is called the absolute permeability and is expressed in Henry m^{-1}.

The absolute permeability of another material can be expressed relative to the permeability of free space by

$$\mu = \mu_r \mu_0,$$

where μ_0 = the permeability of free space; its value is $4\pi \times 10^{-7}$ Henry m^{-1} and it is a constant.

μ_r = the dimensionless quantity called relative permeability; its value depends on the fluid material.

- Like other forms of flowmeter, the electromagnetic flowmeter requires a minimum length of straight pipework upstream from the meter in order to guarantee accuracy of measurement. No pipe bending or pipe fitting is usually allowed for five pipe diameters upstream.

7.11.2 Disadvantages of the electromagnetic flowmeter

- The electromagnetic flowmeter requires a minimum conductivity of 10 µmho. cm^{-1}. Therefore it is not suitable for the measurement of gaseous flows and liquid hydrocarbons.
- It is suitable for the measurement of flowing slurries provided that the liquid phase has adequate conductivity.
- The instrument is expensive in terms of initial purchase cost. One of the reasons for its high cost is the need for the careful individual calibration of each instrument during manufacture, as there are considerable variations in the properties of the magnetic materials used.

7.11.3 Field excitation

There are three possible ways of energizing the magnetic field coils, namely:
- DC,
- AC at 50 Hz, and
- pulsating DC.

The three common difficulties faced by DC excitation are:
- the polarization effect,
- the electrochemical effect, and
- the thermoelectric effect.

All these influence the DC output voltage.

⇒ These effects can be overcome by energizing the field coils with alternating current at 50 Hz.

⇒ For many years, AC excitation was the most common type of excitation. It signals can easily be amplified by a high-gain AC amplifier

The major disadvantage of AC energized field coils is that the output voltage is subject to a 50 Hz interference voltage generated by the transformer action of a loop consisting of the signal leads and the fluid path. The problems of DC and AC magnetic fields can be solved by energizing the coils with pulsating DC or interrupted DC, which is direct current that is pulsed at a fixed interval. The pulsed magnetic field and the induced or output voltage of the flowmeter are shown in figure 7.26.

The output voltage is

$$e = e_2 - \frac{(e_1 + e_3)}{2} - e_4 - \frac{(e_3 + e_5)}{2}.$$

Here, the DC field is switched in a square-wave fashion between the some value and zero at frequency of 3–6 Hz. When the field is zero, any nonzero output from the flowmeter is considered to be an error, called the zero error. By storing the zero error and subtracting it from the output obtained when the field is next applied, the error can be removed. (The principle is similar to the autozeroing of a digital multimeter, in which the input is momentarily grounded and the nonzero output voltage charges a capacitor. This capacitor voltage can be utilized later to correct the output for an actual reading.) In practice, the zero error correction is done several times a second.

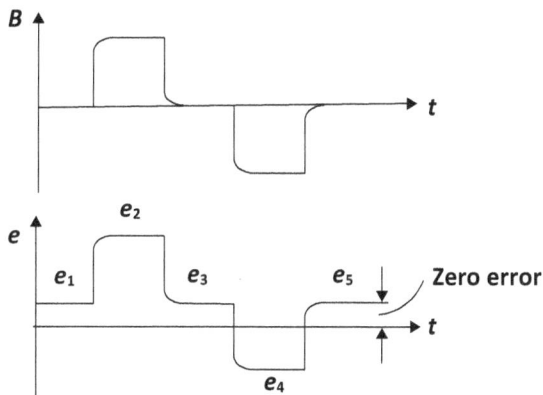

Figure 7.26. The magnetic flux produced by field coils and the output voltage of the electromagnetic flowmeter.

7.11.4 The use of an electromagnetic flowmeter with a metal pipe

To measure the flow of liquid that has high conductivity, a metal pipe is used without an insulating liner. For example, if the liquid is mercury, a stainless steel pipe will be used. Here, the conductivity of mercury is much higher than that of stainless steel. In this case, the electrode can be installed on the outside surface of the pipe.

7.12 The ultrasonic flowmeter

An ultrasonic signal is short burst of sine waves whose frequency is greater than the audible range of frequencies, which is 20 KHz. The typical frequency of ultrasonic waves is 10 MHz. The ultrasonic measurement of volume flow rate is a noninvasive method. There are two types of ultrasonic flowmeters, namely Doppler shift meters and transit time meters. Both methods depend on the transmission and reception of acoustic energy. Piezoelectric crystals are used for both functions. In a transmitter, electrical energy in the form of a short burst of high-frequency voltage is applied to the crystal, causing it to vibrate. The crystal is in contact with the fluid and the vibration is communicated to the fluid and propagated through it. The vibration reaches the receiving crystal which produces an electric signal as an output.

7.12.1 The Doppler shift ultrasonic flowmeter

7.12.1.1 The operational principle of of the Doppler shift ultrasonic flowmeter
A fundamental requirement of this instrument is the presence of scattering elements within the flowing fluid. Doppler meters will not work unless sufficient reflecting particles and/or air bubbles are present. This flowmeter usually employs a clamp-on configuration, as shown in figure 7.27.

The transmitter propagates an ultrasonic wave whose frequency ranges from 0.5 to 10 MHz in the fluid, which is flowing with a uniform velocity V. Particles or bubbles moving with the same velocity reflect some of the energy to the receiver. The reflecting elements cause a frequency shift between the transmitted and reflected

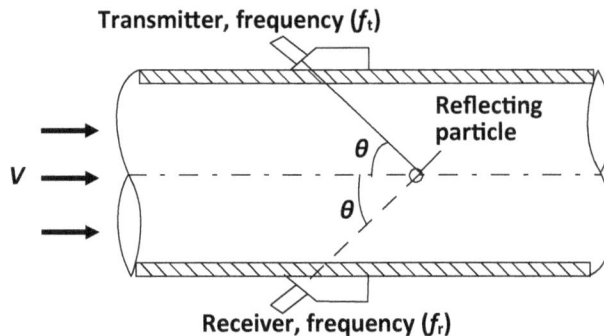

Figure 7.27. Doppler shift ultrasonic flowmeter

ultrasonic energy. The frequency is measured and it is calibrated in terms of the flow velocity.

The flow velocity (V) is given by

$$V = \frac{c(f_t - f_r)}{2f_t \cos \theta},$$

where f_t = the frequency of the transmitted ultrasonic wave,

f_r = the frequency of the received ultrasonic wave,

c = the velocity of sound in the fluid being measured, and

θ = the angle of incidence and that of the reflected ultrasonic wave relative to the axis of the flow in the pipe.

Both the transmitter and receiver are made using piezoelectric oscillator technology. The Doppler shift flowmeter is relatively inexpensive. Its measurement accuracy depends on the flow profile, the constancy of the pipe wall thickness, the number and size of the reflecting particles, and the accuracy with which the velocity of sound in the fluid is known. Accurate measurement can only be achieved by carefully calibrating the instrument for each particular flow measurement application. Recently, a Doppler shift ultrasonic flowmeter has been developed in which the transmitter and receiver are flush with the inner surface of the pipe. This avoids the problem of variable pipe thickness.

However, we must note following points while measuring the fluid velocity using a shift flowmeter with a clamp-on design:

- The dependence on c, the velocity of sound in the fluid, requires compensating changes in $\cos\theta$.
- For such a design, θ is the transducer wedge angle and c is the propagation velocity of the wave in the wedge material.

7.12.2 The transit time ultrasonic flowmeter

The transit time ultrasonic flowmeter is designed to measure the volume flow rate of a clean liquid. It consists of a pair of ultrasonic transducers mounted along an axis aligned at an angle θ with respect to the fluid flow axis. A transit time ultrasonic flowmeter is shown in figure 7.28. Each transducer consists of a transmitter–receiver pair. The transmitter emits ultrasonic energy which travels across to the receiver on the other side of the pipe. The fluid flowing through the pipe causes a difference between the transit time of the beam traveling upstream and the transit time of the beam traveling downstream; measuring the difference in the time of travel gives the flow velocity. Typically, the time difference is 100 ns for a total transit time of 100 µs.

7.12.3 Methods used to measure the time shift

There are three different ways to measure the time shift, namely:

- direct measurement,
- conversion to a phase change, and
- conversion to a frequency change.

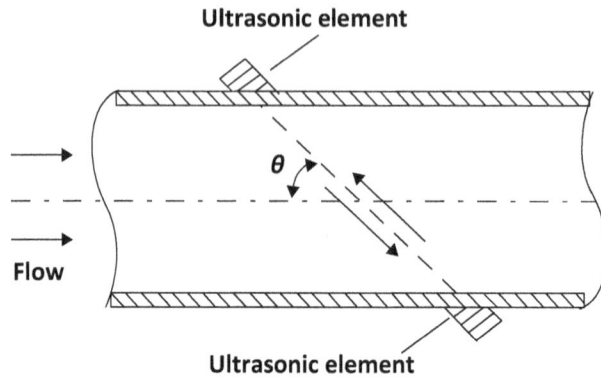

Figure 7.28. Transit time ultrasonic flowmeter.

The third method is attractive, since it does not need to measure the speed of sound in the measured fluid. This method also multiplexes the transmitting and receiving functions so that the same transducer can be used both as transmitter and receiver.

The forward and reverse transit times across the pipe t_f and t_r are given by

$$t_f = \frac{l}{c + v \cos \theta}, \quad t_r = \frac{l}{c - v \cos \theta},$$

where c = the velocity of sound in the fluid,
v = the velocity of the fluid,
l = the distance between the ultrasonic transmitter and receiver, and
θ = the angle between the ultrasonic beam and axis of the fluid flow.
The time difference δt is given by

$$\delta t = t_r - t_f = \frac{2vl \cos \theta}{c^2 - v^2 \cos^2 \theta}.$$

However, in practice, the receipt of a pulse is used to trigger the transmission of next ultrasonic energy pulses. Thus, the frequencies of the forward and return pulse trains are given by

$$F_f = \frac{1}{t_f} = \frac{c + v\cos\theta}{l}$$

$$F_r = \frac{1}{t_r} = \frac{c - v \cos \theta}{l}.$$

If the two frequency signals are now multiplied together, the resulting beat frequency is given by

$$\delta f = F_f - F_r = \frac{2v\cos\theta}{l}$$

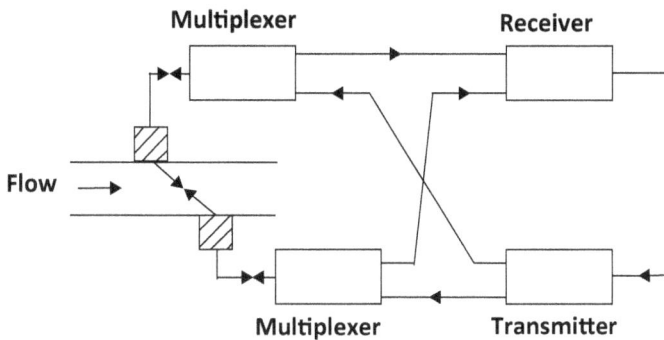

Figure 7.29. The transit time measurement system.

$$\therefore V = \frac{l\delta f}{2\cos\theta}.$$

The two frequencies F_f and F_r are mixed in such a way that the received signal contains the beat frequencies that are their sum and difference frequency. A fast Fourier transform (FFT) analysis of this output will lead us to the peak corresponding to the difference frequency, which is a measure of the flow velocity.

The transit time measurement scheme is shown in figure 7.29. If the pipe diameter is large, the transit time flowmeter is preferred over the Doppler shift flowmeter because the transit time is sufficiently large to be measured with reasonable accuracy, namely ±0.5%. The transit time flowmeter costs more than a Doppler shift flowmeter because of greater complexity of the signal conditioning circuit needed to make accurate transit time measurements.

7.12.4 The advantages of the ultrasonic flowmeter

- The fluid flowing through the pipe does not necessarily have to be a conductive fluid.
- It is particularly useful for measuring the flows of corrosive fluids and slurries.
- A further advantage over other flowmeters is that the instrument is externally clamped to the existing pipe rather than being inserted as an integral part of the flow line.
- Unlike other flowmeters, the ultrasonic flowmeter can be installed without cutting or breaking the pipe. Therefore, it has an enormous cost advantage.
- Its clamp-on mode of operation has a significant safety advantage, as it avoids the possibility that personnel installing flowmeters may come into contact with hazardous materials such as toxic, radioactive, or flammable fluids.
- Any contamination of the fluid being measured can easily be avoided.
- Ultrasonic flowmeters have mainly been used for liquids but are now being applied for gases and streams.

7.12.5 The disadvantages of the ultrasonic flowmeter

- The ultrasonic flowmeter is sensitive to the velocity profile of the flow.
- Ultrasonic flowmeters give different readings for axisystemetic profiles of different shapes but with identical average velocities.
- If we assume the meter coefficient to be one for a uniform profile, this drops to 0.75 for a laminar flow and it varies between 0.93 and 0.96 for turbulent flows.

7.13 The hot-wire anemometer

The principle of operation of the hot-wire anemometer is as follows:

The hot-wire anemometer is based on the principle of heat transfer by convection between a resistive wire and a flowing fluid. The wire is kept at a temperature much higher than the surrounding temperature, hence the name 'hot wire.' It is capable of measuring the average velocity and the turbulent flow of the fluid.

7.13.1 The hot-wire element

The hot-wire filament is usually a fine wire of platinum or tungsten.

Its typical dimensions are:
- diameter 5–300 µm
- length 1–10 mm

Care needs to be taken so that the loss of heat energy due to conduction or radiation is minimized. The element is thus jacketed and placed inside the pipe with the help of a fork-type support, as shown in figure 7.30.

At equilibrium, the heat balance equation can be written as

$$I_0^2 R_{W0} = U(v_0)A(T_{W0} - T_F) \tag{7.17}$$

(i.e. the heat generated in the wire = the convective loss of heat from its surface),

where

I = the current flowing through the resistance wire (A),

R_{W0} = the resistance of the wire (ohms),

A = the heat transfer area (m^2),

T_{W0} = the temperature of the wire (°C),

T_F = the temperature of the fluid (°C), and

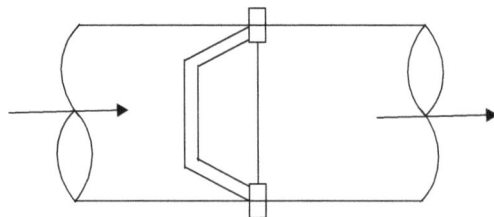

Figure 7.30. A hot-wire anemometer.

$U(v_0)$ = the convective heat transfer coefficient (W m^{-2} °C^{-1})

$$= a + b\sqrt{v}.$$

Here, $a, b = f(d, k, \rho, \eta)$, where
d = the sensor dimensions (m),
k = the thermal conductivity of the fluid (W m^{-1} °C^{-1}),
ρ = the fluid density (kg m^{-3}),
η = the fluid viscosity (Pa s).

7.13.2 Dynamic characteristics

Taking into account the dynamics of the system, we can write:
the rate of heat input − the rate of heat dissipation = the rate of rise of heat energy.
We can write:

$$I^2 R_W - U(v)A[T_W - T_F] = MC\frac{dT_W}{dt}, \tag{7.18}$$

where M = the mass of the heating element and
C = specific heat.
At equilibrium or in the steady state, equation (7.17) applies.
For an incremental change, we can write
$I \rightarrow I_0 + \Delta I$
$T_W \rightarrow T_{W0} + \Delta T$
$R_W \rightarrow R_{W0} + \Delta R_W$
$U(v) \rightarrow U(v_0) + \sigma \Delta v,$
where

$$\sigma = \left.\frac{\partial U}{\partial v}\right|_{\substack{v=v_0 \\ T_W=T_{W0}}}.$$

Substituting the new expressions for I, T_W, R_W and $U(v)$ into equation (7.18), we get

$$(I_0 + \Delta I)^2(R_{W0} + \Delta R_W) - \{U(v_0) + \sigma \Delta V\}A(T_{W0} + \Delta T - T_F) = MC\frac{d}{dt}(T_{W0} + \Delta T).$$

Simplifying, we get

$$[2I_0\Delta I R_{W0} + I_0^2\Delta R_W] - U(v_0)A\Delta T - \sigma \Delta v A[T_{W0} - T_F] = MC\frac{d}{dt}(\Delta T).$$

Putting

$$\Delta T = \frac{\Delta R_W}{K_T},$$

we have

$$\Delta R_T = \frac{K_I}{1 + s\tau_v}\Delta I - \frac{K_v}{1 + s\tau_v}\Delta v,$$

7-30

where

$$\tau_v = \frac{MC}{[U(v_0)A - I_0^2 K_T]}$$

$$K_v = \frac{K_T \sigma A [T_{W0} - T_F]}{[U(v_0)A - I_0^2 K_T]}$$

$$K_I = \frac{2K_T I_0 R_{W0}}{[U(v_0)A - I_0^2 K_T]}$$

A block diagram representation of a hot wire anemometer system is given in figure 7.31.

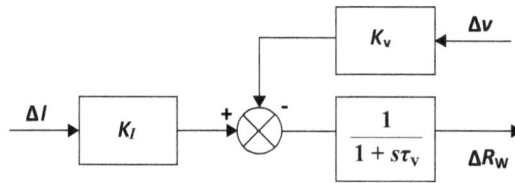

Figure 7.31. Block diagram representation of a hot wire anemometer system.

7.13.3 Constant current and constant temperature hot-wire anemometers

There are two types of arrangements for hot-wire anemometers:
- the constant current type
- the constant temperature type (or constant R_W type).

A constant current hot-wire anemometer with a measuring bridge circuit is shown in figure 7.32.

In the constant current hot-wire anemometer, the current is kept fixed at a certain value. The resistances R_1, R_2, and R_3 are of the same order as R_W. The value of R_S is high. This type of hot-wire anemometer is generally used to measure steady flows or small fluctuations in velocity; it is not good for turbulent flow measurements. A constant temperature hot wire anemometer is shown in figure 7.33. This keeps the resistance R_W constant by incorporating feedback.

As the velocity increases, R_W decreases, thereby creating an unbalanced voltage. The increase in the current brings the resistance back to its initial value. The feedback increases bandwidth and therefore this device is also suitable for turbulent flow measurements.

A block diagram representation of the constant temperature type is shown in figure 7.34.

Analysis of the block diagram of figure 7.34 gives

$$\frac{\Delta e_0}{\Delta v} = \frac{K_v K_B K_A R_0}{\tau_v s + (1 + K_B K_A K_I)} = \frac{K'}{\tau_v' s + 1}$$

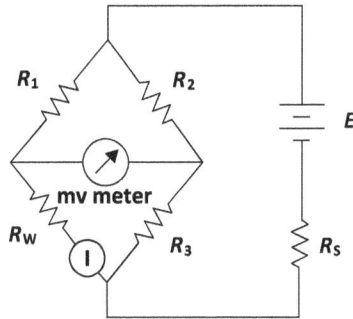

Figure 7.32. Circuit diagram for a constant current hot wire anemometer.

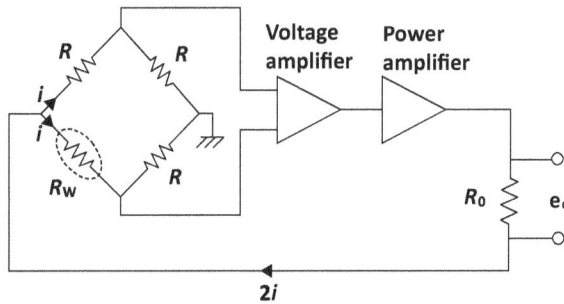

Figure 7.33. Circuit diagram for a constant temperature hot-wire anemometer.

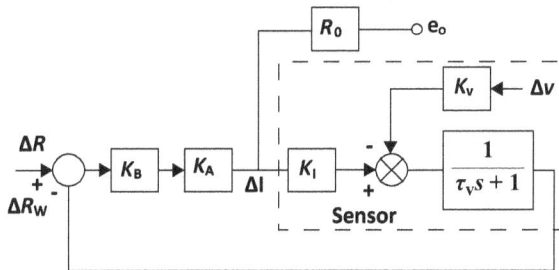

Figure 7.34. A block diagram representation of the constant temperature type.

where,

$$K' = \frac{K_{\mathrm{v}}K_{\mathrm{B}}K_{\mathrm{A}}R_0}{1 + K_{\mathrm{B}}K_{\mathrm{A}}K_{\mathrm{I}}}$$

$$\tau'_v = \frac{\tau_v}{1 + K_{\mathrm{B}}K_{\mathrm{A}}K_{\mathrm{I}}}$$

Thus, we see that the time constant of the system reduces and the system becomes faster.

7.13.4 A comparison of constant current and constant temperature measurements

The disadvantages of constant current hot wire anemometer are as follows:
- The current has to be kept large; a sudden drop in fluid velocity may burn out the wire.
- For dynamic measurements, separate compensating networks are required.

The constant temperature type removes the above problems by introducing a feedback loop. However, it has other problems, namely instability and drift. For small velocity changes, noise has to be taken care of separately; otherwise, the measurements are erroneous.

Problem 7.1. Calculate the flow rate of water flowing through a pipe of diameter 0.1 m using an orifice with a hole diameter of 0.05 m. The differential head read over a mercury column manometer for vena contracta taps is 0.25 m. The flowing water temperature is 35 °C. The manometer temperature is 24 °C. The standard temperature at which the data are derived is 15 °C.

The following data are given:
- $\rho_{35\ °C} = 997$ kg m$^{-3.}$
- $\rho_{24\ °C} = 994$ kg m$^{-3.}$
- $\rho_{15\ °C} = 999.8$ kg m$^{-3.}$
- $\mu_{35\ °C} = 0.06$ poise. (1 poise $= 0.1$ N s m^{-2}).

Problem 7.2. Hydrogen flows in a pipe of diameter 0.05 m with a mass flow rate of 9.6×10^4 kg m^{-2} and a temperature of 20 °C. If, at the given temperature and pressure, the flow rate is 0.493 m^3 min^{-1} for a manometer height of 55 cm of mercury, calculate the orifice diameter.
- The specific heat $\gamma = 1.4$.
- $\mu = 0.003$ poise.
- $R = 28.81$ m/°C.
- The rational expansion factor $Y = 1-[0.41 + 0.35\ \beta^4]\ (\Delta p/\gamma p_1)$.
- $\beta =$ the pipe diameter ratio.

Problem 7.3. A Venturi meter is to be used to measure the flow rate of water in a pipe of diameter D $= 0.20$ m. The maximum flow rate is 2136 m^3 min^{-1}. Venturis with throat diameters of 0.10, 0.12, and 0.14 m are available.
- (a) Choose the most suitable Venturi meter, assuming the differential pressure at maximum flow is 918 kg m^{-2}.
- (b) Calculate the accurate value of the differential pressure developed across the chosen Venturi at the maximum flow rate.

C_d (the discharge coefficient) $= 0.990-0.02\ (\frac{d}{D})^4$.

Problem 7.4. A Venturi tube is to be used to measure a maximum flow rate of water of 3.7859 kg s^{-1} or 0.003 7859 m^3 s^{-1}. The Reynolds number of the flow at the throat is to be at least 10^5. Determine the size of the Venturi and the maximum range of the differential pressure.

The following data are given:

- $\rho_{water} = 1000$ kg m^{-3}.
- $\mu = 0.0116$ poise $= 0.0116 \times 0.1 = 0.001\ 16$ Pa s.

See appendix I for additional data.

Problem 7.5. A measurement of the velocity profile of a pipe is needed prior to the installation of a permanent flowmeter. The mean velocity of a high-pressure incompressible fluid through the pipe was measured using a Pitot tube. At the maximum flow rate, the mean differential pressure was found to be 310 Pa.

(a) Calculate the mean velocity of the gas at maximum flow rate.
(b) Calculate the maximum mass flow rate.
(c) Calculate the Reynolds number at the maximum flow rate.

The following data are given:

- The diameter of the pipe = 0.18 m.
- The density of the fluid = 6 kg m^{-3}.
- The viscosity of the gas = 5.0×10^{-3} Pa s.

Problem 7.6. A turbine flowmeter consisting of a rotor with six blades is suspended in a fluid stream, and the rotational axis of the rotor is parallel to the direction of flow. The blades rotate at an angular velocity of ω rad s^{-1}, where $\omega = 4.2 \times 10^4\ Q$.

If Q m^3 s^{-1} is the volume flow rate of the fluid, the total flux ϕ_T linked by the coil of the magnetic transducer is given by

$$\phi_T = 4.637 + 0.92 \cos(6\theta) \text{ milliwebers.}$$

Here, θ is the angle between the blade assembly and the transducer. The flowmeter can measure flow rates ranging between 0.22×10^{-3} m^3 s^{-1} and 3.74×10^{-3} m^3 s^{-1}.

Find the amplitude and frequency of the transducer output signals at the minimum and maximum flow rates.

Problem 7.7. A Doppler shift ultrasonic flowmeter is shown in figure 7.35 with its signal conditioning circuit. Two piezoelectric crystals, each of which has a natural frequency of 2.4 MHz, are used as the transmitter and receiver. The transmitting crystal directs an ultrasonic wave into the pipe which makes an angle of 45° with the flow. Calculate the received frequency as well as the cutoff frequency of the low-pass

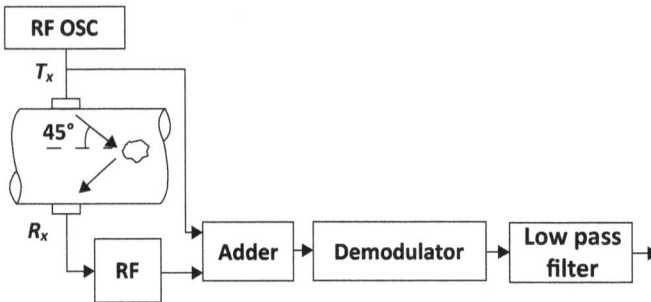

Figure 7.35. The Doppler shift ultrasonic flowmeter.

filter. The velocity of the flowing fluid = 10 m s^{-1} and the velocity of sound in the fluid = 10^3 m s^{-1}.

Problem 7.8. A transit time ultrasonic flowmeter was used to measure velocity of a fluid flowing in a pipe, and it was found that the zero flow transit time was 1.2 ms; however, when there was flow, the differential transit time was 115 μs. The angle between the line connecting the transmitter to the receiver and the direction of flow of the fluid was 30°. Find the velocity of the fluid. By how much does the transit time change for a ±2% change in the velocity of sound? The velocity of sound in the gas is 500 m s^{-1}.

Problem 7.9. In an experiment, an electromagnetic flowmeter was used to measure the average flow rate of a liquid flowing in a cylindrical pipe. The flux density of the magnetic field applied had a peak value of 1.4 Tesla (Wb m^{-2}). The output from the electrodes was fed to an amplifier with a gain of ten and an input impedance of 2.2 MΩ. The internal resistance developed due to the fluid between the electrodes was found to be 200 kΩ.
    (a) Determine the velocity of the liquid when the peak-to-peak output voltage of the amplifier was found to be 4 V.
    (b) Find the percentage change in the reading of the amplifier for a 10% increase in the conductivity of the flowing fluid.

The diameter of the pipe = 0.1 m.

Problem 7.10. The operational principle of the vortex flowmeter is based on the natural phenomenon of vortex shedding. If a circular cylinder of diameter d meters is installed as a bluff body in a pipe of diameter D meters, then the frequency f (in Hz) of vortex shedding is given by

$$\frac{f}{Q} = \frac{4S}{\pi D^3} \frac{1}{\frac{d}{D}\left[1 - 1.4\frac{d}{D}\right]},$$

where Q = the volume flow rate of the fluid (m^3 s^{-1}).

S = the Strouhal number (0.2 in this case).

The Strouhal number is a dimensionless quantity and it is relatively constant ($0.20 \leqslant S \leqslant 0.21$) over the range of Reynolds numbers from 300 to 150 000.

Calculate the correct cylinder diameter for a 0.15 m pipe carrying water flows of between 0.1 and 1.32 m^3 s^{-1}. Find the maximum vortex shedding frequency.

Problem 7.11. In problem 7.10, $L = 0.15$ m and $V = 0.5$ m s^{-1}. The fluctuation in frequency is up to 110 Hz. State whether the above flowmeter, as shown in figure 7.36, is suitable for this application.

Problem 7.12. A schematic of a mass flowmeter or Coriolis flowmeter is shown in figure 7.37.

Assume we have a U-tube ABCD of length 1.2 m rotating with an angular velocity of 4π rad s^{-1}. Water flows in at a velocity of 1.5 m s^{-1}. Calculate the force exerted on each of the limbs AB and CD. The diameter of the U-tube is 0.15 m.

Show that the torque developed is directly proportional to the mass flow rate.

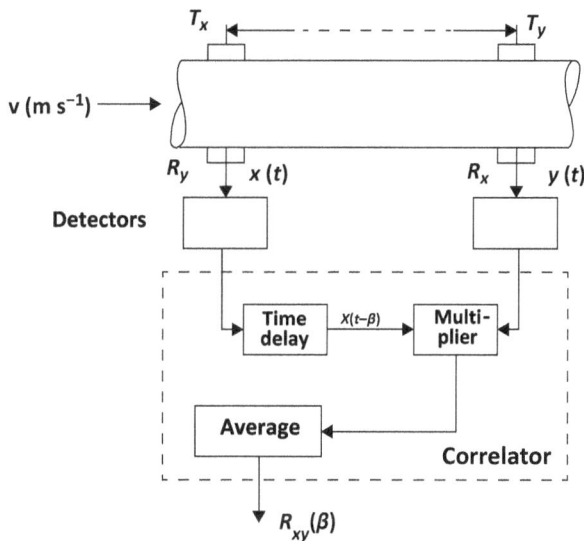

Figure 7.36.

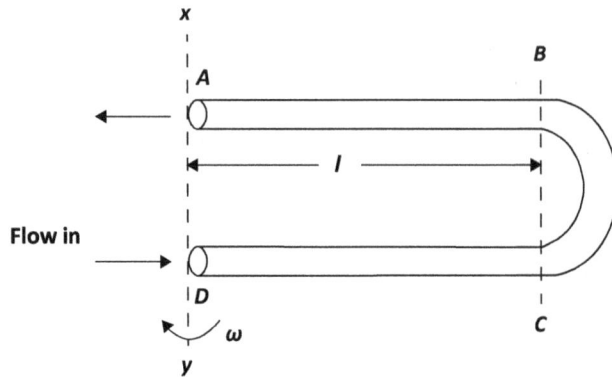

Figure 7.37.

Problem 7.13. An electromagnetic flowmeter is used to measure the average flow rate of a liquid in a pipe of diameter 50 mm. The flux density in the liquid has a peak value of 0.1 T. The output from the flowmeter is fed to an amplifier that has a gain of 1000 and an input impedance of 1 MΩ. If the impedance of the liquid between the electrodes is 200 kΩ, then what is the peak value of the amplifier output voltage for an average flow velocity of 20 mm s^{-1}?

Problem 7.14. A tungsten constant current hot-wire anemometer has the following specifications:
- its resistance at 0 °C is 10 Ω,
- the surface area of the wire is 10^{-4} m^2,
- the linear temperature coefficient of resistance of the wire is 4.8×10^{-3} °C^{-1},
- the convective heat transfer coefficient is 25.2 W m^{-2} °C^{-1},
- the flowing air temperature is 30 °C,
- the hot-wire current is 100 mA, and
- the mass specific heat product is 2.5×10^{-5} J °C^{-1}.

(a) What is the resistance of the wire under steady-state flow conditions?
(b) Calculate the value of the thermal time constant of the hot wire under steady-state flow conditions.

IOP Publishing

Fundamentals of Industrial Instrumentation (Second Edition)

Alok Barua

Chapter 8

The flapper nozzle system

Learning objectives:
- The operational principle of the flapper nozzle.
- Displacement measurement.
- The reverse-acting pneumatic relay.
- The current-to-pressure (I–P) converter.
- The differential pressure transmitter.

8.1 Introduction

The flapper nozzle is one of the most important and versatile devices. It finds its application as a displacement measuring device, in differential pressure transmission, and as an I–P converter. The basic principle, which is the movement of a flapper that depends on the air pressure produced by a nozzle, remains the same in each application.

8.2 The application of the flapper nozzle as a displacement measuring device

The flapper nozzle system is used for the measurement of small displacements. A small movement of the flapper can result in a large change in the output pressure, and hence it is treated as a displacement-to-pressure converter. Such transducers are designed for use in both hydraulic and pneumatic power supplies. The pneumatic systems are operated within an air pressure range of 100–200 KN m^{-2}. A flapper nozzle system is shown in figure 8.1.

The output pressure of the nozzle's rear chamber is measured by a pressure sensor. The flapper nozzle system shown in figure 8.1 consists of a chamber of small volume connected to a constant-pressure source on one side and vented to the atmosphere through a nozzle on the other side. The flapper held in front of the nozzle is used to regulate the rate of air flow bleeding out through the nozzle into the atmosphere. When the flapper is held tightly against the nozzle, no air leaks

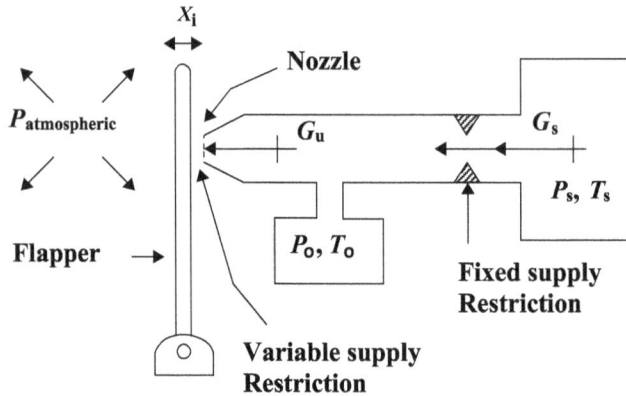

Figure 8.1. A flapper nozzle system. $P_{atmospheric}$ = atmospheric pressure; x_i = the displacement of the flapper; G_u = the outlet mass flow; G_s = the inlet mass flow; P_s, T_s = the pressure of the supply air and its temperature, respectively; P_o = the pressure output.

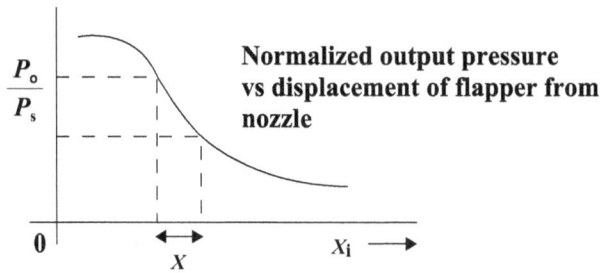

Figure 8.2. Flapper nozzle characteristics.

out and ultimately the output pressure (P_o) reaches the value of the supply pressure (P_s). When the flapper is positioned too far away, it has no effect or control on the output flow rate and the flow rate is governed by the system's size and configuration. As the system is designed to measure displacements, the supply-side restrictor and nozzle are so designed that the output pressure never falls to the level of the atmospheric pressure, even with the flapper positioned far away.

The flapper nozzle characteristics are shown in figure 8.2. The characteristics show that a small change in displacement produces a large output change. Parameters estimated using the experimental curves are shown in figures 8.3–8.5.

In the neighborhood of the operating point, we can take a linear approximation:

$$G_s = G_s(p_o) \cong G_{s,o} + \left. \frac{dG_s}{dp_o} \right|_{p_o = p_{o,o}} (p_o - p_{o,o})$$

$$= G_{s,o} + K_{sf} \, p_{o,p},$$

Supply side restrictor

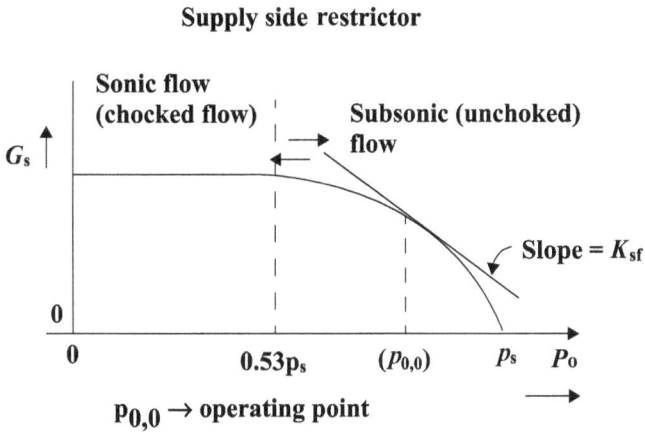

Figure 8.3. Parameters estimated using experimental curves for the supply-side restrictor.

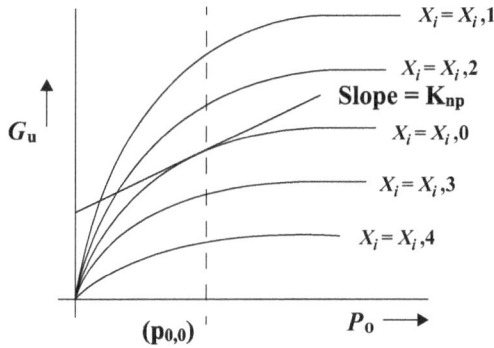

Figure 8.4. Parameters estimated using experimental curves for the load-side restrictor.

where

$G_{s,o}$ = the value of G_s at the equilibrium operating point,

$p_{0,0}$ = the value of p_o at the equilibrium operating point,

$P_{o,p}$ = a small change in p_o from $p_{0,0}$, and

K_{sf} = the value of $\frac{dG_s}{dp_o}$ at the operating point.

Also,

$$G_{u,s} = G_u(p_0,x_i) = G_{u,0} + K_{np}p_{0,p} + K_{nx}x_{i,p}.$$

where $G_{u,s}$ = the value of G_u at the equilibrium operating point and $x_{i,p}$ = a small departure from the equilibrium operating point ($x_{i,0}$).

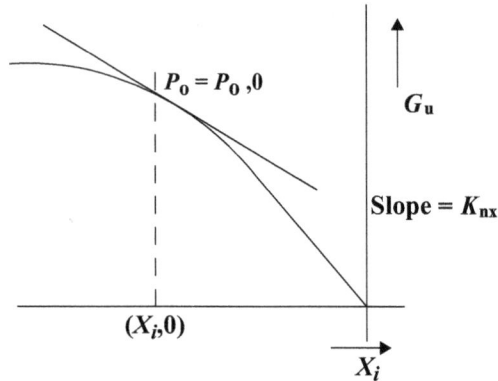

Figure 8.5. An analytical method for finding K_{nx} and K_{np}.

Therefore,

$$G_u = G_{u,0} + \left.\frac{\partial G_u}{\partial p_0}\right|_{\substack{x_{i,0} \\ p_{0,0}}} (p_0 - p_{0,0}) + \left.\frac{\partial G_u}{\partial x_i}\right|_{\substack{x_{i,0} \\ p_{0,0}}} (x_i - x_{i,0})$$

We know that

mass in − mass out = additional mass stored.

For the perfect gas law, we know that $p_0 V = M_p R T_0$. Therefore,

$$\left[G_{s,0} + K_{sf} p_{0,p} \right] dt - \left[G_{u,0} + K_{np} p_{0,p} + K_{nx} x_{i,p} \right] dt$$
$$= dM_p = \frac{V}{RT_0} dp_0$$

Let us assume that $G_{s,0} = G_{u,0}$; then

$$\frac{V}{RT_0} \frac{dp_{0,p}}{dt} + (K_{np} - K_{sf}) p_{0,p} + K_{nx} x_{i,p} = 0$$

Let $K = \dfrac{-K_{nx}}{K_{np} - K_{sf}}$

$$\tau = \frac{V}{RT_0\left(K_{np} - K_{sf}\right)}$$

or $(\tau s + 1) p_{0,p} = K x_{i,p}$

or $\dfrac{p_{0,p}}{x_{i,p}} = \dfrac{K}{\tau s + 1}$

Thus we see that the flapper nozzle is a first-order system. It is also a reverse kind of transducer. It is a very fast-acting pneumatic amplifier with good resolution. It can be used in precision measurement.

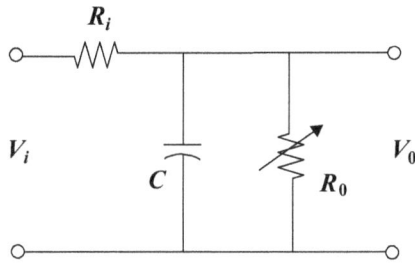

Figure 8.6. Electrical equivalent circuit for a flapper nozzle system. R_0 = the variable restriction; R_i = the fixed restriction; C = the capacity of the system (volume).

Construction:
- Nozzle diameter = 1/32 inch $\approx$ 0.8 mm.
- Volume = 1(inch)^3 $\approx$ 16 cc.
- Supply orifice diameter = 1/64 inch $\approx$ 0.4 mm (fixed-side restriction).
- Value of K = 8000 psi/inch.

The electrical equivalent circuit for a flapper nozzle system is shown in figure 8.6.

If the connecting tube between the supply pressure and the flapper nozzle chamber is a capillary tube, then the supply-side restrictor has a constant value. However, the load-side restrictor is variable if it is not connected to a capillary tube.

The transfer function of the electrical system is given by

$$\frac{V_0}{V_i}(s) = \frac{R_0}{R_i + R_0}\left(\frac{1}{1 + sC\frac{R_0 R_i}{R_0 + R_i}}\right)$$

If the supply pressure (V_i), the supply-side restriction (R_i) and the capacitance (C) of the flapper nozzle system are constant, V_0 is a function of R_0.

8.3 Static sensitivity

Let us redraw the flapper nozzle system in figure 8.7. For static sensitivity calculations, the flow through the restriction is assumed to be incompressible. The flow through the fixed restriction is given by

$$f_1 = C_{d_1}\frac{\pi d_1^2}{4}\sqrt{2(P_S - P_O)} \tag{8.1}$$

where C_{d_1} = the discharge coefficient of fixed restriction.

As the nozzle discharges air to the atmosphere, the flow through the nozzle is

$$f2 = C_{d_2}\pi d_2 \bar{x}\sqrt{2P_0}, \tag{8.2}$$

where C_{d_2} = the discharge coefficient of the nozzle and $\bar{x}$ = the distance from the flapper to the nozzle.

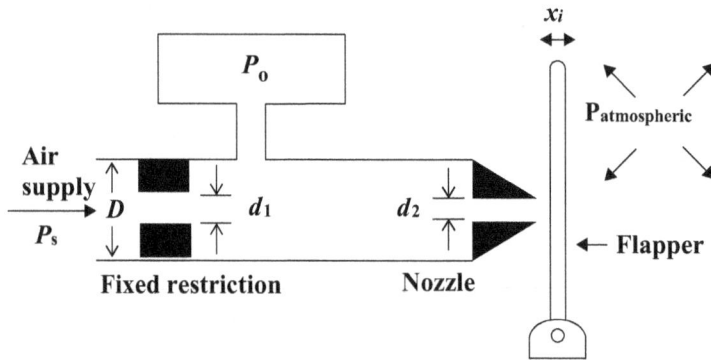

Figure 8.7. A schematic arrangement of the flapper nozzle system.

Assuming now the condition of flow continuity and $C_{d_1} = C_{d_2} = C_d$ using equations (8.1) and (8.2), we get

$$\frac{16\,d_2^2\,\bar{x}^2}{d_1^{\,4}}P_o = P_S - P_o$$

$$\frac{P_o}{P_S} = \frac{1}{1 + 16(d_2\,\bar{x}/d_1^2)^2}.$$

The assumptions made to arrive at equations (8.1) and (8.2) are:
  (i) The velocity of approach is neglected.
  (ii) For small x, the area of the nozzle outlet facing the atmosphere is taken to be $\pi d_2 x$.

The normalized response curve is shown in figure 8.8.

The response slope $(dP_n/d\bar{x}_n)$ is approximately linear for short P_n ranges. The maximum slope or sensitivity is obtained by equating

$$\frac{d^2 P_n}{d\bar{x}_n^{\,2}} = 0$$

such that $\bar{x}_n = 0.144$. This gives $\frac{dP_n}{d\bar{x}_n} = -2.59$ and $P_n = 0.75$.

Note: Negative sign indicates the slope is negative.

When $\bar{x}_n$ is sufficiently large, P_n becomes constant. To avoid this zero-sensitivity portion, the lower and upper limits or bounds are fixed at $P_n = 0.75$ and $P_n = 0.15$. For $P_S = 138$ KN m^{-2} (20 PSI), this causes P_o to lie in between 20.7 KN m^{-2} and 103 KN m^{-2} (3–15 PSI). The upper limit is mainly determined by the pipe thickness; the lower is not kept at zero in order to enable a positive check on the control pressure on the low side and also to avoid any leakage from outside into the system.

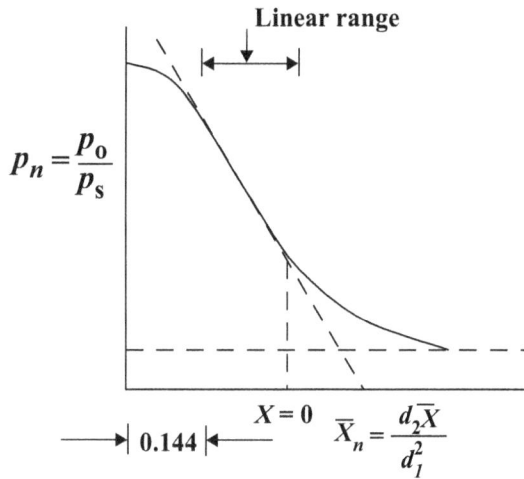

Figure 8.8. The normalized response curve.

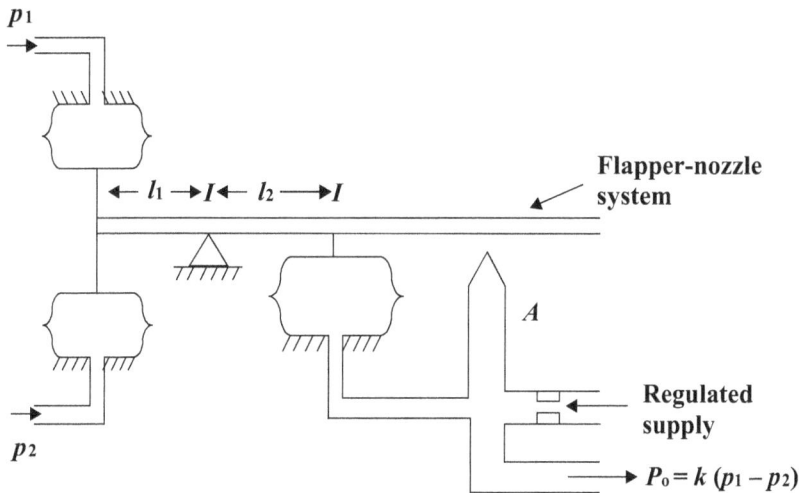

Figure 8.9. A force balance differential pressure transmitter.

8.4 The force balance differential pressure transmitter

The feedback principle is used to measure the differential pressure in a force balance system. Such a pressure transmitter is shown in figure 8.9, and its equivalent block representation is shown in figure 8.10. Here, the differential pressure is converted into single-ended pressure and can be transmitted over a certain distance.

In the schematic diagram of figure 8.9, if $p_1 > p_2$, then the upper bellows exerts more force than the lower one, the flapper moves away from the nozzle, and the

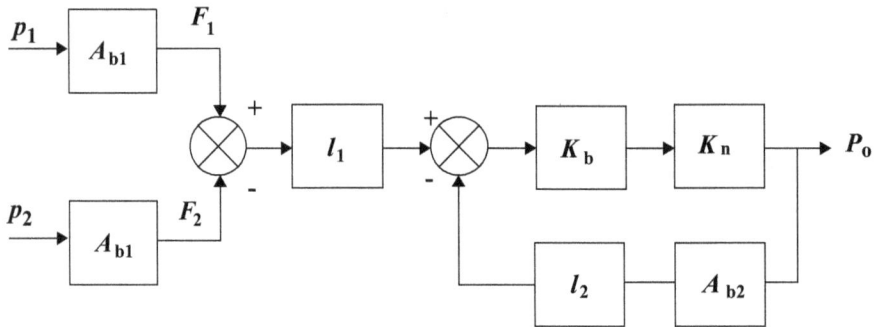

Figure 8.10. A block diagrammatic representation of figure 8.9. A_{b1}, A_{b2} = the areas of the bellows; K_b = bellows linkage compliance; K_n = flapper nozzle gain.

pressure inside tube A falls. The feedback bellow (bigger one) cannot hold the flapper in such a position and brings it down.

The output pressure is given by

$$P_o = \frac{A_{b1}\, l_1}{A_{b2}\, l_2}\,(p_1 - p_2).$$

8.5 A flapper nozzle with an air relay

In order for the flapper nozzle to work satisfactorily, the pressure changes must be small. This is accomplished using an air relay. A control pressure of 23–115 PSI is obtained from the relay for process pressure changes of only 20.7 KN m^{-2} and 103 KN m^{-2} (3 and 15 PSI), and this involves only one sixth of the original movement of the flapper. Here, the flapper position is directly proportional to the pressure change. In other words, the system is strictly operated in the linear region. The schematic arrangement of such a relay is shown in figure 8.11.

The air relay shown in figure 8.11 is reverse-acting type. When the flapper moves away (i.e. due to a larger process pressure) from the nozzle, the pressure p_i falls, the diaphragm comes down; thus, the ball 'd' of the ball valve closes path 'b' more, and the passage between 'a' and 'c' is opened further, giving a larger pressure indication in gauge g_2. The readings in gauges g_1 and g_2 are inversely related. As the output pressure P_O is derived from the supply pressure P_S, the highest reading is limited by this factor. The air relay gives an amplified output for a small change in pressure signal.

8.6 The current-to-pressure (*I*–*P*) transducer

Current-to-pressure transducers (*I*–*P*s) are used primarily in process control to change a 4–20 mA electronic signal produced by an electronic controller into a 20.7 KN m^{-2} to 103 KN m^{-2} (3–15 PSI) pneumatic signal. An effective current-to-pressure converter must incorporate the following features. In the flapper nozzle system, air bleeds through the nozzle quickly when the pressure decreases. It also

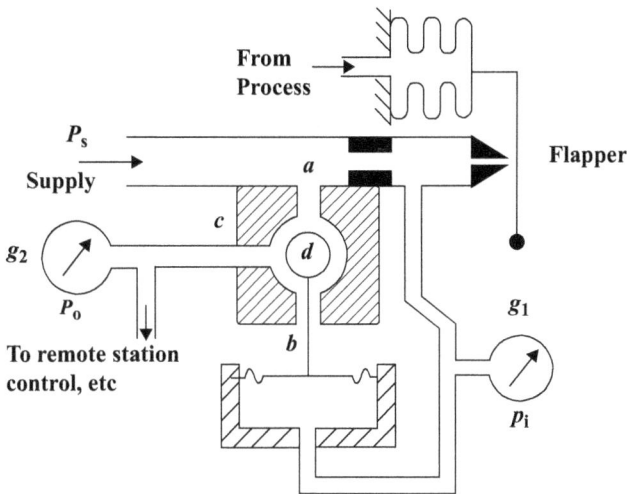

Figure 8.11. A reverse-acting pneumatic relay.

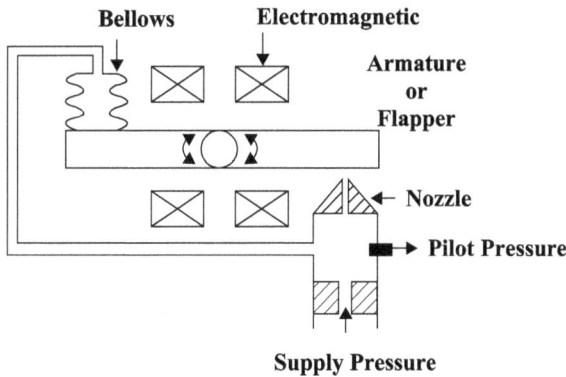

Figure 8.12. The traditional flapper nozzle current-to-pressure converter.

consumes little air from the system. The flapper system is rugged enough to withstand vibration, temperature extremes, and corrosive conditions. All these features make it attractive for the construction of current-to-pressure converters.

The traditional flapper-nozzle-based current-to-pressure converter is shown in figure 8.12.

The input current (4–20 mA) is applied to a coil–armature arrangement that acts on a beam. The beam ('flapper') positions itself against a nozzle that has air flowing through it. The gap between the flapper and the nozzle determines the back pressure (also called pilot pressure) that builds up in the nozzle. A bellows is sometimes connected to the nozzle area to balance the forces on the armature–flapper. The pilot pressure for linear operations is channeled to a pneumatic relay or booster. This booster or relay acts as an amplifier, translating the low pilot pressure into a higher output pressure and capacity. Mechanical *I–P*s that use a flapper nozzle and which do not use electronic feedback sensors face some difficulty in dealing with environmental factors. The flapper is susceptible to vibration and has traditionally forced users to mount the *I–P* separately on a pipe or rack. This requires additional tubing to carry the *I–P* output signal to the valve. This additional cost nullifies the benefits of mounting the *I–P*s together in a common location. The dead time and lag time introduced into the loop by the longer output signal tubing have significant impacts on loop performance. Traditional *I–P*s are also adversely affected by fluctuations in the air supply, downstream tubing leaks, temperature changes, and aging of the magnetic coil within the *I–P*. Periodic calibration checks are required in order to maintain the output of the *I–P* within the desired range. Dirty supply air is a major cause of *I–P* downtime. Mechanical *I–P*s do not have electronic feedback to compensate for partial plugging, but by designing the nozzle opening to be at least 0.04 cm (0.015 inch) in diameter, this problem can be avoided.

8.6.1 New *I–P* converters

Several new concepts have been introduced for current-to-pressure converters. These new concepts have changed the nature of the pilot stage and have incorporated sensor-based electronics. They include some trade-offs, such as trade-offs between efficiency and cost or the improvement of one factor at the expense of of another. One such important device is the piezo-ceramic bender nozzle.

8.6.2 The piezo-ceramic bender nozzle

The piezo-ceramic bender nozzle is shown in figure 8.13. This unit does not use a coil to move the flapper but instead the flapper itself is made of layers of different materials which are laminated together. These different materials flex or bend when a voltage is applied across them. The 4–20 mA input signal to the *I–P* is first

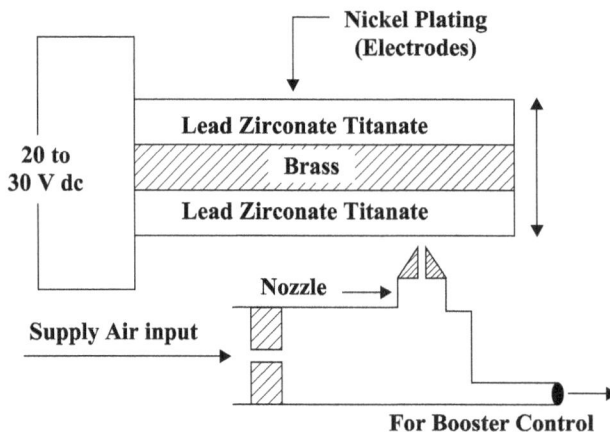

Figure 8.13. The piezo-ceramic bender nozzle.

converted to a voltage in the range of 20–30 V DC. This design tends to be more stable when exposed to vibration than the typical flapper armature.

Some of the drawbacks of this new device are:

(i) The bender does not have very good repeatability and tends to locate itself in different positions for the same input signal. This creep is cumulative and eventually exceeds the adjustment range of the calibration mechanism. An electronic feedback sensor can be combined with the piezo-ceramic bender to temporarily compensate for the creep, but the feedback circuit typically uses much of the power available from the input signal, leaving little to energize the bender.

(ii) The bender cannot balance against the force of the nozzle air unless the nozzle is kept relatively small. Thus, large nozzles must be traded for improved bender control.

(iii) The plugging of small nozzles is the chief reason for I–P field failure.

Chapter 9

Signal conditioning circuits

Learning objectives:
- The design equations for active filters.
- The principle of inductance simulation.
- An metal–oxide–semiconductor (MOS) implementation of a sample and hold circuit.
- Complementary metal-oxide–semiconductor (CMOS) transmission gates.
- A typical application of multiplexing and demultiplexing.

9.1 Active filters

Active filters consist of only operational amplifiers, resistors, and capacitors. Complex roots are achieved by the use of feedback, thereby dispensing with inductors. Many signal conditioning applications require filters of order greater than two. One of the ways to realize higher-order filters is to cascade second-order filters. If the order of the filter is odd, a first order section will be necessary. All modern signal conditioning systems include various types of continuous time filters that the designer has to realize using an appropriate technology. The designer can implement these using active RC, metal–oxide–semiconductor field-effect transistor (MOSFET)-C, switched capacitor, or switched current techniques. The design method used for the MOSFET-C filter most closely resembles the design of an active RC filter. It uses only MOS capacitors, transistors, and op-amps. The only difference is that it replaces the resistors used in active RC circuits by a MOSFET biased in the triode region. Usually, the filters have to share an integrated circuit with other switched or digital systems, so that the ground lines are likely to contain switching transients and are generally noisy. Measuring analog signals relative to ground yields a poor signal-to-noise ratio and low power supply noise rejection. This problem can be remedied by using an op-amp in a fully differential and balanced form. Moreover, this filter is electronically tunable. In the mid 1980s, a monolithic filter with a reduced silicon area was needed. A practical alternative is the switched capacitor (SC) filter. In this technique, a resistor is replaced by a capacitor and four switches that need nonoverlapping clocks. SC filters are characterized by difference equations, in contrast to the differential equations used for continuous time filters. SC filters work in a discrete time domain; therefore, their circuits are analyzed using a z-transform, whereas a Laplace transform is used for continuous time filters. SC circuits became popular for two main reasons: they reduced the silicon area, and the time constant of the integrator can be precisely controlled using capacitor ratios only. Switched current (SI) circuits are similar to switched capacitor circuits except that they process current instead of voltage. As the SI technique offers advantages such as the ability to operate at high frequencies, a wide dynamic range, and a low power supply voltage requirement, they find applications in the implementation of analog filters. SI circuits do not use op-amps, and therefore their high-frequency response is not limited by the gain–bandwidth product of the op-amp. Moreover, SI filters can be implemented using standard digital technology, therefore making them suitable for mixed signal circuit designs. However, in this chapter, we concentrate on design of second-order active RC filters.

Filter type: filters are frequency-selective networks that are categorized as follows:
- low pass (LP);
- band pass (BP);
- high pass (HP);
- band reject (BR);
- low-pass notch (LPN);
- high-pass notch (HPN);
- all pass (AP).

A general biquadratic filter function is shown below:

$$T(s) = K\frac{s^2 + \left(\frac{\omega_z}{Q_z}\right)s + \omega_z^2}{s^2 + \left(\frac{\omega_p}{Q_p}\right)s + \omega_p^2}, \tag{9.1}$$

where
- ω_z = the zero frequency,
- Q_z = zero selectivity,
- ω_p = the pole frequency,
- Q_p = the pole selectivity, and
- K = the gain constant.

The general biquadratic filter function can also be written (with $K = 1$) as:

$$T(s) = \frac{ms^2 + cs + d}{s^2 + as + b}. \tag{9.2}$$

A second-order low-pass filter function is given by the following transfer function:

$$T(s) = \frac{d}{s^2 + as + b}. \tag{9.3}$$

A second-order high-pass filter function is given by

$$T(s) = \frac{ms^2}{s^2 + as + b} \tag{9.4}$$

A second-order band-pass filter function is given by the following transfer function:

$$T(s) = \frac{cs}{s^2 + as + b} \tag{9.5}$$

A second-order band-reject filter function is given by the following transfer function:

$$T(s) = \frac{s^2 + d}{s^2 + as + b} \tag{9.6}$$

If $b = d$, this is called a band reject filter.

If $b < d$, the filter is called a low-pass notch filter.

If $b > d$, the filter is called a high-pass notch filter.

The transfer function of a second order all-pass filter or delay equalizer is given by following equation.

$$T(s) = K\frac{s^2 - as + b}{s^2 + as + b}. \tag{9.7}$$

9.2 The single-amplifier filter

Single amplifier filters can be classified into three major categories:
- positive feedback topology,
- negative feedback topology, and
- enhanced negative feedback topology.

9.2.1 The positive feedback topology

In this case, the RC network is placed in the positive feedback path of the op-amp. However, negative feedback is also introduced to keep the filter stable. A single-amplifier filter based on a positive feedback topology is shown in figure 9.1.

The transfer function of the network can be expressed in terms of the feedforward and feedback transfer functions of the passive RC network.

The feedforward transfer function

$$T_{FF}(s) = \left.\frac{V_3(s)}{V_1(s)}\right|_{V_2(s)=0}. \tag{9.8}$$

The feedback transfer function

$$T_{FB}(s) = \left.\frac{V_3(s)}{V_2(s)}\right|_{V_1(s)=0}. \tag{9.9}$$

From the circuit, we see that

$$V_3(s) = V^+(s) = T_{FF}(s)V_{IN}(s) + T_{FB}(s)V_O(s)$$

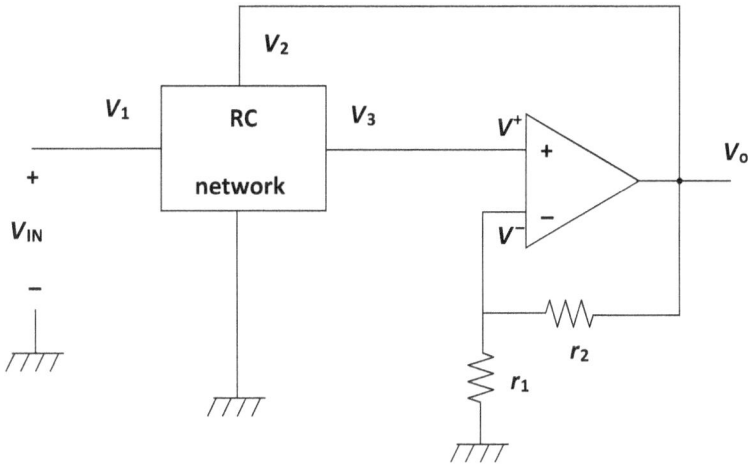

Figure 9.1. A positive feedback topology.

$$V^-(s) = \frac{r_1 V_O(s)}{r_1 + r_2} = \frac{V_O(s)}{k},$$

where $k = 1 + \frac{r_2}{r_1}$

$$V_O(s) = A[V^+(s) - V^-(s)]$$

$$\therefore\ T(s) = \frac{V_O(s)}{V_{IN}(s)} = \frac{k T_{FF}(s)}{1 - k T_{FB}(s) + \frac{k}{A}}. \tag{9.10}$$

For an ideal op-amp,
$A \to \infty.$
Equation (9.10) reduces to

$$\therefore\ T(s) = \frac{k T_{FF}(s)}{1 - k T_{FB}(s)}. \tag{9.11}$$

Now, $T_{FF}(s) = \dfrac{N_{FF}(s)}{D_{FF}(s)}$ and $T_{FB}(s) = \dfrac{N_{FB}(s)}{D_{FB}(s)}.$ \hfill (9.12)

- $N_{FF}(s)$ and $N_{FB}(s)$ are the zeros of the RC network observed at different ports.
- The denominators $D_{FF}(s)$ and $D_{FB}(s)$ are obtained from the nodal determinants of the RC network, and thus we can write $D(s) = D_{FF}(s) = D_{FB}(s)$. Substituting these into equation (9.12) yields

$$T(s) = \frac{k N_{FF}(s)}{D(s) - k N_{FB}(s)} \tag{9.13}$$

Salient features of the positive feedback topology are:
- The zeros of the transfer functions are the zeros of the feedforward RC network, which can be complex.
- The poles of the transfer function are determined by the poles of the RC network, the zeros of the feedback transfer function, and the factor k.

9.2.2 The negative feedback topology

The negative feedback topology is shown in figure 9.2. The RC network is in the negative feedback path of the op-amp, hence the name. As before, the transfer function of the network can be expressed in terms of the feedforward and feedback transfer functions of the passive network.

Thus, we can write.

$$V_3(s) = V^-(s) = T_{FF}(s)V_{IN}(s) + T_{FB}(s)V_O(s) \tag{9.14}$$

$$V_O(s) = A[V^+(s) - V^-(s)] = -AV^-(s). \tag{9.15}$$

Eliminating $V^-(s)$ from (9.14) and (9.15) yields

$$T(s) = \frac{V_O(s)}{V_{IN}(s)} = -\frac{T_{FF}(s)}{T_{FB}(s) + \frac{1}{A}}.$$

Since, for an ideal op-amp, $A \to \infty$

$$\therefore T(s) = -\frac{T_{FF}(s)}{T_{FB}(s)}.$$

As before, we can write

$$T_{FF}(s) = \frac{N_{FF}(s)}{D(s)}; \; T_{FB}(s) = \frac{N_{FB}(s)}{D(s)}$$

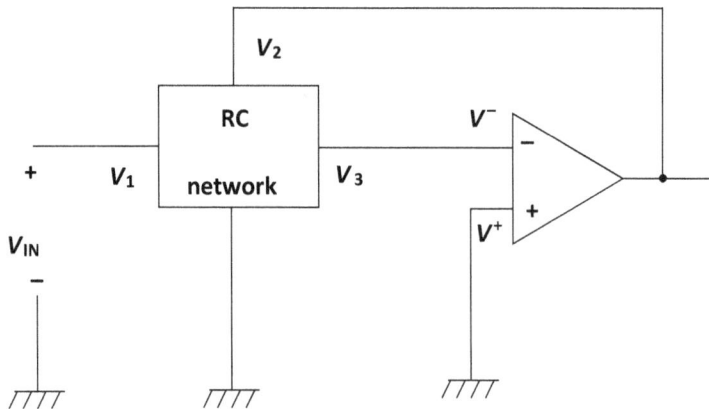

Figure 9.2. A negative feedback topology.

$$\therefore T(s) = -\frac{N_{FF}(s)}{N_{FB}(s)}.$$

The salient features of the negative feedback topology are:
- The zeros of the feedback network determine the poles of the transfer function.
- The zeros of the feedforward network determine the zeros of the transfer function.
- The poles and zeros can be complex; however, for a stable network, the poles should lie on the left-hand side of the s-plane.
- The poles of the RC network do not contribute to the transfer function (assuming that the op-amp is ideal).

9.2.3 The enhanced negative feedback topology

An enhanced negative feedback topology is shown in figure 9.3. This is similar to the negative feedback topology with the only difference that positive feedback is also present. The negative feedback topology is inherently stable; however, the circuit has some interesting features once positive feedback is added.
- From circuit analysis, we get

$$\frac{V_O}{k}(s) = V_{IN}(s)T_{FF}(s) + V_O(s)T_{FB}(s)$$

$$\text{or,} \quad \frac{V_O(s)}{V_{IN}(s)} = -\frac{T_{FF}(s)}{T_{FB}(s) - \frac{1}{k}}, \text{ where } k = 1 + \frac{r_2}{r_1}$$

$$\text{or,} \quad \frac{V_O(s)}{V_{IN}(s)} = \frac{kN_{FF}(s)}{D - kN_{FB}(s)}.$$

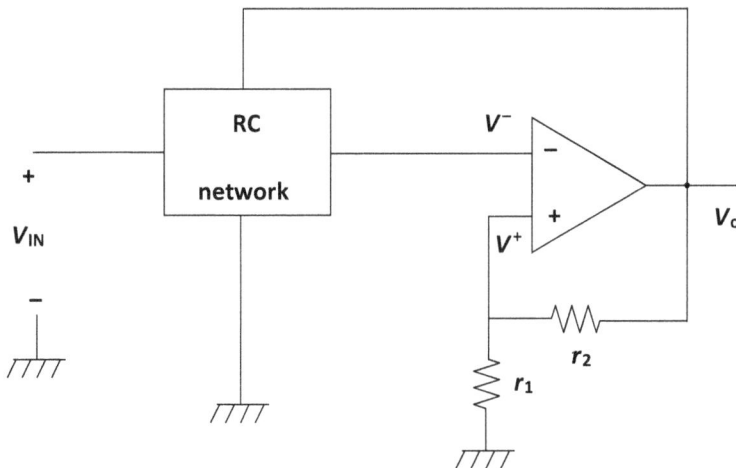

Figure 9.3. A mixed-feedback topology.

9.2.4 Design equations

Among the positive feedback single-amplifier filter topologies, the Sallen and Key filter is popular. For the Sallen and Key filters, there are generally three widely used choices, which are as follows:

1. The equal R and equal C design.
2. The unity gain amplifier design.
3. The Saraga design.

The equal R and C design

- In this design, all the resistors and capacitors are given equal values.
- Its major drawback is that its passive sensitivity is very high.

The unity gain amplifier design

- In this case, the gain of the amplifier is chosen to be one.
- It is found that in this case, the sensitivity is quite low or even zero.
- The main disadvantage of this type of design is is that the capacitor spread is very high.

The Saraga design

For the Saraga design, k is assigned a fixed value of 4/3. By doing so, a good compromise between sensitivity and element spread can be achieved.

9.2.5 The Sallen and Key low-pass filter

The Sallen and Key low-pass filter is shown in figure 9.4. From the circuit, we find that

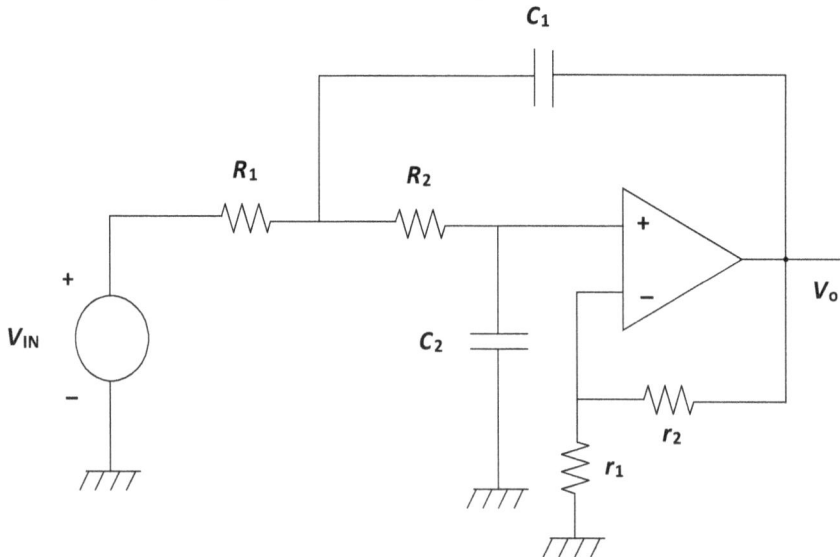

Figure 9.4. Sallen and Key low-pass filter.

$$\frac{V_0}{V_{\text{IN}}}(s) = \frac{k/R_1R_2C_1C_2}{s^2 + s\left(\frac{1}{R_1C_1} + \frac{1}{R_2C_1} + \frac{(1-k)}{R_2C_2}\right) + \frac{1}{R_1R_2C_1C_2}}$$

or $\omega_p = \sqrt{\dfrac{1}{R_1R_2C_1C_2}}$

$$Q_p = \frac{\sqrt{\dfrac{1}{R_1R_2C_1C_2}}}{\dfrac{1}{R_1C_1} + \dfrac{1}{R_2C_1} + \dfrac{(1-k)}{R_2C_2}}$$

$$K = \frac{k}{R_1R_2C_1C_2}.$$

The equal R and equal C design:

The design equations are

$$C_1 = C_2 = 1$$

$$R_1 = R_2 = R = 1/\omega_p$$

$$k = 3 - \frac{1}{Q_p}.$$

The unity gain amplifier design:

The design equations are

$$C_2 = C, \;\; C_1 = 4Q_p^2C$$

$$R_1 = R_2 = R$$

$$CR = \frac{1}{2\omega_pQ_p}.$$

For simplicity, we take

$$C_2 = 1 \;\; \therefore \;\; C_1 = 4Q_p^2$$

$$R_1 = R_2 = \frac{1}{2\omega_pQ_p}$$

The Saraga design

The design equations are:

$$C_2 = 1, \;\; C_1 = \sqrt{3}\,Q_p$$

$$R_1 = \frac{1}{Q_p\omega_p}, \;\; R_2 = \frac{1}{\sqrt{3}\,\omega_p}$$

$$k = \frac{4}{3} = \left(1 + \frac{r_2}{r_1}\right).$$

9.2.6 The Sallen and Key high-pass filter

The Sallen and Key high-pass filter circuit is shown in figure 9.5.
From the circuit, we get

$$\frac{V_0}{V_{IN}}(s) = \frac{ks^2}{s^2 + s\left(\dfrac{1}{R_2C_2} + \dfrac{1}{R_2C_1} + \dfrac{(1-k)}{R_1C_1}\right) + \dfrac{1}{R_1R_2C_1C_2}}.$$

Thus, $\omega_p = \sqrt{\dfrac{1}{R_1R_2C_1C_2}}$

$$Q_p = \frac{\sqrt{\dfrac{1}{R_1R_2C_1C_2}}}{\dfrac{1}{R_2C_2} + \dfrac{1}{R_2C_1} + \dfrac{(1-k)}{R_1C_1}}$$

$$K = k = \left(1 + \frac{r_2}{r_1}\right).$$

The equal R and equal C design:
The design equations are

$$C_1 = C_2 = 1,$$

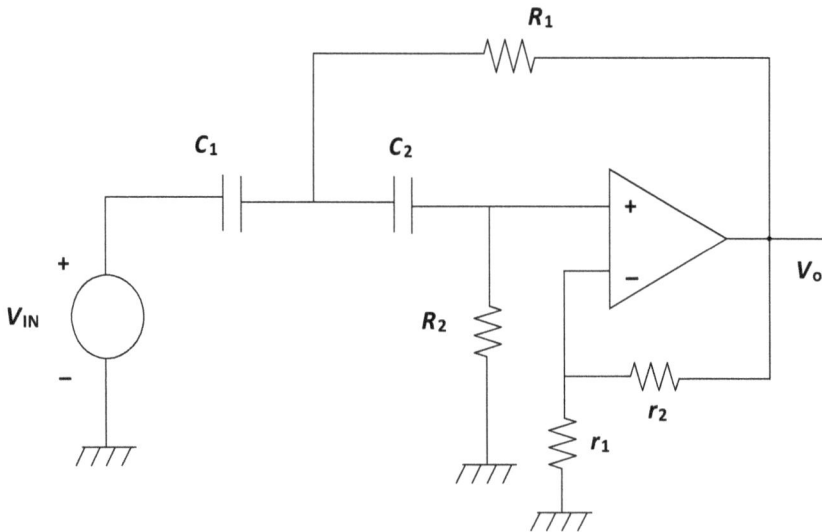

Figure 9.5. Sallen and Key high-pass filter.

$$R_1 = R_2 = 1/\omega_p$$

$$k = 3 - \frac{1}{Q_p}$$

The unity gain amplifier design:
With $k = 1$, the design equations are as follows:

$$C_1 = C_2 = C, \quad R_1 = R, \quad R_2 = 4Q_p^2 R$$

$$CR = \frac{1}{2\omega_p Q_p}.$$

For simplicity, we take

$$C_1 = C_2 = 1$$

$$R_1 = R = \frac{1}{2\omega_p Q_p}$$

$$R_2 = 2\frac{Q_p}{\omega_p}.$$

The Saraga design:
The design equations are:

$$C_1 = Q_p, \quad C_2 = \sqrt{3}$$

$$R_1 = \frac{1}{\sqrt{3}\,\omega_p Q_p}, \quad R_2 = \frac{1}{\omega_p}$$

$$k = 1 + \frac{r_2}{r_1} = \frac{4}{3}$$

9.2.7 The Sallen and Key band-pass filter

The Sallen and Key band-pass filter circuit is shown in figure 9.6.
From circuit analysis, we get

$$\frac{V_O}{V_{IN}}(s) = \frac{ks/R_1C_1}{s^2 + s\left(\dfrac{1}{R_1C_1} + \dfrac{1}{R_3C_2} + \dfrac{1}{R_3C_1} + \dfrac{(1-k)}{R_2C_1}\right) + \dfrac{R_1+R_2}{R_1R_2R_3C_1C_2}}.$$

Thus,

$$\omega_p = \sqrt{\frac{R_1 + R_2}{R_1 R_2 R_3 C_1 C_2}}$$

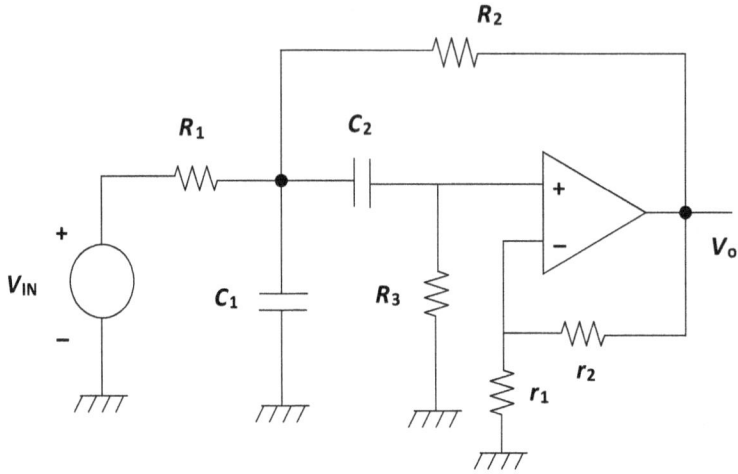

Figure 9.6. A Sallen and Key band-pass filter.

$$Q_{\mathrm{p}} = \frac{\sqrt{\dfrac{R_1 + R_2}{R_1 R_2 R_3 C_1 C_2}}}{\dfrac{1}{R_1 C_1} + \dfrac{1}{R_3 C_2} + \dfrac{1}{R_3 C_1} + \dfrac{(1-k)}{R_2 C_1}}$$

$$K = \frac{k}{R_1 C_1}.$$

The equal R and equal C design:

$$C_1 = C_2 = 1,$$

$$R_1 = R_2 = R_3 = \frac{\sqrt{2}}{\omega_{\mathrm{p}}}$$

$$k = 4 - \frac{\sqrt{2}}{Q_{\mathrm{p}}}.$$

The unity gain amplifier design:
With $k = 1$, the design equations are

$$C_1 = C_2 = C$$

$$R_1 = R_3 = \left(9 Q_{\mathrm{p}}^2 - 1\right) R$$

$$R_2 = R, \quad CR = \frac{3 Q_{\mathrm{p}}}{\left(9 Q_{\mathrm{p}}^2 - 1\right) \omega_{\mathrm{p}}}$$

For simplicity,

$$C_1 = C_2 = 1$$

$$\therefore R_2 = \frac{3Q_p}{\left(9Q_p^2 - 1\right)\omega_p}$$

$$R_1 = R_3 = \frac{3Q_p}{\omega_p}$$

9.3 Negative feedback circuits

These types of circuits have multiple feedback paths and hence they are also known as multiple feedback single-amplifier circuits. They may have two types of structures:
1. Without positive feedback.
2. With positive feedback.

9.3.1 The band-pass circuit without positive feedback

A multiple feedback band-pass circuit is shown in figure 9.7. The circuit is a negative feedback circuit when the RC network is in the negative feedback path. It is important to note that there is no positive feedback in the circuit.

From the circuit, we have

$$\frac{V_O}{V_{IN}}(s) = \frac{-s/R_1C_2}{s^2 + s\left(\dfrac{1}{R_2C_1} + \dfrac{1}{R_2C_2}\right) + \dfrac{1}{R_1R_2C_1C_2}}.$$

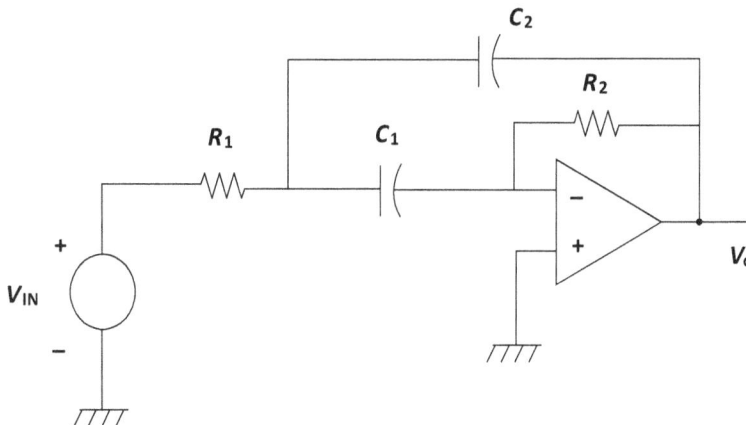

Figure 9.7. A multiple feedback band-pass filter circuit.

We have $\omega_p = \dfrac{1}{\sqrt{R_1 R_2 C_1 C_2}}$

$$Q_p = \frac{\sqrt{\dfrac{1}{R_1 R_2 C_1 C_2}}}{\dfrac{1}{R_2 C_1} + \dfrac{1}{R_2 C_2}}$$

$$K = \frac{1}{R_1 C_2}.$$

The design equation:

Let us take $C_1 = C_2 = 1$.

$$R_2 = 2\frac{Q_p}{\omega_p}; \quad R_1 = \frac{1}{2\omega_p Q_p}$$

9.3.2 The bandpass circuit with positive feedback

Let us introduce positive feedback into the negative feedback circuit of figure 9.7. A mixed-feedback circuit is shown in figure 9.8.

Analyzing this circuit, we have

$$\frac{V_O}{V_{IN}}(s) = \frac{-s/\left[R_1 C_2 \left(1 - \frac{1}{k}\right) \right]}{s^2 + s\left(\dfrac{1}{R_2 C_1} + \dfrac{1}{R_2 C_{2+}} - \dfrac{1}{k-1}\dfrac{1}{R_1 C_1} \right) + \dfrac{1}{R_1 C_1 R_2 C_2}}$$

where $k = 1 + \dfrac{R_B}{R_A}$.

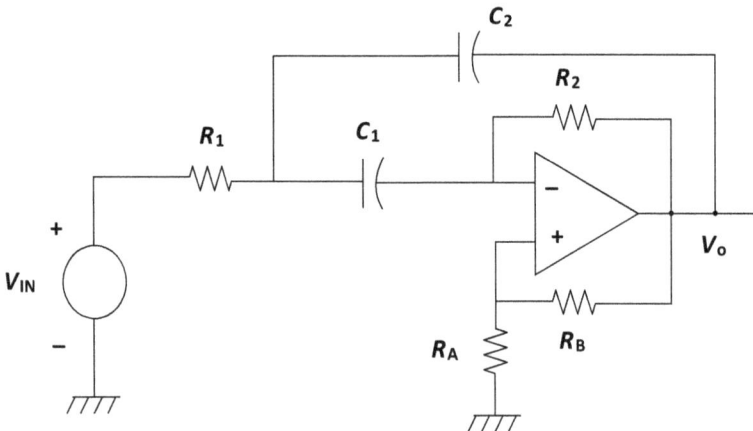

Figure 9.8. A negative feedback band-pass filter with positive feedback.

The pole frequency (ω_p), pole selectivity (Q_p) and gain constant (K) are given by

$$\omega_p = \sqrt{\frac{1}{R_1 R_2 C_1 C_2}}$$

$$Q_p = \frac{\sqrt{\frac{1}{R_1 R_2 C_1 C_2}}}{\frac{1}{R_2 C_1} + \frac{1}{R_2 C_2} - \frac{1}{k-1}\frac{1}{R_1 C_2}}, \quad \text{and}$$

$$K = \frac{-1}{R_1 C_2 \left(1 - \frac{1}{k}\right)}.$$

It is interesting to note that the expression for pole selectivity contains a subtractive term in the denominator. Therefore, the selectivity of the circuit can be made very high by controlling the positive feedback or k without the use of a high component spread. However, the disadvantages of a single-amplifier circuit are that the sensitivities are high, the component spread is large, and orthogonal tuning of the filter parameters is impossible.

9.4 Inductor simulator

An active RC filter can also be designed by simulating the inductance of an LC ladder filter using R, C, and an active device which is essentially an op-amp. Therefore, we shall look at an active RC circuit that has an input impedance of $Z_{in} = sL$, where L is the value of the simulated inductance. An inductance simulation circuit is shown in figure 9.9.

From circuit analysis, we obtain the following.

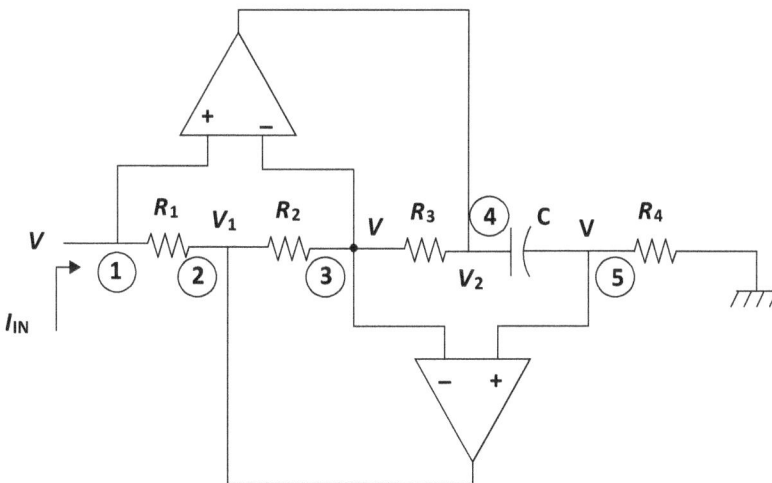

Figure 9.9. An inductance simulation circuit.

Applying Kirchoff's current law (KCL) at node 5,

$$(V_2 - V)sC = \frac{V}{R_4}. \tag{9.16}$$

Applying KCL at node 3,

$$\frac{V_1 - V}{R_2} = \frac{V - V_2}{R_3}$$

$$\text{or } \frac{V - V_1}{R_2} = \frac{V_2 - V}{R_3}.$$

Substituting $(V_2 - V)$ from (9.16) yields

$$\frac{V - V_1}{R_2} = \frac{V}{sR_3R_4C}. \tag{9.17}$$

Applying KCL at node 1,

$$I_{\text{IN}} = \frac{V - V_1}{R_1} = \frac{VR_2}{sR_4R_3R_1C}$$

$$\text{or } \frac{V}{I_{\text{IN}}} = \frac{sR_4R_3R_1C}{R_2}.$$

Therefore, the input impedance of the circuit can be written as

$$\text{or } Z_{\text{IN}} = \frac{sR_4R_3R_1C}{R_2}.$$

Thus, $L_{\text{eq}} = \frac{R_4R_3R_1C}{R_2}.$

Thus, the two-amplifier structure of figure 9.9 simulates an inductance.

Low-pass, high-pass, and band pass filters using the two-amplifier structure are shown in figures 9.10, 9.11 and 9.12, respectively.

9.5 The low-pass filter

From circuit analysis,

$$\frac{V_O}{V_{\text{IN}}}(s) = \frac{(1/R_1R_2)(1/R_4 + 1/R_5)(R_5/C_1C_2)}{s^2 + (s/R_3C_2) + (R_5/R_1R_2R_4C_1C_2)}. \tag{9.18}$$

The design equations are:

$$R_1 = R_2 = R_4 = R_5 = R;$$
$$R_3 = Q_pR$$
$$C_1 = C_2 = C$$
$$CR = \frac{1}{\omega_p}.$$

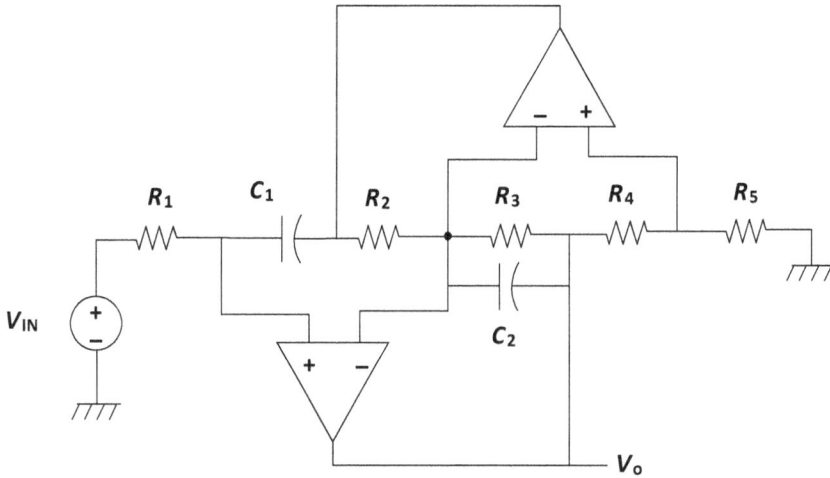

Figure 9.10. A low-pass filter.

9.6 The high-pass filter

From circuit analysis,

$$\frac{V_O}{V_{IN}}(s) = \frac{s^2\,[1 + (R_4/R_5)]}{s^2 + (s/\,R_1C_1) + (R_4/R_2R_3R_5C_1C_2)} \tag{9.19}$$

The design equations are:

$$\begin{aligned}
R_1 &= Q_p R \\
R_2 &= R_3 = R_4 = R_5 = R; \\
C_1 &= C_2 = C \\
CR &= \frac{1}{\omega_p}.
\end{aligned}$$

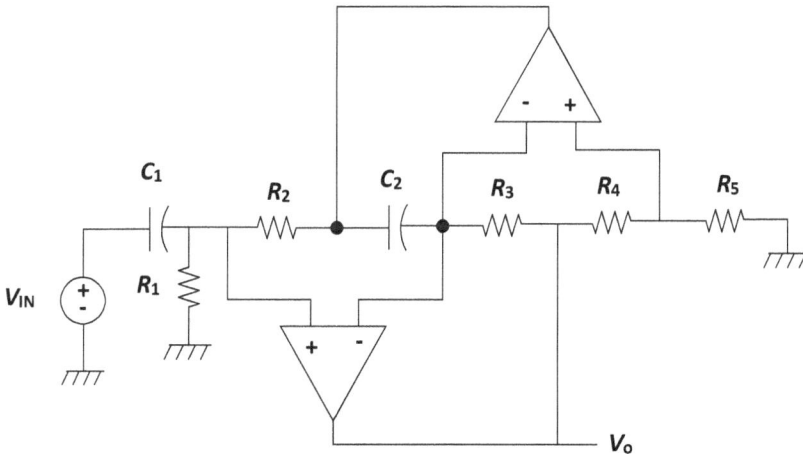

Figure 9.11. A high-pass filter.

9.7 The band-pass filter

From circuit analysis,

$$\frac{V_O}{V_{IN}}(s) = \frac{[s\,\{1 + (R_4/R_5)\}]\,/C_1R_1}{s^2 + (s/R_1C_1) + (R_4/C_1C_2R_2R_3R_5)}. \tag{9.20}$$

Design equations are:

$$
\begin{aligned}
R_1 &= Q_pR \\
R_2 &= R_3 = R_4 = R_5 = R; \\
C_1 &= C_2 = C \\
CR &= \frac{1}{\omega_p}.
\end{aligned}
$$

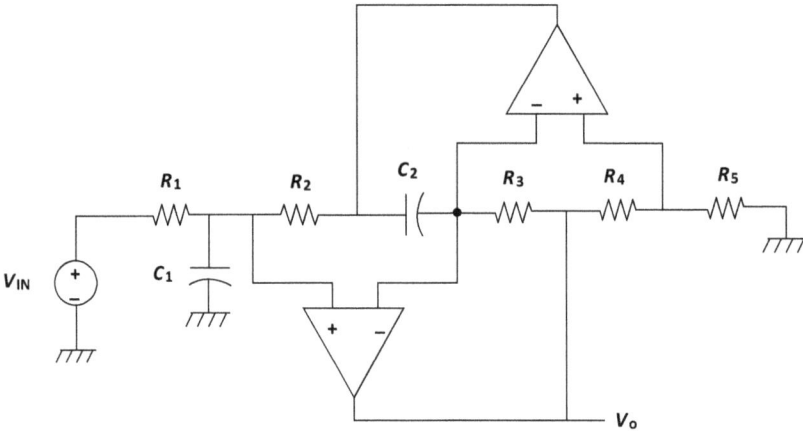

Figure 9.12. A band-pass filter.

9.8 The state variable filter

A state variable filter with a triple-amplifier structure is shown in figure 9.13. Here, the input signal is fed forward via resistors R_5 and R_6.

From circuit analysis,

$$\frac{V_O}{V_{IN}}(s) = T(s) = -\frac{R_8}{R_6}\,\frac{s^2 + s\left(\dfrac{1}{R_1C_1} - \dfrac{1}{R_4C_1}\dfrac{R_6}{R_7}\right) + \dfrac{R_6}{R_7}\dfrac{1}{R_3R_5C_1C_2}}{s^2 + s\,\dfrac{1}{R_1C_1} + \dfrac{R_8}{R_7}\dfrac{1}{R_2R_3C_1C_2}}. \tag{9.21}$$

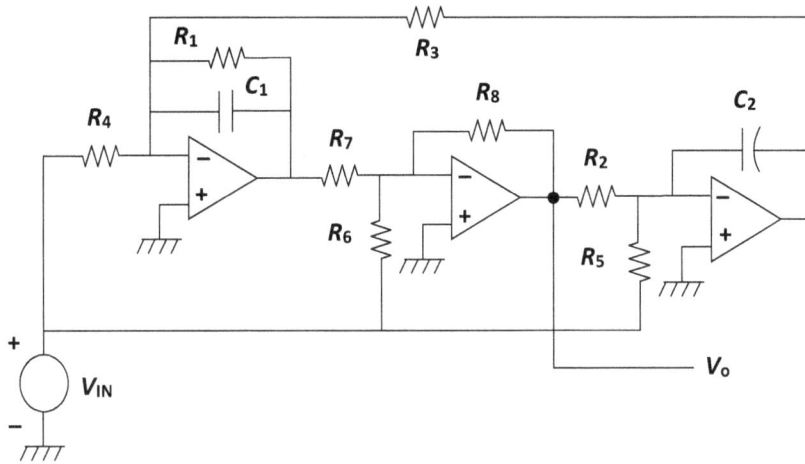

Figure 9.13. A three-amplifier feedforward structure.

The above equation can also be written as

$$T(s) = K \frac{(s^2 + cs + d)}{(s^2 + as + b)}.$$

(9.22)

The design equations are:

$$C_1 = C_2 = 1$$

$$R_2 = R_3 = R_7 = R_8 = R = \frac{1}{\sqrt{b}}$$

$$R_1 = \frac{1}{a}, \quad R_4 = \frac{1}{K(a-c)}, \quad R_5 = \frac{\sqrt{b}}{Kd}, \quad R_6 = \frac{1}{K\sqrt{b}}.$$

The design equations given above yield positive element values for $a \geqslant c$. As is evident from equation (9.21), tight component tolerances are required to realize high-pass and band-reject functions. Equation (9.22) can also be written as

$$T(s) = K \frac{s^2 + \frac{\omega_z}{Q_z} s + \omega_z^2}{s^2 + \frac{\omega_p}{Q_p} s + \omega_p^2}.$$

(9.23)

Thus, for orthogonal tuning, we can use the following elements as shown in the table below, and they need to be applied sequentially as given in table 9.1.

9.9 The sample and hold circuit

A sample and hold circuit samples an input signal and holds its last sampled value until the input signal is sampled again. This type of circuit is very useful for interfacing analog signal sources to analog-to-digital converters. A circuit schematic of a sample and hold circuit is shown in figure 9.14.

Table 9.1. The required variation of resistive elements for orthogonal tuning of filter parameters.

Parameter	Element
K	R_8
ω_z	R_3
ω_p	R_2
Q_p	R_1
Q_z	R_4

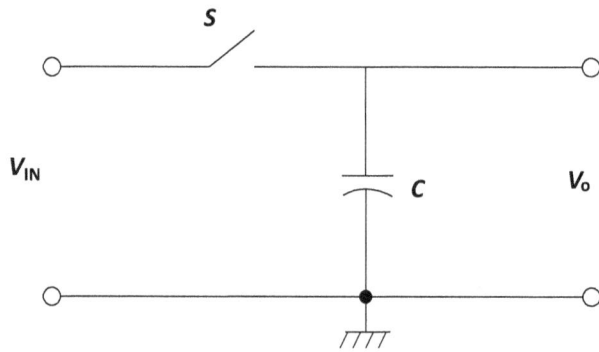

Figure 9.14. A simple sample and hold circuit.

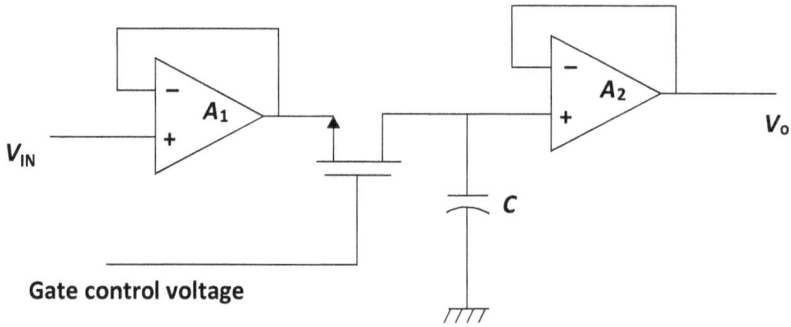

Figure 9.15. A practical sample and hold system.

The voltage across capacitor C follows the input signal during time T_s when switch S is closed. The capacitor holds the charge or the voltage signal (V_i) attained at the end of the interval T_s when the switch S is open. The switch may be a bipolar transistor, a MOSFET controlled by voltage gating (a signal voltage), or a CMOS transmission gate. One of the simplest practical sample and hold circuits is shown in figure 9.15.

The input buffer amplifier (A_1) has a low output impedance that allows the capacitor to charge quickly. In contrast, the high input impedance of the output amplifier (A_2) maintains the charge on C over a longer period of time when switch S is open.

A positive voltage at the gate of the N-type metal–oxide–semiconductor (NMOS) turns on switch S. The capacitor C charges to the instantaneous value of the input voltage with a time constant of $(r_O + r_{DS}(ON))C$, where r_O = the very small output resistance of the input op-amp A_1 and $r_{DS}(ON)$ = the ON resistance of the MOSFET. The switch is turned off when the voltage at the gate of the NMOS is removed and capacitor C is isolated from any load through the op-amp A_2 and thus holds the voltage impressed on it. It is recommended that this circuit should use a capacitor with a polycarbonate, polyethylene, polystyrene, Mylar, or Teflon dielectric. Most capacitors other than those mentioned above have two major problems, namely:

(i) A polarization phenomenon which causes the stored voltage to decay with a time constant of several seconds. In effect, this is the leakage resistance of the capacitor.

(ii) Dielectric absorption, which causes a capacitor to 'remember' a fraction of its previous charge.

Even if polarization and absorption do not occur, the OFF current and the bias current of the op-amp flow through C. For the maximum input bias current and a capacitor of $0.1\ \mu F$, the drift rate during the hold period is less than $10\ mV\ s^{-1}$. There are two more factors which influence the operation of the circuit, namely:

(i) The aperture time (typically less than 100 ns), which is the delay between the time at which the voltage is applied to the gate of the switch and the actual time at which the switch closes.

(ii) The acquisition time, which is the time that the capacitor takes to change from one level of holding voltage to the new value of the input voltage after the switch is closed.

9.10 The logarithmic amplifier

The logarithmic amplifier is used when it is desired that the output voltage should be proportional to the logarithm of the input voltage. A logarithmic amplifier which uses an op-amp and a differential amplifier is shown in figure 9.16. For analysis purposes, we assume $i_{B_1} < <i_{C_1}$ and $i_{B_2} < <i_{c_2}$.

$$V_x = V_{BE_2} - V_{BE_1} = V_T\ \ln i_{c_2} - V_T\ \ln i_{C_1}$$
$$= - V_T\ \ln \frac{i_{C_1}}{i_{C_2}}. \tag{9.24}$$

Neglecting V_x compared to the reference voltage V_R gives

$$i_{C_2} = V_R/R_2 \text{ and } i_{C_1} = \frac{V_{IN}}{R_1}. \tag{9.25}$$

$$\text{Again,} \quad V_O = \frac{V_x(R_3 + R_4)}{R_3}. \tag{9.26}$$

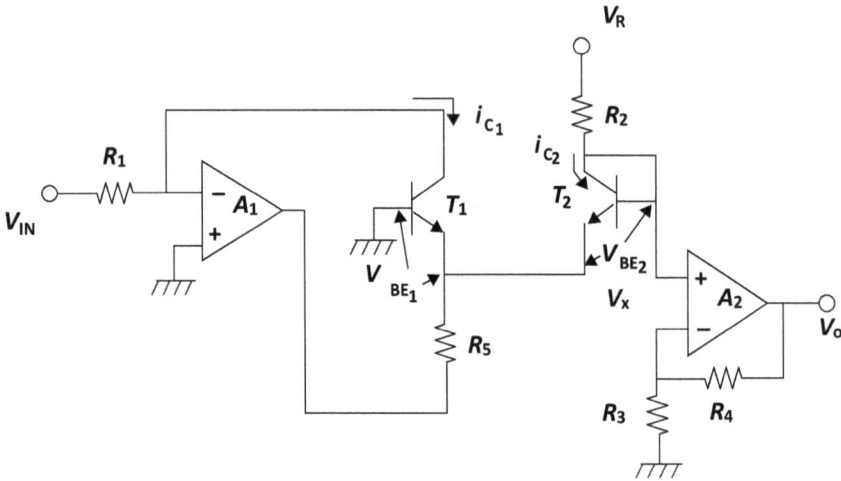

Figure 9.16. A logarithmic amplifier.

Combining equations (9.24), (9.25), and (9.26) yields

$$V_O = -V_T \left(\frac{R_3 + R_4}{R_3} \right) \left[ln\frac{V_{IN}}{R_1}\frac{R_2}{V_R} \right],$$

where i_{B_1}, i_{B_2}, i_{C_1} and i_{C_2} are the bias currents and collector currents of transistors T_1 and T_2, respectively. V_T is the voltage equivalent of temperature.

9.11 The antilogarithmic amplifier

The antilogarithmic (antilog) amplifier is used when it is desired that the output voltage should be proportional to the antilog or exponential of the input voltage. An antilog amplifier is shown in figure 9.17. The reference voltage V_R passes a constant current i_{C_1} through the transistor T_1. However, i_{C_2} depends on the input voltage V_{IN}.

The input of the second op-amp is at virtual ground. Therefore

$$-V_x = V_{BE_1} - B_{VE_2}.$$

Again,

$$i_{C_1} = \frac{V_R}{R_2} \text{ and } i_{C_2} = \frac{V_O}{R_1}.$$

Moreover,

$$-V_x = \frac{R_3 V_{IN}}{R_3 + R_4} = V_T ln\frac{i_{C_1}}{i_{C_2}}$$

After simplification, we obtain

$$V_O = \frac{R_1 V_R}{R_2} \exp\left\{ \frac{-V_{IN}}{V_T} \cdot \frac{R_3}{R_3 + R_4} \right\}.$$

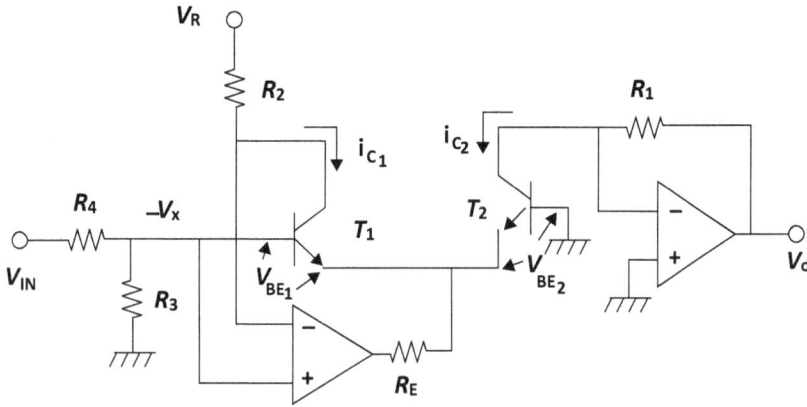

Figure 9.17. An antilog amplifier.

9.12 The analog switch

An analog switch is a voltage-controlled device that connects an analog signal to a circuit as and when desired. The analog switches are used in analog-to-digital and digital-to-analog converters, switched capacitor filters, multiplexing, demultiplexing, and data acquisition systems.

A complementary MOSFET switch or CMOS switch (a combination of P-type metal–oxide–semiconductor (PMOS) and NMOS) provides all the desired features of an analog switch. This circuit is commonly known as a transmission gate. A CMOS transmission gate is shown in figure 9.18. The ON resistance of a metal–oxide–semiconductor (MOS) device is a nonlinear function of the voltage drop across the device i.e. $(V_i - V_O)$. As V_i increases, the resistance of T_1 increases and that of T_2 decreases. Thus, the ON resistance of CMOS is linear and independent of V_i and is valid for both positive and negative V_i.

The desired features of an analog switch are as follows:
- The OFF resistance of the switch should be very high.
- It should offer minimum resistance when it is closed.

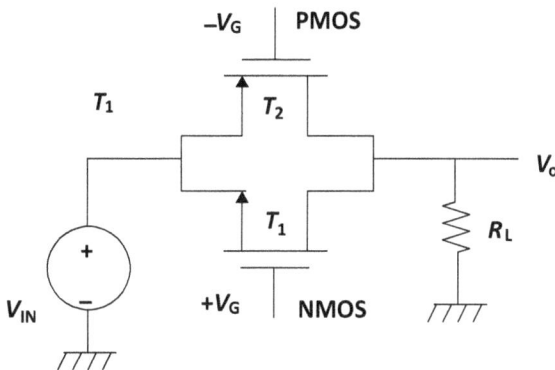

Figure 9.18. A CMOS transmission gate.

- The ON resistance of the switch should be independent of the input voltage, otherwise the output signal will be nonlinearly distorted.
- The input voltage can be positive or negative, therefore the switch should be able to conduct in both directions. Hence, it must be a bidirectional switch.
- The transmission gate is ON when the control signal (V_C) is high.

9.13 Analog multiplexers and demultiplexers

A multiplexer selects one of several sources and transmits the signal over a single transmission line. Each analog signal occupies its own time slot. At the end of the transmission line, each signal must be separated from the others and placed in an individual output channel. The reverse process is called demultiplexing. A four-channel multiplexer and demultiplexer is shown in figure 9.19. The signals flow in and out of the channel through the analog switches or transmission gates. The control signals are obtained from a 2–4 line decoder, which must be synchronized at the sending and receiving ends. The process of multiplexing and demultiplexing saves on the cost of the channels, since all the signals from different sources are transmitted using a single channel . The symbol of CMOS transmission gate is shown in figure 9.20.

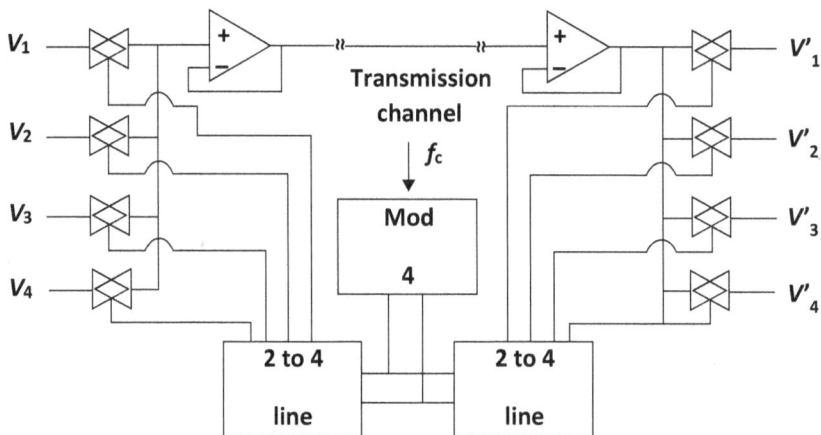

Figure 9.19. A four-channel multiplexer and demultiplexer.

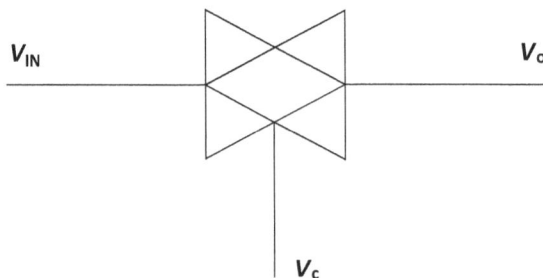

Figure 9.20. The CMOS transmission gate symbol.

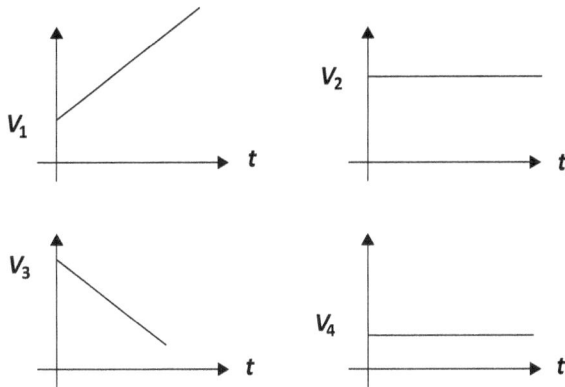

Figure 9.21. Various signal inputs provided to the multiplexer.

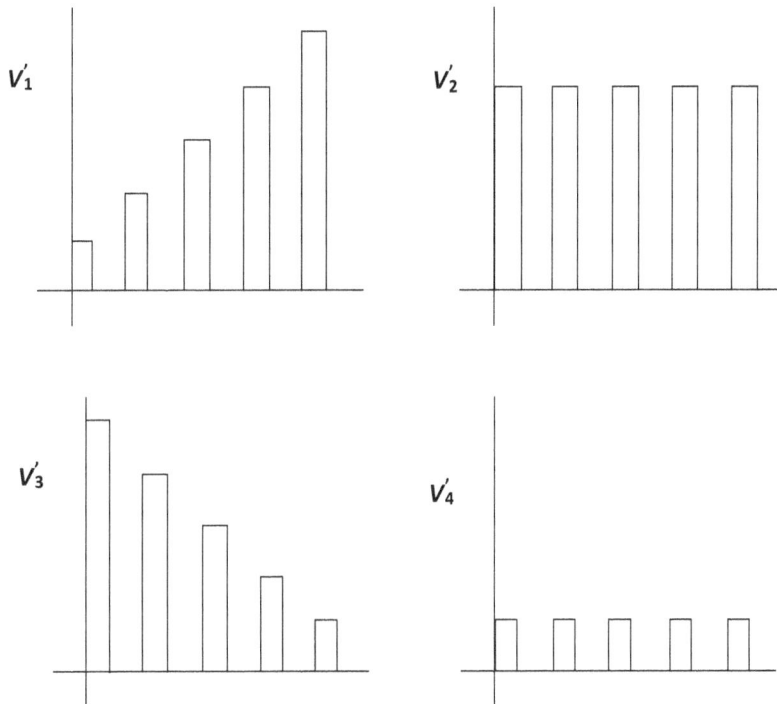

Figure 9.22. Signals at the output of the demultiplexer.

Using the circuitry discussed above, one can save the cost of many cables even though the number of switches increases by this method. Four different voltage inputs and outputs are shown in figures 9.21 and 9.22, respectively.

On passing the output of the demultiplexer through a low-pass filter, we retrieve the original signal.

The National Semiconductor 4016 and 4066 devices and the Motorola Corporation MC 14066 are analog switches. These devices are quad switches on a single chip. They are bidirectional analog switches that are accommodated in a 14-pin DIP (dual in-line package). The ON resistance of analog switches is 100–600 Ω. The input signals must lie between V_{DD} and V_{SS}, where $V_{DD} = +5$ V and $V_{SS} = -5$ V.

Problem 9.1. An all-pass filter has the transfer function

$$G(s) = \frac{s^2 - 3s + 9}{s^2 + 3s + 9}.$$

Calculate the group delay (in seconds) of the filter at the frequency $\omega = 5$ rad s^{-1}.

Problem 9.2. Find the transfer function of a second-order band-pass filter that has a centre frequency of 1000 rad s^{-1}, a selectivity of 100, and a gain of 0 dB at the centre frequency.

Problem 9.3. Synthesize the following band-pass transfer function using a Sallen and Key filter design with equal R and C.

$$T(s) = \frac{400s}{s^2 + 4s + 500}$$

Problem 9.4. Synthesize a second-order low-pass filter that has a pole frequency (ω_p) of 2 kHz and a pole selectivity (Q_p) of 10 using the Saraga version of the Sallen and Key filter. The filter circuit is redrawn in figure 9.23.

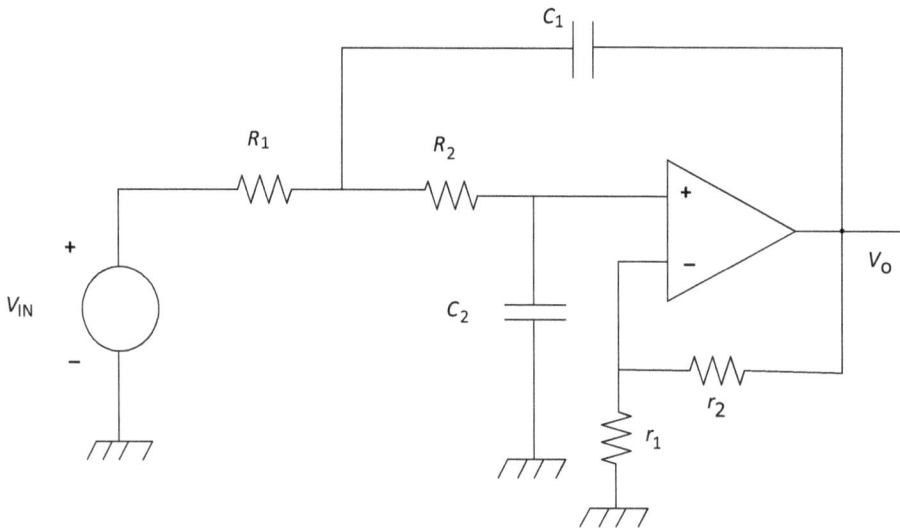

Figure 9.23.

The equations for the Saraga design are:

$$C_2 = 1, \quad C_1 = \sqrt{3}\, Q_p$$

$$R_1 = \frac{1}{Q_p \omega_p}, \quad R_2 = \frac{1}{\sqrt{3}\, \omega_p}$$

$$k = \frac{4}{3} = \left(1 + \frac{r_2}{r_1}\right).$$

IOP Publishing

Fundamentals of Industrial Instrumentation (Second Edition)

Alok Barua

Chapter 10

Piezoelectric sensors

Learning objectives:
- The piezoelectric phenomenon.
- Piezoelectric materials.
- Piezoelectric transducers.
- The equivalent circuit for the crystal.
- The charge amplifier.
- Limitations of piezoelectric transducers.
- Accelerometers.
- Unimorphs and bimorphs.
- Pyroelectricity.

10.1 Introduction

When some solid materials are deformed, they generate an internal electric charge. The effect, known as the piezoelectric effect, is reversible. Sensors which work on the principle of electromechanical energy conversion are called piezoelectric sensors. If a force is applied to a piezoelectric crystal which is placed in between two electrode plates, a charge is developed across the plates that can be calibrated in terms of the force or acceleration applied to the crystals. The main application areas for piezoelectric sensors are measurements of acceleration and force. Piezoelectric crystals are also used to generate ultrasonic signals, since the effect is reversible.

10.2 The piezoelectric phenomenon

The piezoelectric effect was discovered in certain solid crystalline dielectric substances, such as quartz. Under stress-free conditions, when force is applied in the longitudinal direction, the placement of atoms in a piezoelectric crystal is as shown in figure 10.1. Charge is developed on the two faces A and B (figure 10.2).

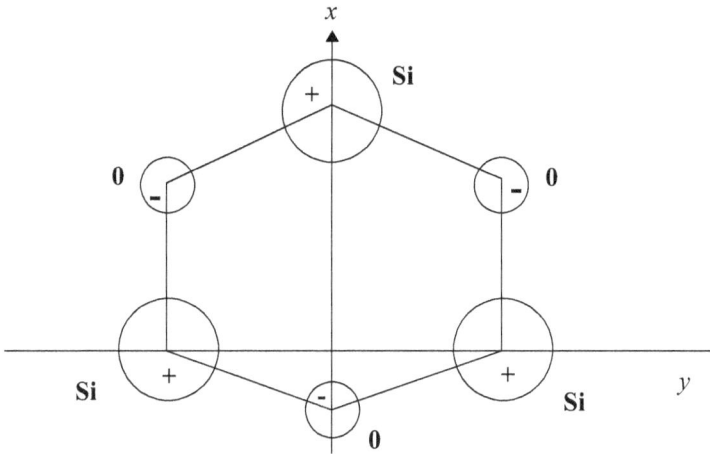

Figure 10.1. The arrangement of atoms in a piezoelectric crystal.

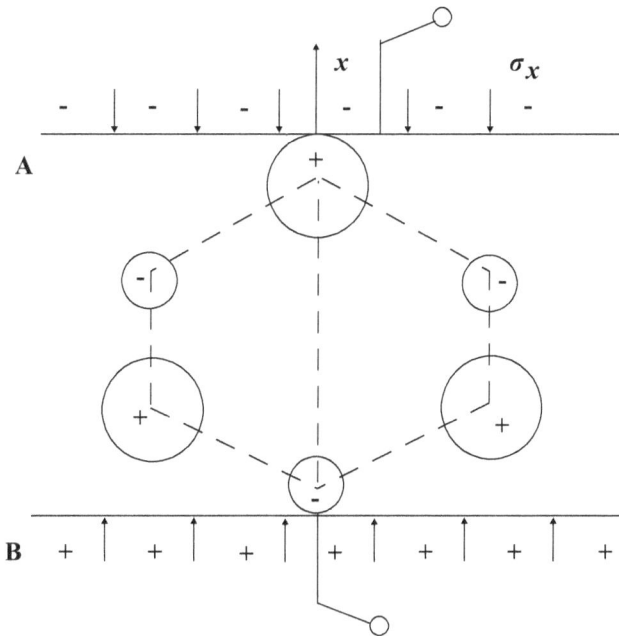

Figure 10.2. The development of charge in the longitudinal direction.

This is known as the longitudinal effect. When force is applied in the transverse direction (figure 10.3), charge is developed on the two faces A and B, which is known as the transverse effect.

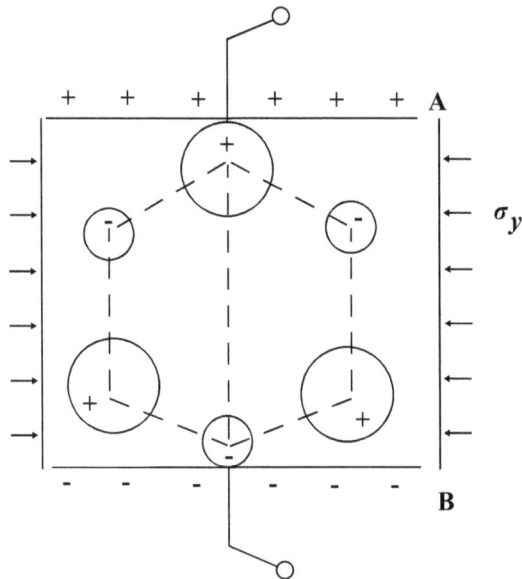

Figure 10.3. The development of charge in the transverse direction.

10.3 Piezoelectric materials

The materials that have a prominent piezoelectric effect fall into three main groups:
- Natural
    - quartz, Rochelle salt.
- Synthetic
    - lithium sulfate, ammonium dihydrogen phosphate.
- Polarized ferroelectric crystals
    - barium titanate, lead zirconate titanate.

Because of their natural asymmetric structure, crystalline materials other than ferroelectric crystals exhibit the effect without further processing. However, ferroelectric crystals need to undergo a certain process. Typically, ferroelectric materials must be polarized by applying a strong electric field to the material while it is heated to a temperature above the Curie point of the material. They are then slowly cooled down while the field remains intact. When the externally applied field is removed, they have a residual polarization which allows them to show the piezoelectric effect.

10.4 Piezoelectric transducers

When a force is applied, the crystal deforms and negative and positive charges are displaced from each other within the crystal. This displacement of charges causes charges of opposite polarity to appear on the two opposite surfaces of the crystal. When the surfaces are in contact with a metal electrode, the charge developed can be measured and calibrated in terms of the force applied. The electrodes become the plates of a capacitor since the piezoelectric material is basically an insulator. Therefore, when the force is applied to a piezoelectric transducer, it can be considered to be a charge generator and a capacitor.

We now define and explain the main characteristics of these crystals. Different double subscripted physical constants describe the phenomenon that occurs within the crystal. The first subscript refers to the direction of the electrical effect and the second to that of the mechanical effect using an axis numbering system. In figure 10.4, directions 1, 2, and 3 represent compression or tension and directions 4, 5, and 6 represent shear. A piezoelectric transducer with electrodes is shown in figure 10.5.

10.4.1 The physical constants of the piezoelectric transducer

Two families of constants, g constants and d constants, are used to describe the piezoelectric effect. These are expressed as g_{ij} and d_{ij}, where i is the direction of the

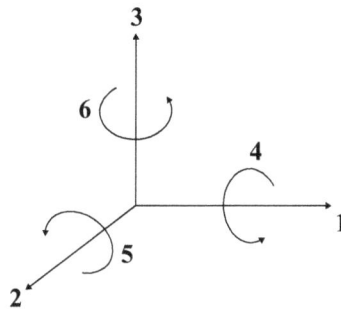

Figure 10.4. Axes representing compression, tension, and shear.

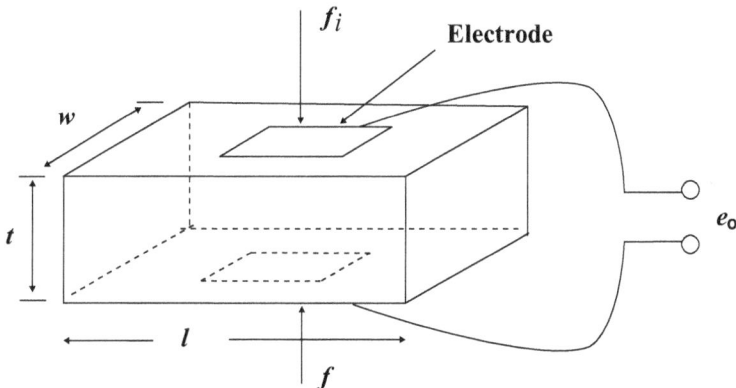

Figure 10.5. A schematic of a piezoelectric transducer and its electrodes.

electric effect and j is the direction of the mechanical effect. The g constant is defined as

$$g_{33} = \frac{\text{field produced in direction 3}}{\text{stress applied in direction 3}}$$

$$= \frac{e_0/t}{f_i/wl}.$$

(10.1)

The d constant is defined as

$$d_{33} = \frac{\text{charge generated in direction 3}}{\text{force applied in direction 3}}$$

$$= Q/f_i.$$

(10.2)

Moreover, $P = [d]\,\sigma$, where P is the polarization, σ is the stress, and $[d]$ is the sensitivity matrix.

10.5 Measuring circuits

The equivalent circuit for a piezoelectric crystal is shown in figure 10.6.

In figure 10.6,

R_{cry} = the leakage resistance of the crystal, which is of the order of 10^{12} Ω,

$C_{cry} = \frac{\epsilon\,wl}{t}$ is the capacitance of the capacitor formed by the two plates placed on the two opposite faces of the crystal, and

q = the charge generated.

The piezoelectric transducer consists of a crystal, an amplifier, and a cable. Such a measurement system is shown in figure 10.7. The crystal's source impedance is very high; therefore, the input impedance of the amplifier should also be extremely high. Since the capacitance of the cable is comparable to the crystal's capacitance, the cable's capacitance should also be considered when drawing an equivalent circuit for

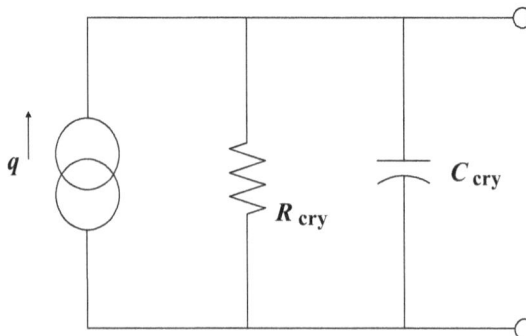

Figure 10.6. Equivalent circuit for a piezoelectric crystal.

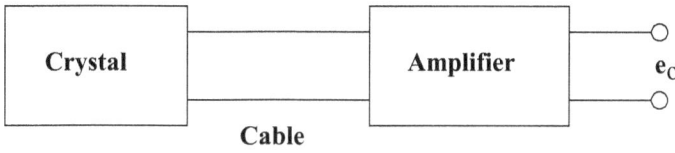

Figure 10.7. Measuring circuit.

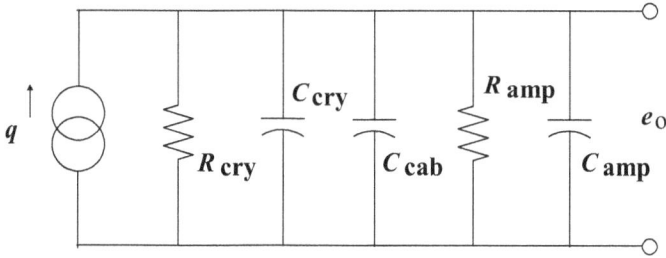

Figure 10.8. Equivalent circuit for a piezoelectric sensor.

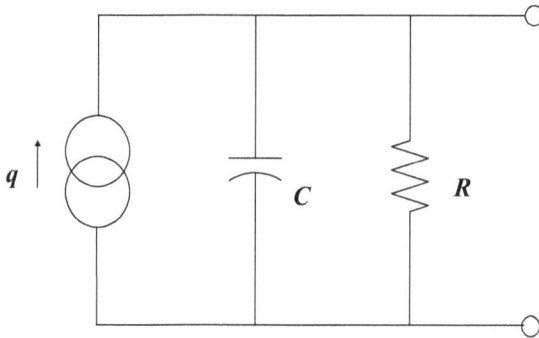

Figure 10.9. A simplified equivalent circuit.

the piezoelectric measuring system as a whole. The complete equivalent circuit is shown in figure 10.8.

A simplified equivalent circuit is shown in figure 10.9.

In figure 10.9, $C = C_{cry} + C_{cab} + C_{amp}$ and $R = R_{cry}||R_{amp}$. The output voltage e_o is given by

$$e_o = iR = \left(\frac{dq}{dt} - C\frac{de_o}{dt} \right)R, \tag{10.3}$$

where i is the current that passes through the resistor R. Equation (10.3) can be rewritten as

$$RC\frac{de_o}{dt} + e_o = R\frac{dq}{dt}$$

$$\therefore \frac{e_o}{q}(s) = \frac{Rs}{RCs + 1}. \tag{10.4}$$

We know that $q = d \times f$ (equation (10.2)), where, 'd', is the d constant of the crystal and f is the force applied to the crystal. Combining equations (10.2) and (10.4) yields

$$\therefore \frac{e_o}{f}(s) = \frac{dRs}{RCs + 1} = \frac{(d/C)\tau s}{1 + \tau s} \tag{10.5}$$

where $\tau = RC$, the time constant of the measurement system.

Equation (10.5) represents a high-pass transfer function. It is not possible to measure any static force with this circuit. The static sensitivity of the sensor system (d/C) changes with a change in any of the values of C_{cab} or C_{amp}. All the problems of measurement can be immediately overcome by using a charge amplifier as shown in figure 10.10.

An analysis of the circuit shown in figure 10.10 gives us
$\frac{dq}{dt} = -\frac{e_o}{R_f} - C_f \frac{de_o}{dt}$, where C_f and R_f are the feedback resistor and the capacitor, respectively,

$$\text{or } R_f C_f \frac{de_o}{dt} + e_o = -R_f \frac{dq}{dt}$$

$$\text{or } \frac{e_o}{q}(s) + e_o = \frac{-R_f s}{1 + R_f C_f s}$$

$$\text{or } \frac{e_o}{f}(s) = \frac{-(d/C_f)\tau_f s}{1 + \tau_f s}, \tag{10.6}$$

where $\tau_f = R_f C_f$.

Equation (10.6) shows that the charge amplifier circuit also cannot measure the static force, but it removes the dependencies of the output voltage on the cable capacitance and the resistances.

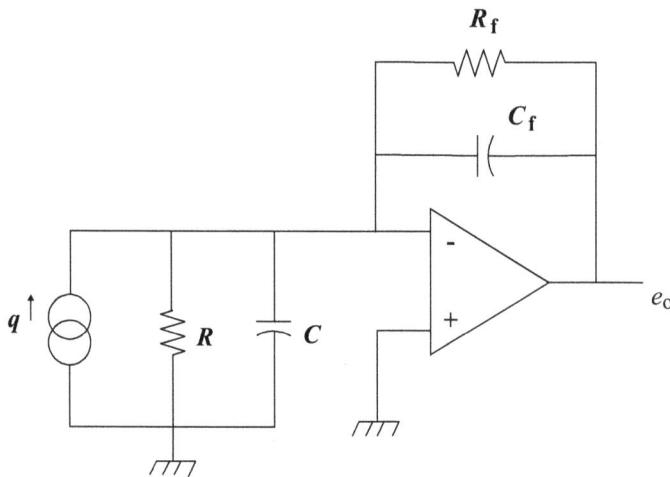

Figure 10.10. Measurement using a charge amplifier.

10.6 Piezoelectric accelerometers

Accelerometers are used for the measurement of acceleration, shock, and vibration. The destructive forces generated by vibration and shock are quantified by measuring acceleration. The piezoelectric accelerometer employs the principles of a seismic transducer by using a piezoelectric element to provide a portion of the spring force. The basic configuration of an accelerometer is shown in figure 10.11; however, the differences with respect to the seismic transducer are in the spring element, the transducer used to sense the relative motion, and the damping. A block diagram of a piezoelectric accelerometer is shown in figure 10.12. The crystal is bonded to a mass in such a way that when the accelerometer is subjected to a force g, the mass compresses the crystal and charge is developed that is measured and calibrated as the g force. The crystal is housed in a sensor body that withstands the environmental conditions. Typically, the body is made of stainless steel welded at various points to prevent the ingress of dust and water. A preload is applied to the piezoelectric element by tightening the nut that holds the mass and the piezoelectric element in place. Preloading is necessary for the sensor to work closer to the linear region of its charge–strain characteristics. The initial preloading produces an output voltage of a certain polarity.

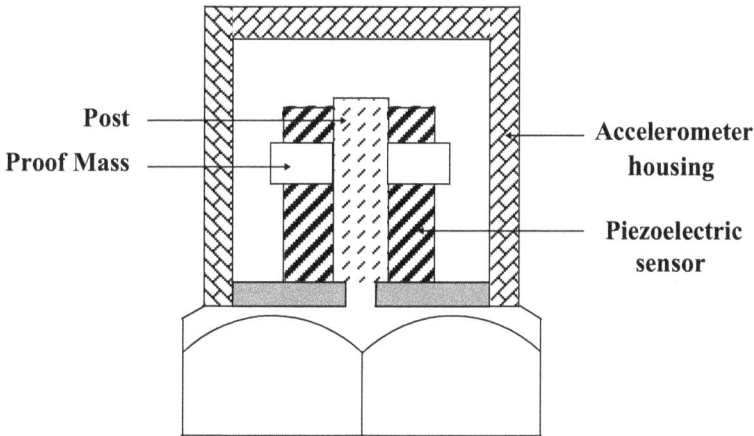

Figure 10.11. The basic configuration of a seismic pickup.

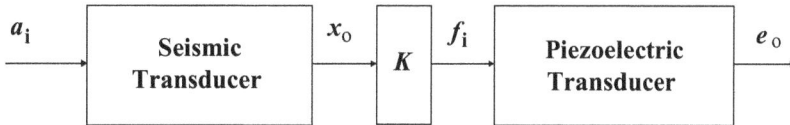

$K \rightarrow$ **Stiffness of the crystal.**

Figure 10.12. A block diagram of a piezoelectric accelerometer.

Considering the seismic transducer as a second-order system, we can write

$$\frac{x_o}{a_i}(s) = \frac{1}{s^2 + 2\xi\omega_n s + \omega_n^2},$$ (10.7)

where ξ and ω_n are the damping ratio and the natural frequency of the system, respectively. For a piezoelectric transducer, we can write

$$\frac{e_o}{f_i}(s) = \frac{(d/c)\tau s}{1 + \tau s},$$ (10.8)

where $\tau = RC$ and d = sensitivity.

Combining equations (10.7) and (10.8) yields

$$\frac{e_o}{a_i}(s) = \frac{e_o}{f_i} \cdot \frac{f_i}{x_o} \cdot \frac{x_o}{a_i}$$

$$= \frac{(d/c)\tau s}{1 + \tau s} \cdot K \cdot \frac{1}{s^2 + 2\xi\omega_n s + \omega_n^2}$$

or, $\dfrac{e_o}{f_{input}}(s) = \dfrac{1}{M}\dfrac{e_o}{a_i}(s)$

$$= \frac{(d/c)\tau s}{1 + \tau s} \cdot \left(\frac{K}{M}\right) \cdot \frac{1}{s^2 + 2\xi\omega_n s + \omega_n^2}$$

$$= \frac{(d/c)\tau s}{1 + \tau s} \frac{\omega_n^2}{s^2 + 2\xi\omega_n s + \omega_n^2}$$

The response curve of a typical piezoelectric accelerometer is shown in figure 10.13. Such curves provides guidance when selecting a sensor for dynamic measurement.

The magnitude of the deformations due to forces and the piezoelectric voltages developed are relatively small. The maximum deformation of a single element is in the order of fractions of a micron. Mechanical amplification is often required, which can be achieved by various arrangements of piezo-ceramic materials such as unimorphs, bimorphs, and stacks. Voltage amplification can be achieved using a high-gain direct-coupled charge amplifier.

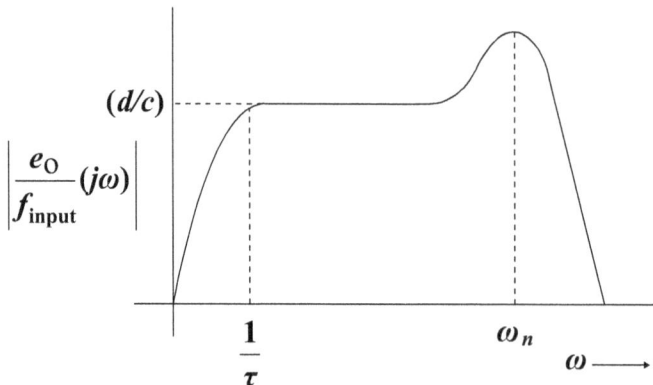

Figure 10.13. The response curve of a typical piezoelectric accelerometer.

10.7 Unimorphs

A unimorph is made by bonding a thin piece of piezo-ceramic material to an inactive substrate. Applying a force to the piezo-ceramic leads to the deformation of the entire structure. High-specification unimorphs are used for hydrophones, sensors, and actuators. A unimorph is shown in figure 10.14.

10.8 Bimorphs

A bimorph is made by bonding two pieces of piezo-ceramic material together so that the differential changes in the lengths of the two pieces can produce relatively large movements. This element consists of two transverse expander plates placed together face to face in such a manner that if a voltage is applied to the electrodes, it causes the plates to deform in opposite directions, resulting in a bending action. This process is reversible. The displacement of a bimorph in response to an applied voltage is many times greater than the corresponding displacement of a single plate and is typically in excess of 10 μm per volt. A bimorph is shown in figure 10.15.

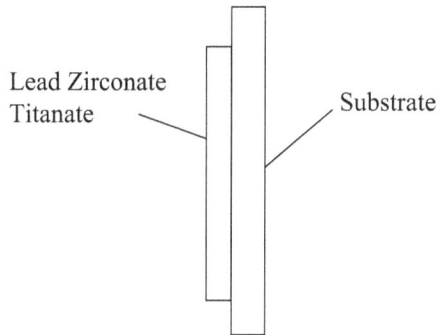

Lead Zirconate Titanate

Substrate

Figure 10.14. A unimorph.

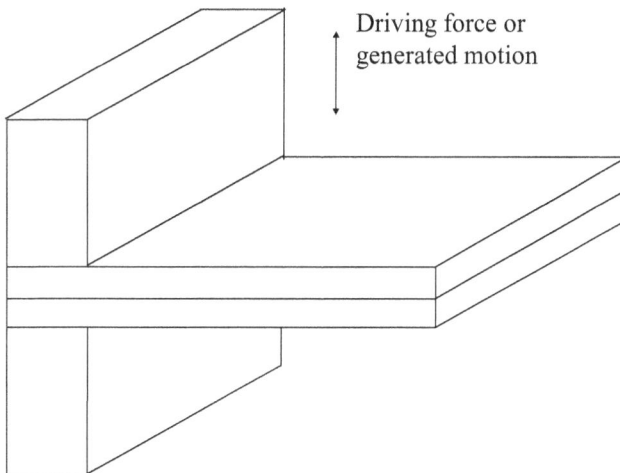

Driving force or generated motion

Figure 10.15. A bimorph.

10.9 Actuator stacks

Stacks consist of several piezoelectric sensors connected mechanically in series and electrically in parallel. The deformation of each transducer element adds to the total displacement. Essentially, the displacement of the whole stack is equal to the sum of the displacements produced by the individual sensors.

Stacks are typically required for measurements that require large displacements (usually between 5 and 180 μm). An actuator stack is shown in figure 10.16.

10.10 Sandwich piezoelectric transducers

These are used for low-frequency applications. It is not possible to manufacture single blocks of ceramic whose resonant frequency lies below 100 kHz. Therefore, a composite half-wave resonator is used consisting of two or more piezo-ceramic rings sandwiched between metal layers. A schematic of a sandwich transducer is shown in figure 10.17. This is a typical sandwich transducer which is used to achieve low frequencies at high drive levels.

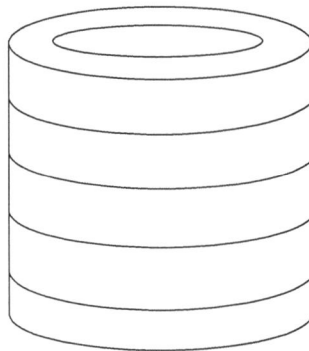

Figure 10.16. An actuator stack

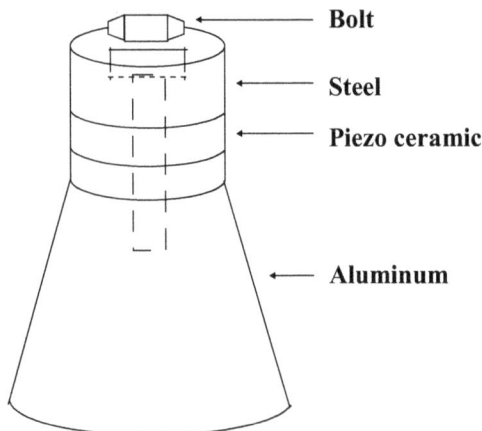

Figure 10.17. A schematic of a sandwich piezoelectric transducer.

10.11 Pyroelectricity

Piezoelectric materials are also pyroelectric. Their properties depend on temperature. Hence, a discussion of piezoelectricity would be incomplete without describing the pyroelectric effect. The pyroelectric effect is exhibited by synthetic ferroelectric ceramics which also show piezoelectric properties. Pyroelectric materials are substances that produce an electric charge when they are exposed to a temperature change. When their temperature is increased, a voltage develops that has the same orientation as the polarization voltage. When their temperature is decreased, a voltage develops that has an opposite polarity to the polarization voltage, which creates a depolarizing field with the potential to degrade the state of polarization of that part.

The maximum electric field which arises due to a temperature shift is

$$E_{(Pyro)} = \frac{\alpha(\Delta T)}{K_3 \, \varepsilon_0},$$

where

$E_{(Pyro)}$ = the induced electric field V m^{-1},
α = the pyroelectric co-efficient Coulomb °C^{-1} m^{-2},
ΔT = the temperature difference °C,
K_3 = the dielectric constant, and
ε_0 = the dielectric permittivity of free space = 8.9×10^{-12} F m^{-1}.

Note: For lead zirconate titanate piezo ceramic, α is typically 400×10^{-6} Coulomb °C^{-1} m^{-2}.

10.12 Limitations of piezoelectric materials

Each piezoelectric material has a particular operating range in terms of temperature, voltage, and applied stress. The characteristics or composition of the materials determines these limits. Using the material outside these ranges may cause partial or total depolarization of the material, which leads to the reduction or loss of its piezoelectric properties.

10.12.1 Temperature limitation of piezoelectric ceramics

As the temperature of operation increases, the performance of a piezoelectric material deteriorates until complete and permanent depolarization occurs at a temperature that is called the material's Curie temperature. Each ceramic material has its own Curie point, which is the maximum temperature of operation for any piezoelectric ceramic. The manufacturer usually documents this temperature. Moreover, the high-temperature limit rapidly decreases with greater continuous operation or exposure. At higher temperatures, the ageing process of sensors happens fast. Hence, the performance of a piezoelectric material decreases, and the safe stress level is lowered. Frequent sensor calibration is suggested for this type of use.

10.12.2 Voltage limitation

A piezoelectric ceramic can be depolarized by a strong electric field whose polarity is opposite to the original voltage used during polarization. The limit on the field strength depends on the type of material, the duration of the application, and the operating temperature. The typical voltage limit is between 500 and 1000 V mm^{-1} for continuous operation.

10.12.3 Mechanical stress limitation

A piezoelectric ceramic becomes depolarized if it is subjected to high mechanical stress. The limit on the applied stress is dependent on the type of ceramic material and the duration of the stress. However, for dynamic stress measurements, the limit is not low and materials with higher energy outputs (high g constants) can be used.

10.12.4 Power limitation

Ultrasonic transducers (both transmitters and receivers) are basically piezoelectric sensors, since they are reversible in nature. The acoustic power handling capacity of an ultrasonic transducer is limited by various factors, namely the dynamic mechanical strength of the ceramic, reduction in efficiency due to dielectric losses, reduction in efficiency due to mechanical losses, depolarization of the ceramic due to electric fields, depolarization of the ceramic due to temperature increases, and instability resulting from the positive feedback between dielectric losses and internal heating. Here, frequent calibration of transducers is also advised.

Problem 10.1. A piezoelectric sensor that has a sensitivity of 2.0 pC N^{-1}, a capacitance of 1600 pF, and a leakage resistance of 10^{12} Ω is connected to a charge amplifier as shown in figure 10.18. If a force of $0.1 \sin 10t$ N is applied to the sensor, what is the amplitude of the charge amplifier output?

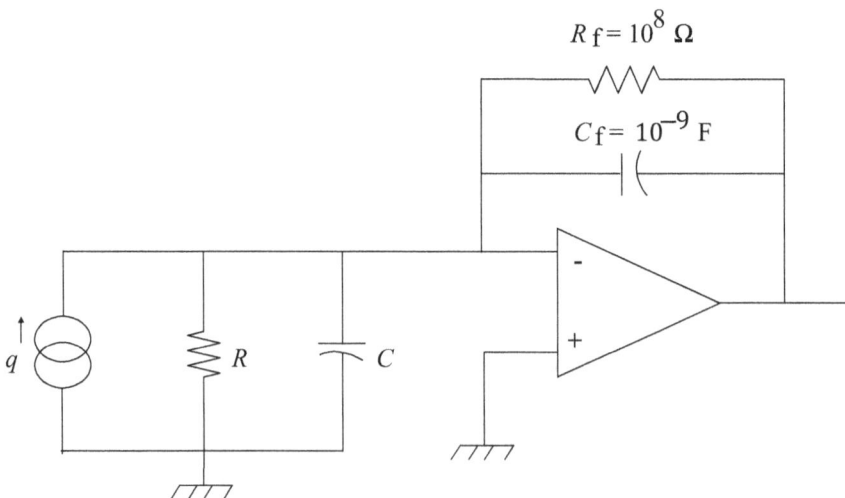

Figure 10.18. Measurement using charge amplifier.

Problem 10.2.   A displacement signal of 10^{-8} sin 100t meter is applied to a quartz crystal of dimensions $10 \times 10 \times 2$ mm. Find the voltage generated. The crystal data are as follows:

- charge sensitivity $(d) = 2\text{pC N}^{-1}$,
- Young's modulus $(E) = 8.6 \times 10^{10}$ N m^{-2}, and
- permittivity $(\varepsilon) = 42 \times 10^{-12}$ F m^{-1}.

Problem 10.3.   A quartz clock employs a ceramic crystal with a nominal resonant frequency of 32.768 kHz. The clock loses 30.32 s in every 23 days. What is the actual resonant frequency of the crystal?

IOP Publishing

Fundamentals of Industrial Instrumentation (Second Edition)

Alok Barua

Chapter 11

Ultrasonic sensors

Learning objectives:
- The ultrasonic principle.
- Transmitters and receivers.
- The equivalent circuit for an ultrasonic transmitter.
- The characteristics of the transmitter.
- Why it is easier to launch ultrasonic waves in metals or liquids than in the air.

11.1 Introduction

Ultrasonic measurement systems consist of three elements: a transmitter, the transmission medium, and the receiver. Commonly used ultrasonic sensors are made with piezoelectric sensing elements for both the transmitter and the receiver. A schematic of an ultrasonic transmitter and an ultrasonic receiver is shown in figure 11.1.

11.2 Analysis

The piezoelectric effect is reversible, i.e. if we apply a force the an ultrasonic sensor, charge is developed on particular faces of the crystal and vice versa. The ultrasonic transmitter uses the inverse piezoelectric effect, i.e. if a voltage (V) is applied to the transmitter, the crystal undergoes deformation (X) in a particular direction. If we apply an AC input to the crystal, it vibrates, and the vibration of the crystal is transmitted through the medium from one end to the other. The particle displacement creates an accompanying pressure which is picked up by the receiver. The receiver uses the direct piezoelectric effect and converts the force (F) into the corresponding voltage (V).

For the transmitter, when a voltage V is applied to the crystal, the displacement x can be written as

$$x = d \times V,$$

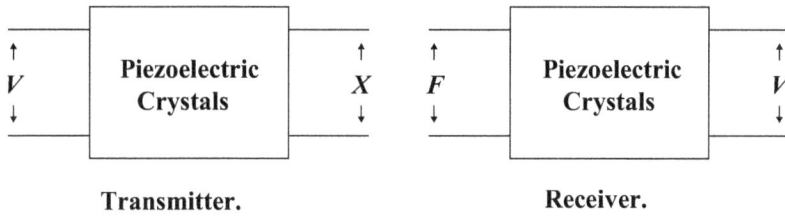

Figure 11.1. A schematic of an ultrasonic transmitter and an ultrasonic receiver.

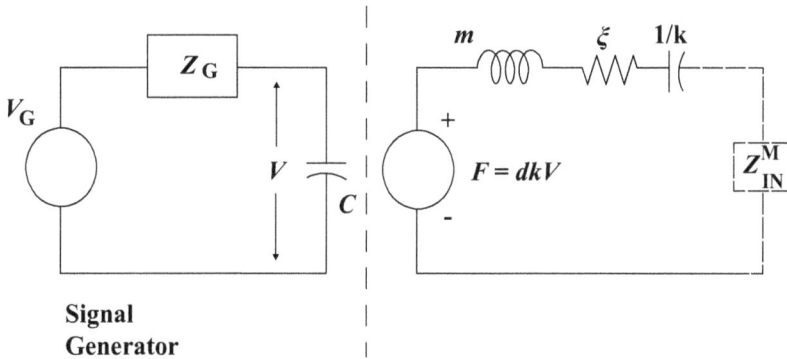

Figure 11.2. Equivalent circuit of a transmitter. Z_G = the output impedance of the signal source; m = the mass of the crystal; ξ = the damping coefficient of the crystal; K = the spring constant; Z_{IN}^M = the input impedance of the medium; x = the velocity of ultrasonic waves in the medium.

where d $(=K/k)$ is the charge sensitivity to force and its unit is C N^{-1}. Here, k is the stiffness of the crystal.

For the receiver, the net charge developed can be expressed as

$$q = d \times F,$$

where F is the applied force. The performance characteristic d is the same in both cases.

Moreover,

$$F = x \times k$$
$$= d \times V \times k.$$

11.3 The equivalent circuit for the transmitter

Figure 11.2 illustrates the equivalent circuit for a transmitter
Ideally,

$$Z_G = 0; \quad Z_{IN}^M = 0.$$

The ideal equivalent circuit for a transmitter is shown in figure 11.3, and its equivalent circuit with mass (m), damping coefficient (ξ), and the inverse spring constant ($1/K$) reflected on the primary side is shown in figure 11.4. The transformer ratio is $1{:}dk$, where $L_1 = m/(dk)^2$, $R_1 = \xi/(dk)^2$, and

$$C_1 = d^2 k.$$

The overall impedance = $H(s)$.

$$\text{or,} \quad \frac{1}{H(s)} = sC + \frac{1}{R_1 + sL_1 + \frac{1}{sC_1}}.$$

Therefore,

$$H(j\omega) = \frac{\omega R_1 C_1 - j(1 - \omega^2 L_1 C_1)}{\omega[(C + C_1) - \omega^2 L_1 C C_1] + j\omega^2 C C_1 R_1}.$$

It is evident that there are two natural frequencies:

ω_n (the series natural frequency) $= \dfrac{1}{\sqrt{L_1 C_1}}$

ω_p (the parallel resonant frequency) $= \sqrt{\dfrac{(C + C_1)}{L_1 C C_1}}$.

Magnitude and phase plots for the transmitter are shown in figures 11.5 and 11.6, respectively.

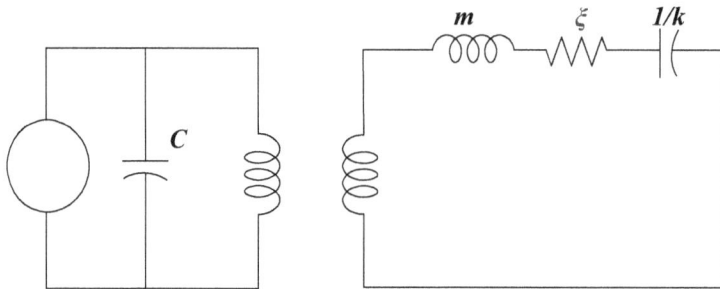

Figure 11.3. The ideal equivalent circuit for the transmitter.

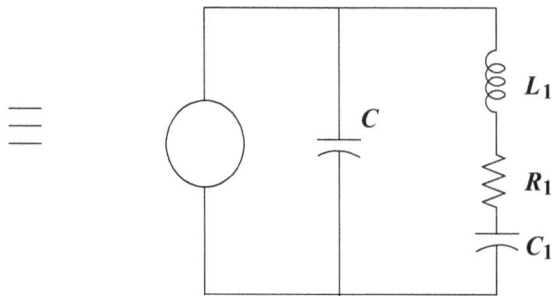

Figure 11.4. The equivalent circuit for a transmitter with m, ξ, and $1/k$ reflected on the primary side.

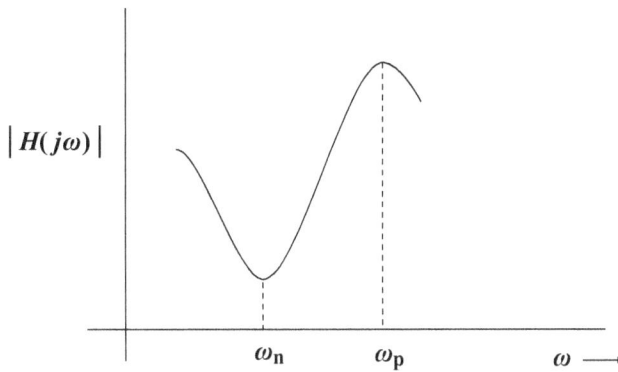

Figure 11.5. A magnitude versus frequency plot for the transmitter.

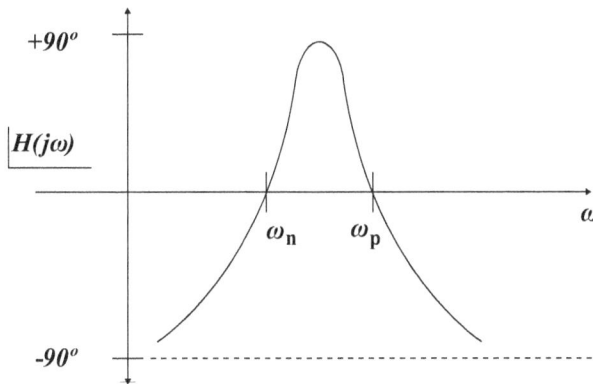

Figure 11.6. A phase plot for the transmitter.

At $\omega = \omega_n$, the magnitude is minimized, and it is maximized at $\omega = \omega_p$. The system becomes resistive at $\omega = \omega_n$ and $\omega = \omega_p$ if we assume that $R_1 = 0$. If $R_1 \neq 0$, the above plot moves toward the right-hand side. Therefore, the circuit behaves as an inductor in the frequency range between ω_n and ω_p.

11.4 The transmission of ultrasound

If P = pressure or stress,

$$\dot{x} \; = \; u = \text{velocity.}$$

The characteristic impedance of the medium

$$Z = P/u.$$

The power intensity of an ultrasonic wave

$$W = P \times u.$$

11.4.1 The average power intensity

A schematic of the transmission of an ultrasonic signal between two media is shown in figure 11.7.

The power intensity is given by

$$W = \frac{1}{\lambda} \int_0^\lambda W(z)dz$$

WI is lower than the power intensity of the signal generated by the crystal at the launch time due to transmission losses in medium 1.

We define α_R = the reflection coefficient = $\frac{WR}{WI} = \frac{(Z_2 - Z_1)^2}{(Z_2 + Z_1)^2}$

and α_T = the transmission coefficient = $\frac{WT}{WI} = \frac{4Z_1 Z_2}{(Z_2 + Z_1)^2}$.

Thus, $\alpha_R + \alpha_T = 1$.

If $Z_2 \gg Z_1$ then most of the incident power intensity is reflected back. For example, if medium 1 is a liquid or a metal and medium 2 is air, then such a situation will arise.

11.4.2 The characteristic impedances of selected materials

Quartz	$= 1.5 \times 10^7$
Barium titanate	$= 2.5 \times 10^7$
Polymer (PVDF)	$= 0.4 \times 10^7$
Steel	$= 4.7 \times 10^7$
Aluminum	$= 1.7 \times 10^7$
Bone	$= 0.8 \times 10^7$
Water	$= 0.15 \times 10^7$
Air	$= 430$

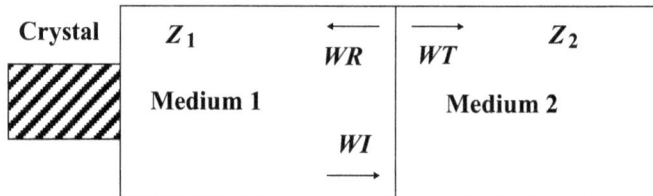

Figure 11.7. The transmission of an ultrasonic signal between two media. Z_1 = the characteristic impedance of medium 1; Z_2 = the characteristic impedance of medium 2; WI = the incident power intensity in medium 1; WR = the reflected power intensity in medium 1; WT = the transmitted power intensity in medium 2.

The following table shows α_R and α_T for different interfaces.

	α_R	α_T
Quartz/steel	0.27	0.73
Quartz/water	0.67	0.33
Quartz/air	1.00	1.1×10^{-4}

Thus we can say that air is a poor choice for the transmission of ultrasound signals, as the characteristic impedance of air is much higher than those of liquids or metals. Polyvinylidene difluoride (PVDF) is used for biomedical applications.

11.5 Measuring ultrasound

In this section, a pulse echo technique for the measurement of ultrasound will be discussed. Such a measurement scheme is shown in figure 11.8. A piezoelectric crystal is used as a transmitter/receiver which is attached to medium 1. The characteristic impedances of media 1 and 2 must be widely different. A pulse is generated by the pulse generator. Initially, the crystal acts as a transmitter and it sends out a pulse of width T_W.

Most of the pulse energy is reflected at the boundary of medium 1 and 2 because of the wide difference between the characteristic impedances of the two media. The crystal now acts as a receiver and it receives the reflected pulse. The total time taken by the reflected pulse to travel is

$$T_T = \frac{2L}{C},$$

where $L =$ the distance from the interface between the two media to the crystal and $C =$ the velocity of sound in medium 1.

The reflected wave of the ultrasonic signal is shown in figure 11.9. The subsequent reflected pulses are of reduced amplitude.

The repetition rate T_R should be long enough that all the reflected pulses of interest are observed before sending the second pulse. The transmit time T_T should be large compared to the pulse width T_W to avoid overlap between the outgoing pulse and the incoming or reflected pulses.

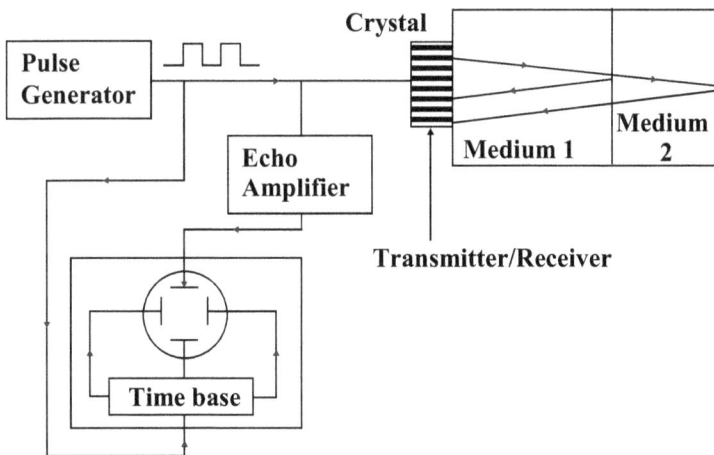

Figure 11.8. An ultrasound measurement scheme.

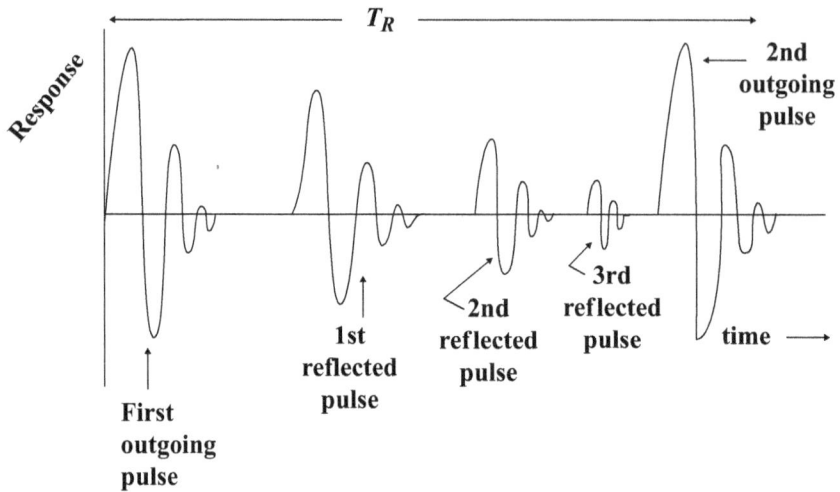

Figure 11.9. The reflected wave of the ultrasonic signal.

11.5.1 Applications

The method discussed above can be used for the following cases of measurement of level and crack detection.

(a) Level measurement.

It should be mentioned that the transmitter/receiver must be placed at the bottom of the liquid container and not at the top. If it is placed at the top, due to presence of air in between the position of the ultrasonic sensor and the top surface of the liquid, no waves can propagate through the liquid, thus giving the wrong measurement. If the sensor is placed at the bottom, then most of the pulse energy is reflected back from the junction of the liquid medium and the air. The reflected ultrasonic signal is received by the crystal when it works as a receiver. A schematic of the measurement of liquid levels is shown in figure 11.10.

(b) Crack detection.

Here, a crack or gap acts as the second medium and the ultrasonic wave is reflected from the junction of solid and crack, which helps us to detect where the crack has occurred by measuring the travel time between the transmitted signal and the received signal (figure 11.11). Ultrasonic sensors are used to measure liquid flows, which was discussed in detail in chapter 7.

11.5.2 Advantages

- It is easy to launch and focus the beam of an ultrasonic signal, as the diffraction of these waves is small due to their short wavelength. They do not disperse.

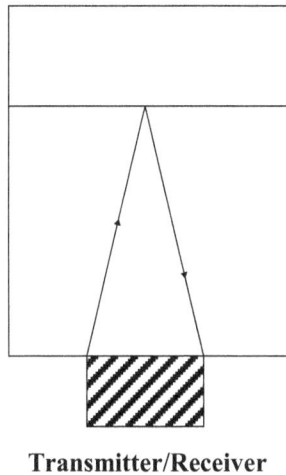

Transmitter/Receiver

Figure 11.10. A schematic of an ultrasonic level measurement system.

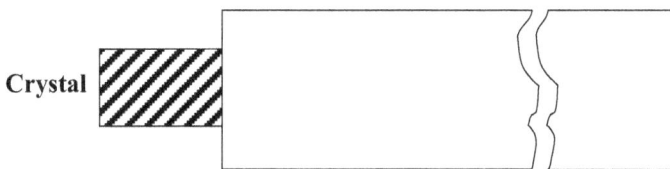

Figure 11.11. A schematic of a crack detection method.

- Ultrasonic signals can easily pass through metals and liquids. This helps because it makes it possible to install the measurement system outside the item to be measured, and it has led to the development of noninvasive sensors.

11.6 Special applications

Ultrasonic sensors have wide application in biomedical instrumentation and some such applications are discussed in this section (figure 11.12).

11.6.1 Determining blood pressure

This measurement system utilizes a Doppler shift technique that detects the motion of the blood vessel walls at various states of occlusion. Two ultrasonic crystals (8 MHz) are used, one as a transmitter and other as a receiver. The signal trasmitted by the ultrasonic transmitter is focused on the vessel wall and the blood. The reflected signal is detected by the ultrasonic receiver and calibrated. The difference between the transmitted and received signals is in the frequency range of 40–500 Hz and is proportional to the velocity of the wall motion and the velocity of the blood. When the cuff pressure is above the diastolic pressure but below the systolic pressure, the vessel opens and closes for each heartbeat, as the pressure in the artery also oscillates below and above the external cuff pressure. The opening and closing of the vessel/

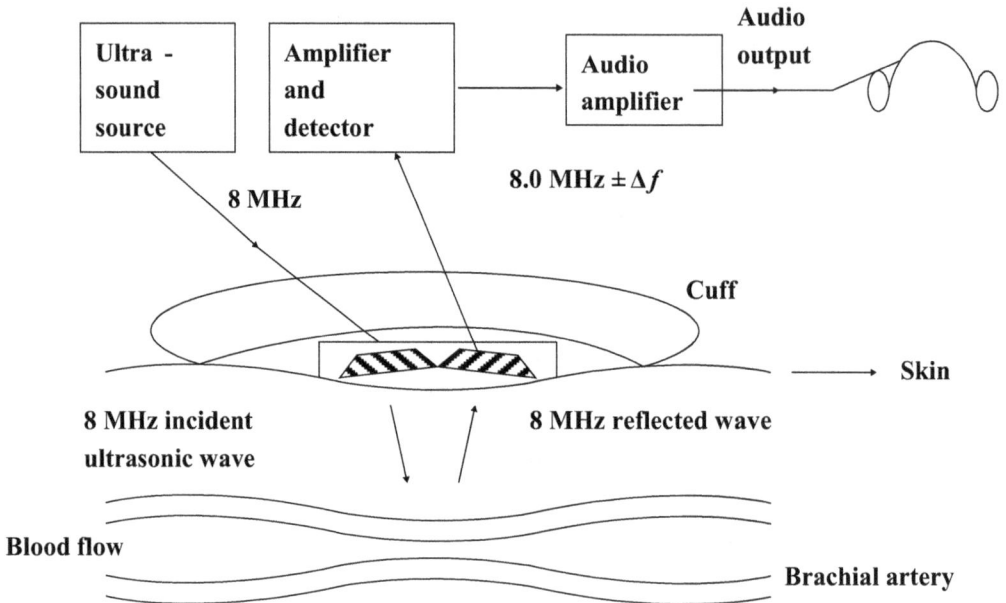

Figure 11.12. Determination of blood pressure by an ultrasonic technique.

artery are determined by the ultrasonic system. As the pressure is increased further, the time between opening and closing decreases and the measurement indicates the systolic pressure. On the other hand, when the pressure in the cuff is decreased, the diastolic pressure is recorded. The advantages of the system are that it can be used on children and newborn babies and it is immune to highly noisy environments. Its main disadvantage is that movements of the body cause changes in the ultrasonic path between the sensor and the blood vessel, thus giving false outputs.

11.6.2 The blood flow meter

The measurement of the flow of liquid in a pipe is discussed in chapter 7. The same principle is applied here for the measurement of blood flow in an artery. As we know, two types of measurement are possible using ultrasonic sensors: one is the measurement of transit time and the other is the Doppler shift method.

Figure 11.13 illustrates the various ultrasonic transducer configurations. For case (I), the transit time probe technique requires two ultrasonic transducers facing each other. The path length between them is L and they are inclined to the tube/artery axis at an angle θ. The shaded region represents an acoustic pulse. For a transmitter/ receiver probe as shown in (II), both transducers are placed on the same side. They can be placed on the skin of the human body. The beam intersection region is shaded. For case (III), a lens can be used to focus and narrow the beam. Case (IV) is used for pulse operations. In this case, the transducer is loaded by a backing material which is a mixture of tungsten powder in epoxy. The shaded region indicates a single time of range gating.

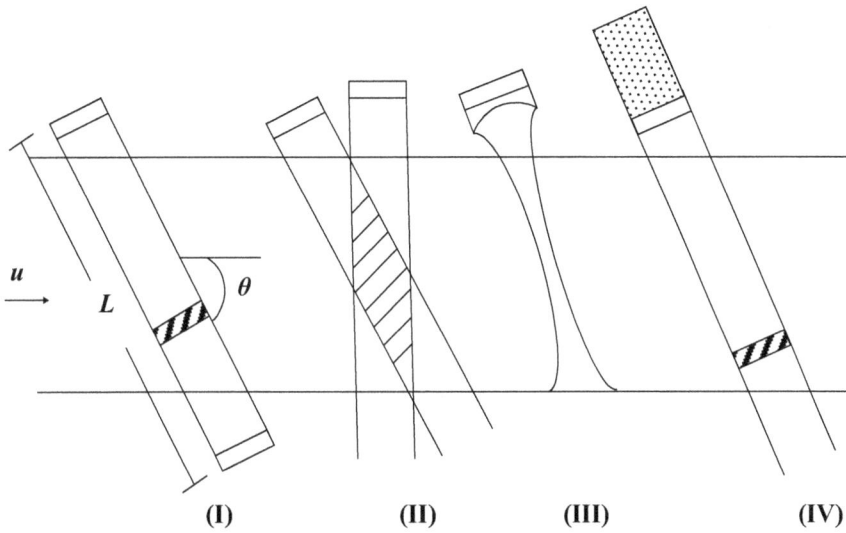

Figure 11.13. Transit time ultrasonic transducer configurations.

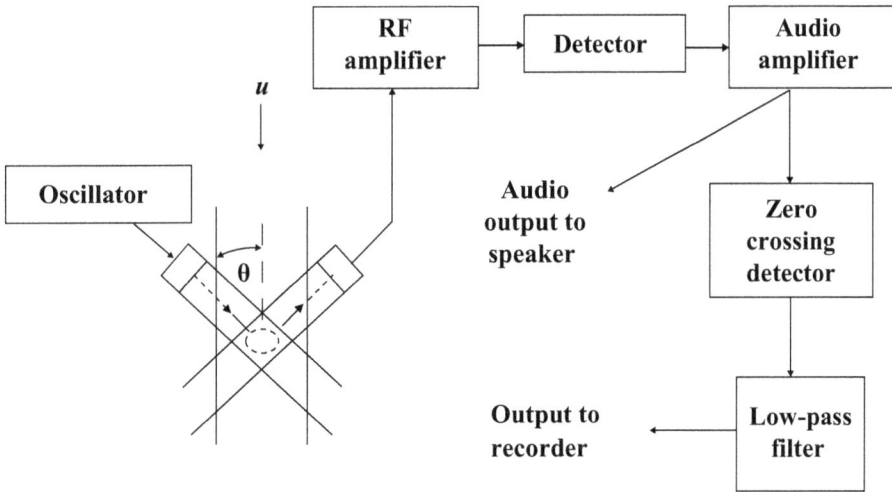

Figure 11.14. Doppler shift ultrasonic blood flow meter.

A Doppler shift blood flow meter is shown in figure 11.14. For the Doppler flow meter, we know that

$$\frac{f_d}{f_o} = \frac{u}{c},$$

where

- f_d = the Doppler frequency shift,
- f_o = the source frequency,
- u = the target velocity, and
- c = the velocity of sound in blood.

For blood flow meters, the blood cells act as the particulate matters which form reflecting particles. In the arrangement as shown in the figure, a frequency shift occurs twice, once between the transmitting source and the moving cell and again between the moving cell and the receiver:

$$\frac{f_d}{f_o} = \frac{2u}{c}$$

as $c \gg u$.

Taking the angle θ, which is the angle between the ultrasonic wave and the axis of the blood artery,

$$f_d = \frac{2f_o u \cos \theta}{c}$$

As shown in the figure, the ultrasonic waves are transmitted to the blood cell, which reflects the Doppler shifted wave to the receiver. The amplified radio frequency (RF) signal along with the carrier signal are detected and demodulated to separate the signal from the carrier frequency and thereby produce an audio-frequency signal.

Problem 11.1. An ultrasonic sensor is used to find cracks or gaps inside a metal. Once transmitted, the ultrasound wave is reflected back to the transmitter/receiver if there is a crack or gap in the metal. An ultrasonic transmitter uses a frequency of 330 kHz. When testing a specimen, a reflected wave is recorded 0.05 ms after the transmitted pulse. If the velocity of sound in the test object is 6.0 km s^{-1}, at what depth is the crack located?

Problem 11.2. An ultrasonic beam that has a frequency of 1 MHz and an intensity of 0.5 W cm^{-2} passes through a layer of soft tissue of thickness t with an attenuation coefficient of 1.18 cm^{-1}. The ratio of the output power to the input power is $\frac{1}{e^2}$. What is the thickness of the tissue?

IOP Publishing

Fundamentals of Industrial Instrumentation (Second Edition)

Alok Barua

Chapter 12

The measurement of magnetic fields

Learning objectives:
- The construction of a search coil.
- The Hall-effect transducer.
- The relations between various parameters such as the magnetic field strength, flux density, the output voltage of the coil, the total flux through the loop, etc. for a magnetometer search coil.
- The way in which parameters such as flux density, the dimensions of the sensor, its Hall coefficient, current density, Hall voltage, etc. are related in Hall-effect transducers.
- The use of Hall-effect transducers to measure current, flux density, power, etc.

12.1 The measurement of magnetic fields using search coils

A schematic of a search coil used to measure magnetic fields is shown in figure 12.1. It consists of a flat coil with N turns which is placed in the magnetic field as shown in figure 12.1. Let us assume that the length of the coil is L and its cross-sectional area is A.

The magnetic field strength and magnetic flux density are H and B, respectively. They act in the direction shown. H and B are related by the following expression:

$$B = \mu H,$$

where
- B = the magnetic flux density in Wb m^{-2},
- H = the magnetic field strength in A m^{-1}, and
- μ = the magnetic permeability of free space, $4\pi \times 10^{-7}$ H m^{-1}.

The voltage output (E) of the coil is given by

$$E = NA \cos \alpha \frac{\mathrm{d}B}{\mathrm{d}t},$$

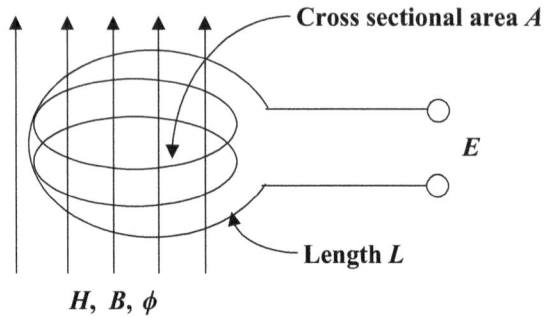

Figure 12.1. A magnetometer search coil.

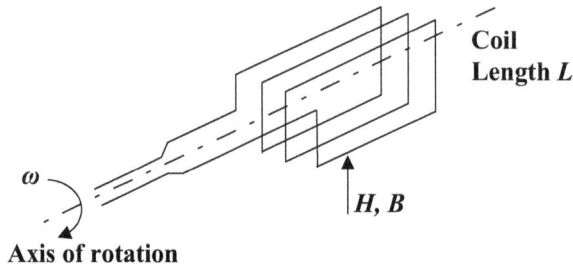

Figure 12.2. A rotating search coil used for the measurement of a magnetic field.

where A is cross-sectional area of the coil in m², α is the angle formed between the direction of the magnetic field and the perpendicular line to the plane of the coil, and N is the number of turns of the coil.

The total flux through the loop is

$$\phi = A \cos \alpha \, B,$$

So that

$$E = N\frac{d\phi}{dt}.$$

The voltage output of the transducer is proportional to the rate of change of the magnetic field. If a stationary coil is placed in a steady magnetic field which is not time varying, then there no output voltage is produced across the two terminals of the coil. The search coil is thus a transducer that converts a magnetic field into a voltage. To facilitate the measurement of a steady magnetic field, it is therefore necessary to introduce movement of the search coil. A schematic of a rotating search coil in a magnetic field is shown in figure 12.2.

The root mean square (rms) voltage output of such a device is

$$E_{\text{rms}} = \frac{1}{\sqrt{2}} N A B \omega,$$

where ω is the angular velocity of rotation of the coil.

The dimensions of the coil must be measured accurately to obtain accurate measurements of magnetic fields. The coil should be small enough that the magnetic field is constant over its area.

12.2 The Hall effect

The Hall effect is responsible for the voltage developed (known as the Hall voltage) on opposite sides of a thin sheet of a conductor or a semiconductor material (called a 'Hall bar') through which a current is flowing and a magnetic field is applied perpendicular to the Hall element. The ratio of the voltage developed to the magnitude of current flowing is known as the *Hall resistance* and is a characteristic of the materials used in transducers. A schematic of a Hall-effect sensor is shown in figure 12.3.

12.2.1 Hall-effect transducers

A semiconductor plate of thickness t is installed as shown in the figure so that, due to the applied voltage, a current I passes through the material. When a magnetic field of flux density B is applied to the plate in a direction perpendicular to the surface of the plate, a voltage E_H is generated, as shown. This voltage is called the Hall voltage and is given by

$$E_H = K_H \frac{IB}{t}$$

where I is in amperes, B is in Gauss, and t is in centimeters.

The proportionality constant (K_H) is called the Hall coefficient and has units of volt centimeters per ampere gauss. Hall-effect devices produce a very low level of signal voltage that has to be amplified. Hall-effect devices contain both the sensor and a high-gain integrated circuit amplifier in a single package.

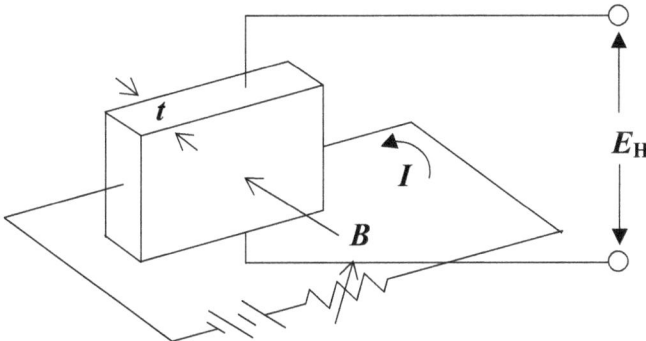

Figure 12.3. A schematic of a Hall-effect transducer.

12.2.2 The advantages of Hall-effect transducers

Hall-effect transducers are immune to dust, dirt, mud, and water. These characteristics are appropriate and suitable for the manufacture of position sensors compared to alternative means such as optical and electromechanical sensing methods.

12.2.3 The measurement of current by Hall-effect transducers

If both AC and DC are to be measured, sensors based on the Hall effect can be used. In a typical 'open-loop' sensor (figure 12.4), the current-carrying conductor is passed through a 'hole' in a gapped ferrite core which is used to concentrate the magnetic field. The Hall sensor is placed in the gap. As we know, the Hall sensor is a magneto-sensitive semiconductor that provides an output voltage proportional to the product of its current (which is held constant) and the component of the magnetic field that is perpendicular to its surface. If we keep the current constant, then the output voltage is directly proportional to the magnetic field. Moreover, this field is proportional to the current being measured. Therefore, the device output voltage can be calibrated in terms of the current that is to be measured. The Hall current sensor has a range of 0–350 A and a frequency response from 0 to 1 kHz.

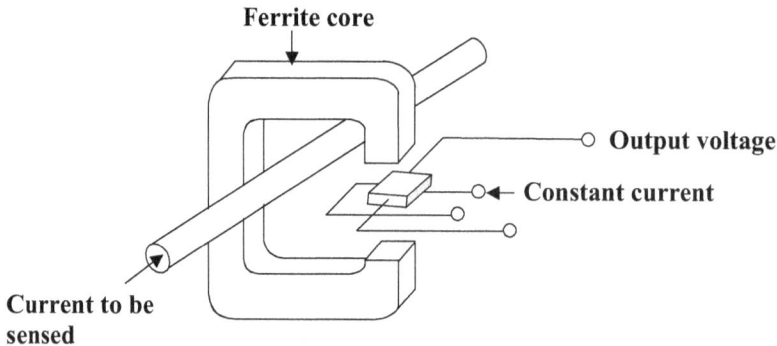

Figure 12.4. A current transducer based on the Hall effect.

12.2.4 Commercially available Hall-effect current transducers

An electromagnetic field is produced by the flow of current in the coil of an electromagnet. It is thus possible to create a noncontact current sensor in which the current-carrying conductor of the electromagnet is threaded through a hole in the sensing device. The device has three terminals. It is shown in figure 12.5.

Hall-effect current sensors are available with built-in integrated circuit amplifiers. The opening is typically 8 mm. The zero current output voltage is midway between the supply voltages, which lie in between 4 and 8 V. The nonzero current output is proportional to the voltage supplied and is linear to 60 A for this particular device. Among the three terminals, the sensor voltage is available across two terminals and at the third terminal, the voltage is proportional to the current being measured. This has several advantages; no resistance or shunt needs to be inserted in the primary

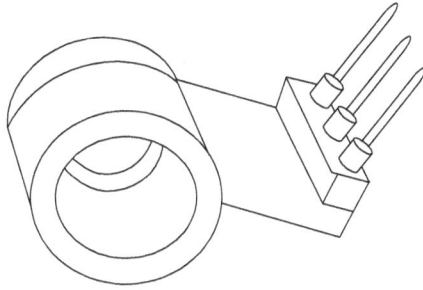

Figure 12.5. A current transducer.

circuit. Moreover, the voltage present on the line or conductor to be sensed is not connected to the transducer. This characteristic improves the safety of measuring equipment. The commercially available Hall-effect current sensors have a wide variety of ranges.

12.2.5 Split ring clamp-on current transducers

The split ring clamp-on current transducer is a noninvasive technique for current measurement that does not require the current-carrying conductor to be cut. It is a variation of the ring transducer (which has a permanent hole). Instead, it uses a split sensor which is clamped onto the line rather than having the line threaded through it, enabling the device to be included in test equipment that is not permanently installed on the current-carrying conductor to be tested. This helps the user to install the sensor at any portion of the current-carrying conductor without disrupting the supply line.

12.2.6 Analog multiplication

It is observed that for the current transducer, the output is proportional to both the applied magnetic field or flux density and the current passing through the sensor or the applied sensor voltage. If the magnetic field is generated by a solenoid, the sensor output is proportional to the product of the current passing through the solenoid and the sensor voltage or current.

12.2.7 Power measurement

As we know, direct electrical power is the product of current and applied voltage. By measuring the current through a load and using the device's supply voltage as the sensor voltage, it is possible to determine the power flowing through a device. With additional circuitry, the device may be used for alternating current power.

12.2.8 Position and motion sensing

Hall-effect devices can be used for motion sensing and also as motion limit switches or for position sensing with enhanced reliability in extreme environments. As there are no moving parts involved within the transducer or magnet, its life expectancy or mean time between failure is far better than those of electromechanical switches.

Problem 12.1. A search coil has 10 turns and a cross-sectional area of 10 cm². It rotates at a constant speed of 100 r.p.m. The output voltage is 80 mV. Calculate the magnetic field strength.

Problem 12.2. A magnetometer search coil has a nominal area of 1 cm² and 100 turns. The rotational speed is nominally 180 r.p.m. Calculate the voltage output when the coil is placed in a magnetic field of 1 Wb m^{-2}.

Problem 12.3. A germanium crystal that has dimensions of 6 × 6 mm² and a thickness of 3 mm is used for the measurement of flux density using a Hall-effect transducer. When the Hall field and the Lorentz force balance each other, it is observed that the current density in the crystal is 0.3 A mm^{-2} and the Hall voltage developed is −0.35 V. Find the value of the flux density and that of the electron velocity. The following data is given: the Hall coefficient for the germanium crystal = −8 × 10^{-3} Vm A^{-1} Wb m^{-2}.

Problem 12.4. The distributed (self) capacitance of a low-loss coil is found to be 820 pF when measured using a Q meter. Resonance of the coil occurred at an angular frequency of 10^6 rad s^{-1} and a capacitance of 9.18 nF. What is the inductance of the coil?

IOP Publishing

Fundamentals of Industrial Instrumentation (Second Edition)

Alok Barua

Chapter 13

Optoelectronic sensors

Learning objectives:
- The principle of photoconductivity.
- Some applications of the photodiode and the photoresistor in measurement systems.
- The way in which light travels in optical fibres.
- The difference between monomode and multimode fibres.
- Fibre optic switches.
- Intrinsic fibre optic sensors.
- Extrinsic fibre optic sensors.

13.1 Photoconductivity

The conductivity of a material is proportional to the concentration of charge carriers present in it.

The current density in A m^{-2} can be expressed as follows:

$$I = (n\mu_n + p\mu_p)eE,$$

where
- n = the magnitude of the free electron concentration,
- p = the magnitude of the hole concentration,
- σ = conductivity,
- μ_n = electron mobility cm^2/V-sec, and
- μ_p = the hole mobility cm^2/V-sec.

The conductivity $\sigma = (n\mu_n + p\mu_p)e$, where E = the electric field and e = the electronic charge.

For an intrinsic semiconductor, $n = p = n_i$ = the intrinsic concentration in cm^{-3}. The radiant energy that falls on the semiconductor causes covalent bonds to break, creating hole–electron pairs. This in excess of those generated thermally. The resistance of the material decreases due to the increase in current carriers. Hence, such a device is called a photoresistor or a photoconductor. For a light intensity change of 100 ft-candle (1 ft-candle = 10.76 lux), 1 lumen (lm) m^{-2} = 1 lux and 1 lumen (lm) ft^{-2} = 10.76 lux. The minimum photon energy required for intrinsic excitation is the forbidden-gap energy E_G (in electron volts) of the semiconductor material.

The threshold wavelength of incident light $= \lambda_C = \frac{1.24}{E_G}$ μ or μm.

For Si, E_G = 1.1 eV, therefore λ_C = 1.13 μ.
For Ge, E_G = 0.72 eV, therefore λ_C = 1.73 μ.

13.2 Photocurrent

The steady-state current is given by

$$i = \frac{eP_r\tau}{T_t},$$

where

- P_r = the rate at which carriers are produced by light,
- τ = the average lifetime of newly created carriers before they recombine, and
- T_t = the average transit time taken by carriers to reach the ohmic contact.

One of the most commonly used commercial photoconductivity cells is CdS (cadmium sulfide) with antimony as a pentavalent impurity. Its resistance varies from 2 MΩ to 10 Ω for dark to bright light, respectively. CdS has a high dissipation capability and excellent sensitivity in the visible spectrum. Its power dissipation is more than 300 mW. A selenium cell is sensitive throughout the visible range and particularly toward the blue end of the spectrum.

13.3 The semiconductor photodiode

If a reverse-biased p–n junction is illuminated at the junction, the current varies almost linearly with the light flux. This effect is utilized in the semiconductor photodiode. The reverse saturation current I_o in a p–n diode is proportional to the concentrations p_{no} and n_{po} of the minority carriers in the n and p regions, respectively.

The voltage–current characteristic is given by

$$I = I_s + I_o\left(1 - e^{V/\eta V_T}\right),$$

where I_s is the short-circuit current, which is proportional to the light intensity current under a large reverse bias (V is positive for forward bias and negative for reverse bias):

$$I \approx I_s + I_o.$$

The parameter (η) is unity for germanium and two for silicon. V_T is the volt equivalent of temperature. The equivalent circuit for a reverse-biased photodiode is shown in figure 13.1.

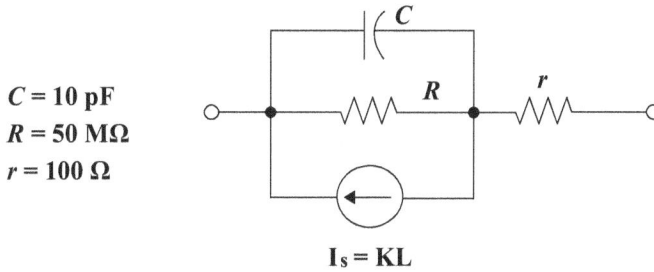

Figure 13.1. The equivalent circuit for a reverse-biased photodiode.

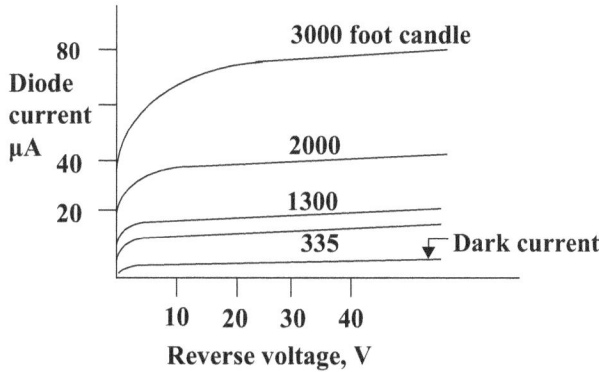

Figure 13.2. Photodiode characteristics.

The symbol L represents the light flux in lumens (lm) and K is the proportionality constant in the range of 10–50 mA lm^{-1}. The reverse saturation current I_o doubles for every 10 °C increase. Its typical sensitivity is 10 mA lm^{-1}. The lumen is defined so that 1 W of light at $\lambda = 0.55$ μ can produce 680 lm. Photodiode characteristics are shown in figure 13.2.

An n–p–n phototransistor provides electrical amplification but has a larger time constant. However, it is more sensitive than the photodiode, although its basic principle of operation is same as that of the photodiode.

13.3.1 Some typical applications

A digital revolution counter can be made using a photodiode. A schematic for such a counter is shown in figure 13.3.

The light source and the photodetector are aligned and there is a hole in the surface of the disc which is mounted of the shaft of the rotating body. For each complete revolution of the disc, one pulse arrives an input to the photodetector. This signal has an irregular shape and it is therefore shaped or made rectangular by a pulse shaper or Schmitt trigger. The output of the Schmitt trigger is now a well-shaped digital signal which is to be processed further. The integrated circuit (IC) chips used for this purpose are a decade counter (7490), a four-bit latch (7475), a

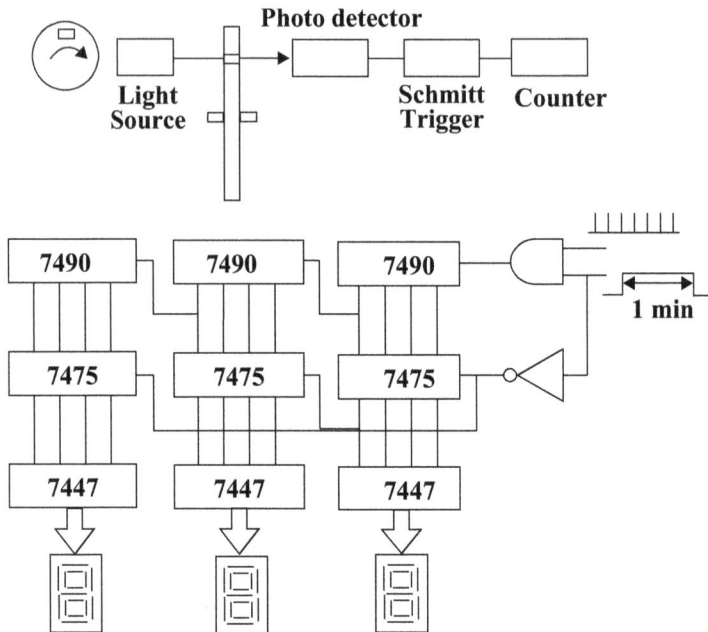

Figure 13.3. A circuit schematic of a revolution counter.

binary-coded decimal (BCD) to seven-segment decoder (7447), and a seven-segment display. All the above are transistor–transistor logic (TTL) chips. The clock input of the first counter is provided by the output of the AND gate, which has two inputs, one from the output of the Schmitt trigger and other is the clock, which has a mark or ON time of one minute. All the ICs are based on trailing-end logic. At the end of 1 min, the clock latch IC holds and displays the reading until a new pulse arrives. This design has the limitation of the lack of a zero reference and cannot sense the direction of rotation of the shaft. A phase-sensitive technique has to be implemented to make the system sensitive to the direction of rotation.

13.3.2 The measurement of steel strips in a hot rolling mill

13.3.2.1 Correlation function
The autocorrelation function of a signal is closely related to its power density spectrum. The autocorrelation function of a waveform is defined as

$$R_{xx} = \lim_{T \to \infty} \frac{1}{T} \int_0^T x(t)x(t-\tau)dt.$$

The waveform $x(t)$ is multiplied by a delayed version of itself $x(t-\tau)$ and the product is averaged over T seconds. The cross-correlation function between two identical waveforms $x(t)$ and $y(t)$ is given by

$$R_{xy}(t) = \overline{x(t-\tau)y(t)} = \lim_{T \to \infty} \frac{1}{T} \int_0^T x(t-\tau)y(t)dt.$$

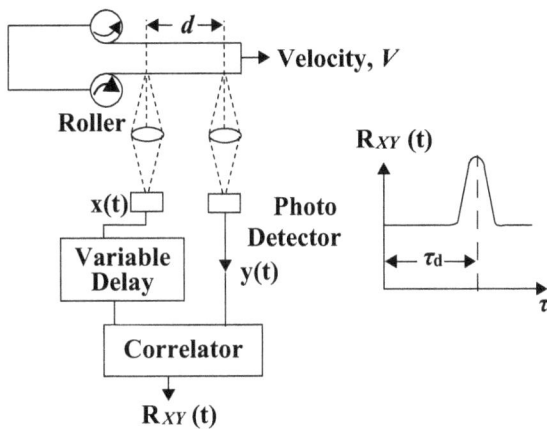

Figure 13.4. A schematic of the measurement of rolled strip velocity in a hot rolling mill.

This cross-correlation function is used to perform noncontact velocity measurements of rolled metal sheets in hot rolling mills.

Measuring the velocity of rolled steel or sheet material is difficult when it is red hot and its temperature is extremely high. A basic schematic of correlation measurement is shown in figure 13.4. The fluctuations of intensity of the light reflected from the metal surface are transformed into electrical signals by means of two photodetectors mounted at a distance d apart. The cross-correlation coefficient reaches a maximum when the variable delay equals the time for a point on the steel sheet to travel between two points of measurement.

The sheet velocity, $V = \dfrac{d}{\tau_d}$

13.4 The transmission of light in optical fibres

Fibre optics is the technology of using light to transmit information. Light has a number of advantages over electricity as a medium for transmitting signals. It is immune to noise created by neighboring electromagnetic fields, its attenuation over a given transmission distance is much lower than that of electric cables, and it is also intrinsically safe. Fibre optic technology is used in two major ways. First, the fibre optic cable itself can be used directly as a sensor in which the variable being measured causes some measurable change in the characteristics of the light trans-mitted by the cable. Second, fibre optic cables can be used as a transmission medium for data. This latter use of fibre optic cables for transmitting information can be further divided into three separate application areas.

First, relatively short fibre optic cables are used as part of various instruments to transmit light from conventional sensors to a more convenient location for processing, often in situations where space is very scarce at the point of measure-ment. Second, longer fibre optic cables are used to connect remote instruments to controllers in instrumentation networks. Third, longer links are still used for data transmission systems in telephone and computer networks. These three application classes have different requirements, particularly in terms of bandwidth, and tend to use different types of fibre optic cables.

13.4.1 The principle of the transmission of light in two media of different refractive indices

Figure 13.5 shows materials of refractive indices n_1 and n_2. Light entering from the external medium (of refractive index n_0) at an angle of α_0 is refracted at an angle of α_1 in the core. When the light meets the core–cladding boundary, part of it is reflected at an angle of β_1 back into the core and part is refracted into the cladding at an angle of β_2. Here, α_1 and α_0 are related by Snell's law according to the following equation:

$$n_0 \sin \alpha_0 = n_1 \sin \alpha_1. \tag{13.1}$$

Similarly, β_1 and β_2 are related by

$$n_1 \sin \beta_1 = n_2 \sin \beta_2.$$

The light which enters the cladding is eventually lost and contributes to the attenuation of the transmitted signal in the cable. However, equation (13.1) shows how this loss can be prevented. If $\beta_2 = 90°$, then the refracted ray travels along the boundary between the core and cladding, and if $\beta_2 > 90°$, all of the beam is reflected back into the core due to total internal reflection. The case in which $\beta_2 = 90°$, corresponding to incident light at an angle of α_c, is therefore the critical angle at which total internal reflection occurs at the core–cladding boundary.

The condition for this is that $\sin \beta_2 = 1$.

Putting $\sin \beta_2 = 1$ into equation (13.1) yields

$$\frac{n_1 \sin \beta_1}{n_2} = 1.$$

Thus:

$$\sin \beta_1 = n_2/n_1.$$

From figure 13.5, we can write $\cos \alpha_1 = \sin \beta_1$.

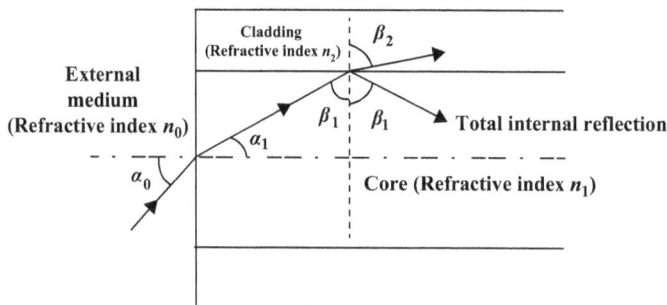

Figure 13.5. The transmission of light through fibre.

Therefore,

$$\sin \alpha_1 = (1 - \cos^2 \alpha_1)^{1/2} = (1 - \sin^2 \beta_1)^{1/2} = [1 - (n_2/n_1)^2]^{1/2}.$$

From equation (13.1), we can write

$$\sin \alpha_c = \sin \alpha_0 = \frac{n_1}{n_0} \sin \alpha_1,$$

where α_c is the critical angle, which will be defined below.

Hence,

$$\sin \alpha_c = \frac{n_1}{n_0} [1 - (n_2/n_1)^2]^{1/2}.$$

Therefore, provided that the angle of incidence of the light into the cable is greater than the critical angle given by $\theta = \sin^{-1} \alpha_c$, all of the light is internally reflected at the core–cladding boundary. Further reflections occur as the light passes down the fibre and it thus travels in a zigzag fashion by total internal reflection to the end of the cable. Although this minimizes attenuation, there is a remaining problem in that the transmission times of the parts of the beam that travel in this zigzag manner are greater than that of light which enters the fibre perpendicular to the face and so travels in a straight line to the other end.

In practice, the rays of light entering the cable are spread over the range given by $\sin^{-1} \alpha_c < \theta < 90°$ and so the transmission times of these separate parts of the beam are distributed over a corresponding range. These differential delay characteristics of the light beam are called the modal dispersion. The practical effect is that a step change in the light intensity at the input end of the cable is received over a finite period of time at the output. It is possible to largely overcome this latter problem in multimode cables by using cables made solely from glass fibres in which the refractive index changes gradually over the cross-section of the core rather than abruptly at the core–cladding interface as in the step-index optical fibre. This special type of cable is known as graded index cable and it progressively bends light incident at less than 90° to its end face rather than reflecting it at the core–cladding boundary.

Although the parts of the beam away from the center of the cable travel further, they also travel faster than the beam passing straight down the center of the cable because the refractive index is lower away from the center. Hence, all parts of the beam are subject to approximately the same propagation delay. In consequence, a step change in the light intensity at the input produces a similar step change of light intensity at the output. An alternative solution is to use a monomode cable. This propagates light using a single mode only, which means that time dispersion of the signal is almost eliminated.

13.5 The components of an optical fibre system

The central part of a fibre optic system is a light-transmitting cable containing at least one glass or plastic fibre (but more often a bundle of them). This is terminated at each end by a transducer or a transmitter and receiver pair as shown in figure 13.6. At the input side, the transducer converts the signal from its electrical form into light.

At the output end, the transducer converts the transmitted light back into an electrical form suitable for use by data recording, manipulation, and display systems. These two transducers are often known as the transmitter and receiver, respectively. The arrangement described above is somewhat modified when the cable is directly used as a sensor. In this case, the light launched into the cable comes directly from a light source and does not originate as an electrical signal. However, the same mechanisms as those described below have to be used to introduce the light into the cable.

13.5.1 Fibre optic cables

Fibre optic cables consist of an inner cylindrical core surrounded by an outer coaxial cylindrical cladding, as shown in figure 13.7. To achieve total internal reflection, the refractive index of the inner material is made greater than that of the outer material, and the relative refractive index affects the transmission characteristics of light along the length of the cable. The amount of light attenuation that occurs as it travels along the cable varies with the wavelength of the light transmitted. This characteristic is very nonlinear, and thus a graph of attenuation against wavelength shows a number of peaks and troughs. The positions of these peaks and troughs vary according to the material used for the fibres. It should be noted that fibre manufacturers rarely mention these nonlinear attenuation characteristics and quote the value of attenuation which occurs at the most favorable wavelength. The user should adhere to the prescribed wavelength of light.

Two forms of cable exist, known as monomode and multimode. Monomode cables have a small core diameter, typically 6 µm, whereas multimode cables have

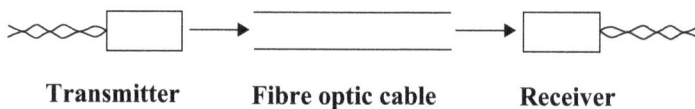

Transmitter   **Fibre optic cable**   **Receiver**

Figure 13.6. Cables, transmitter, and receiver.

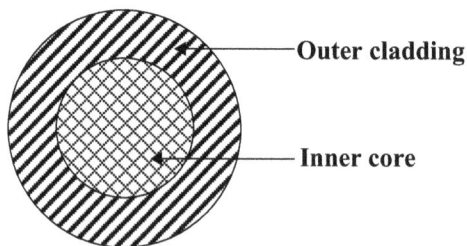

Outer cladding

Inner core

Figure 13.7. A cross-section of a fibre optic cable.

much larger diameter, typically between 50 and 200 μm. Both glass and plastic in different proportions are used to make various forms of cables. One option is to use different types of glass fibres for both the core and the cladding. A second, and cheaper, option is to use a glass fibre core and plastic cladding. This has the additional advantage of being less brittle than the all-glass version. However, all-plastic cables also exist, in which two types of plastic fibres with different refractive indices are used. This is the cheapest form of all optical fibres but it has the disadvantage of having high attenuation characteristics, making it unsuitable for the transmission of light over large distances. Protection is normally given to the cable by enclosing it in the same types of insulating and armoring materials as those used for copper cables. These protect the cable from various hostile operating environments and also from mechanical damage. When suitably protected, fibre optic cables can even withstand flames.

13.5.2 Fibre optic transmitters

A light-emitting diode (LED) is commonly used as the transducer which converts an electrical signal into light and transmits it into the cable. The LED is particularly suitable for this task, as it has an approximately linear relationship between the input current and the light output. The type of LED chosen must closely match the attenuation characteristics of the light path through the cable and the spectral response of the receiving transducer. An important characteristic of the transmitter is the proportion of its power which is launched into the fibre optic cable, and this is more important than its absolute output power. This proportion is maximized by making specially shaped LED transmitters which have a spherical lens incorporated into the chip during manufacture. This sends an approximately parallel beam of light into the cable with a typical diameter of 400 μm. The proportion of light entering the fibre optic cable is also governed by the quality of the end face of the cable and the way it is bonded to the transmitter. A good end face can be produced by either polishing or cleaving. Polishing involves grinding the fibre end down with progressively finer polishing compounds until a surface of the required quality is obtained. The transmitter is then attached by gluing.

The proportion of light transmitted into the cable is also dependent on the proper alignment of the transmitter with the center of the cable. The effect of misalignment depends on the relative diameters of the cable. Figure 13.8 shows the effect on the proportion of power transmitted into the cable for the case in which the cable diameter is greater than beam diameter. This shows that some degree of misalignment can be tolerated unless the beam and cable diameters are equal. The cost of achieving exact alignment of the transmitter and cable is very high, as it requires the LED to be exactly aligned in its housing, the fibre to be exactly aligned in its connector and the housing to be exactly aligned with the connector. Therefore, great cost savings can be achieved wherever some misalignment can be tolerated in the specification of the cable.

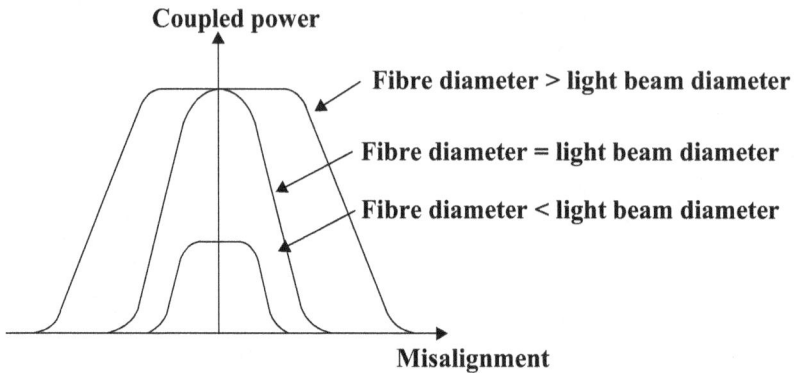

Figure 13.8. The effect of transmitter alignment on the light power transmitted.

13.5.3 Fibre optic receivers

The device used to convert the signal back from light to its electrical form is usually either a p–intrinsic–n (PIN) diode or a phototransistor. Phototransistors have good sensitivity but only low bandwidth. On the other hand, PIN diodes have a much wider bandwidth but lower sensitivity. If both high bandwidth and high sensitivity are required, then special avalanche photodiodes are used, but these increase the cost of the transmitter. The same considerations regarding the losses at the interface between the cable and receiver apply as for the transmitter, and both polishing and cleaving are used to prepare the fibre ends. The output voltages produced by the receiver are very small and amplification is always necessary. A low-noise, high-gain amplifier is the only solution.

13.5.4 The transmission characteristics of fibre optic cables

Monomode step-index fibres have very simple transmission mechanism because the core has a very small diameter and light can only travel down it in a straight line. This type of fibre can sustain only one mode of propagation and requires a coherent laser light source. On the other hand, multimode step-index fibres have quite complicated transmission characteristics because of the relatively large diameter of the core. Many modes can be propagated in multimode fibres; because of their larger core diameter, it is also much easier to launch optical power into the fibre and also to connect two or more fibres together. Moreover, an ordinary LED source can be used to launch the light. Whilst the transmitter is designed to maximize the amount of light which enters the cable in a direction parallel to its length, some light inevitably enters multimode cables at other angles. Light which enters a multimode cable at any angle other than normal to the end face is refracted in the core by total internal reflection. It then travels in a straight line until it meets the boundary between the core and cladding materials. At this boundary, some of the light is reflected back into the core and some is refracted into the cladding. Multimode graded index fibres have a core with a nonuniform refractive index. These fibres are characterized by curved ray paths which offer some advantages but are more expensive than the step-index type.

13.6 Fibre optic sensors

The operational basis of fibre optic sensors is the conversion of the physical quantity measured into a change in one or more parameters of a light beam. The light parameters that can be modulated are one or more of the following:

1. Intensity.
2. Phase.
3. Polarization.
4. Wavelength.
5. Transmission time.

Fibre optic sensors usually incorporate either glass/plastic cables or all-plastic cables. All-glass types are rarely used because of their fragility. Plastic cables have particular advantages for sensor applications because they are cheap and have a relatively large diameter of 0.5–1.0 mm, making connection to the transmitter and receiver easy. The cost of the fibre optic cable itself is insignificant for sensing applications, as the total cost of the sensor is dominated by the cost of the transmitter and receiver. Fibre optic sensors typically have a long life. For example, the life expectancy of reflective fibre optic switches is quoted at 10 million operations. If properly calibrated, the accuracy is also around $\pm 1\%$ of the full-scale reading level for a fibre optic pressure sensor. However, in spite of these obvious merits, industrial usage is currently quite low.

Two major classes of fibre optic sensor exist, namely intrinsic sensors and extrinsic sensors. In intrinsic sensors, the fibre optic cable itself is the sensor, whereas in extrinsic sensors, the fibre optic cable is only used to guide light to/from a conventional sensor.

13.6.1 Intrinsic sensors

Intrinsic sensors can modulate the intensity, phase, polarization, wavelength, or transit time of light. Sensors which modulate light intensity tend to use mainly multimode fibres, but only monomode cables are used to modulate the other light parameters. A particular feature of intrinsic fibre optic sensors is that they can, if required, provide distributed sensing over distances of up to 1 m. Light intensity is the simplest parameter to manipulate in intrinsic sensors because only a simple source and detector are required. The various forms of switches are shown in figures 13.9–13.11. Perhaps the simplest form of these is the shutter switch, as the light path is simply blocked and unblocked as the switch changes state. Modulation of the intensity of transmitted light takes place in various simple forms of proximity, displacement, pressure, pH, and smoke sensors. Some of these are sketched in figures 13.12–13.14.

In proximity and displacement sensors, the amount of reflected light varies with the distance between the fibre ends and a boundary.

In pressure sensors, the refractive index of the fibre, and hence the intensity of light transmitted, varies according to the mechanical deformation of the fibres caused by pressure.

In a pH probe, the amount of light reflected back into the fibre depends on the pH-dependent color of the chemical indicator in the solution around the probe tip.

In a smoke detector, two fibre optic cables placed on both side of a space detect any reduction in the intensity of light transmission between them caused by the presence of smoke.

A simple form of accelerometer can be made by placing a mass subject to acceleration on a multimode fibre. The force applied to the fibre by the mass causes a change in the intensity of light transmitted, thus allowing the acceleration to be

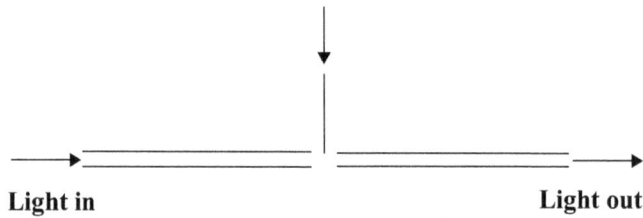

Light in   **Light out**

Figure 13.9. The shutter switch.

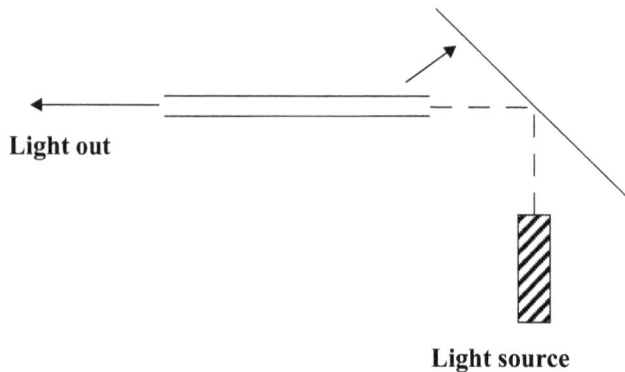

Light out

Light source

Figure 13.10. The reflective switch.

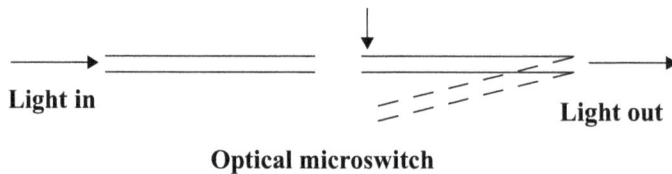

Light in   **Light out**

Optical microswitch

Figure 13.11. Fibre optic switches.

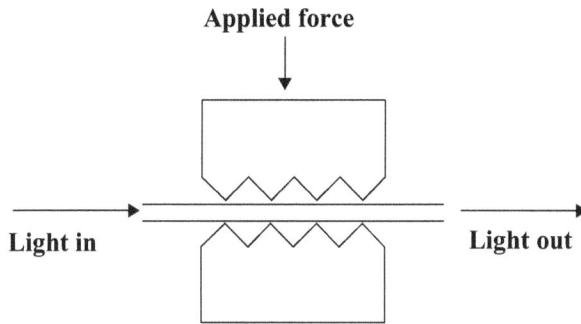

Figure 13.12. A pressure sensor that uses the intensity modulation principle.

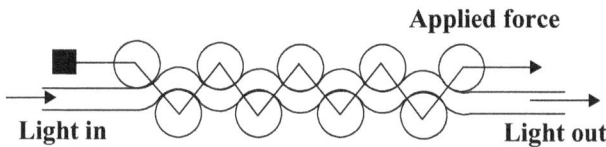

Figure 13.13. A micro bend pressure sensor based on the intensity modulation principle.

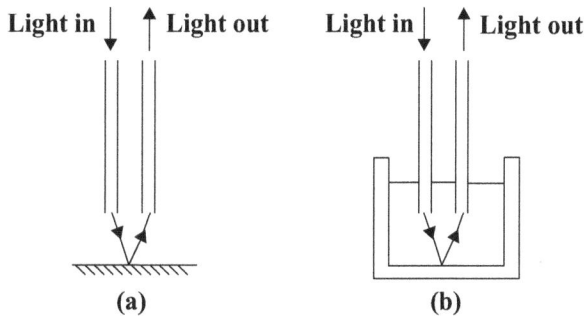

Figure 13.14. Intensity-modulating sensors: (a) a proximity sensor, (b) a pH sensor.

determined. The typical accuracy quoted for this device is ±0.02 g in the measurement range of ±5 g and ±2% in the measurement range up to 100 g. A similar principle is used in probes which measure the internal diameter of tubes. These probes consists of eight strain-gauged cantilever beams which track changes in diameter, giving a measurement resolution of 20 μm.

A slightly more complicated method of effecting light intensity modulation is the variable shutter sensor shown in figure 13.15. It consists of two fixed fibres with two collimating lenses and a variable shutter in between them.

Movement of the shutter changes the intensity of light transmitted between the fibres. This is used to measure the displacement of various devices such as Bourdon tubes, diaphragms, and bimetallic thermometers.

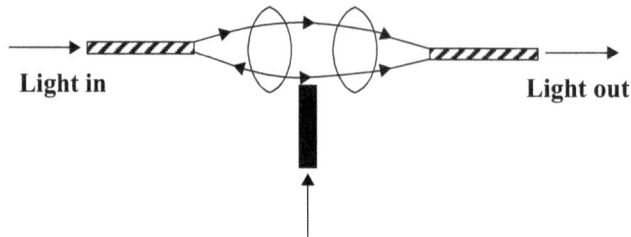

Figure 13.15. The variable shutter sensor.

Yet another type of intrinsic sensor uses a cable in which the core and cladding have similar refractive indices but different temperature coefficients. This type of sensor is used as a temperature sensor. An increase in temperature causes the refractive indices to become closer together and the losses from the core to increase, thus reducing the quantity of light transmitted.

13.6.2 Extrinsic sensors

Extrinsic fibre optic sensors use a fibre optic cable, normally a multimode cable, to transmit modulated light from a conventional sensor. A major feature of extrinsic sensors, which makes them very useful, is their ability to reach places which are otherwise inaccessible.

One example of this is the insertion of fibre optic cables into the jet engines of aircraft to measure temperature. This is achieved by transmitting light into an optical pyrometer located remotely from the engine. Extrinsic fibre optic sensors provide excellent protection of the measurement signals from any electrical or electromagnetic noise. Unfortunately, the output of many forms of conventional sensor is not in an electrical form which can be directly fed to the transmitter. Conversion into the electrical form must therefore take place prior to transmission. For example, in the case of a platinum resistance thermometer, the temperature changes are converted into the unbalanced voltage of a Wheatstone bridge. The unbalanced voltage is modulated and launched into the fibre optic cable through the usual type of transmitter. This complicates the measurement process and means that low-voltage power cables must be routed with the fibre optic cable to the transducer. One particular adverse effect of this is that the advantage of intrinsic safety is lost.

IOP Publishing

Fundamentals of Industrial Instrumentation (Second Edition)

Alok Barua

Chapter 14

The measurement of pH and viscosity

Learning objectives:
- The definition of pH.
- Why is it necessary to measure pH?
- The pH probe.
- Characteristics of the pH amplifier.
- Rotating drum viscosity measurement.
- The capillary flow method.
- The Saybolt viscosimeter.

14.1 An introduction to pH

pH is the measurement of the number of grams of hydrogen ions per litre of a solution. Mathematically, it is expressed as the negative logarithm of the hydrogen ion concentration:

$$pH = -\log_{10}[H^+], \tag{14.1}$$

where $[H^+]$ is the hydrogen ion concentration in the solution.

The value of pH can range from zero to 14, where zero describes extreme acidity and 14 describes extreme alkalinity. Distilled water has a pH of 7. As the pH value changes by one unit, the acidity or alkalinity values change by a factor of ten.

The pH scale is shown in figure 14.1

14.2 Why is pH measurement important?

The measurement of the pH of a solution is necessary to judge whether the liquid is acidic, alkaline, or neutral. Such measurements are necessary in the food and beverage industry, for aquariums, and in agriculture, gardening, color photography, etc. The measurement of the pH of liquids is also necessary in many condensate systems as a check on corrosion, in pharmaceutical and drug production for maximum yield, in food manufacture to reduce spoilage and improve taste, and

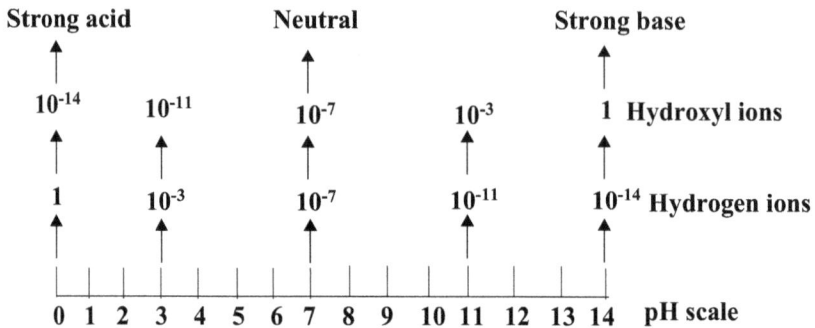

Figure 14.1. The pH scale.

Table 14.1. The pH values of some common substances.

Substance	pH
Sulfuric acid	0.3
Lemon juice	2.3
Vinegar	2.9
Orange juice	4.3
Boric acid	5.0
Milk	6.7
Distilled water	7.0
Blood	7.5
Sea water	7.9
Ammonia	11.3
Bleach	12.5

in innumerable chemical processes. The continuous monitoring of blood pH is essential for the proper treatment of patients suffering from metabolic and respiratory problems. The pH values of some common substances are given in table 14.1.

14.3 The pH probe

The traditional method for indicating the hydrogen ion concentration to use litmus paper, which does not give an exact value, as it is a qualitative measurement. Therefore, for quantitative measurement, electrode-based probes are used.

14.3.1 The functions of the electrode

The pH measurement method requires an electrode to be immersed in the solution. An electrolytic potential is produced at the electrode, which forms an electrolytic half-cell. This is called the *measuring cell*. A second electrode is required to provide a

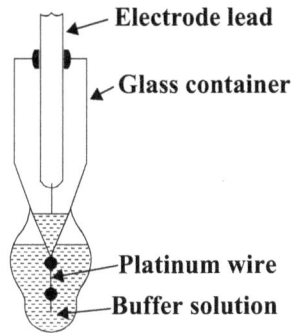

Figure 14.2. A measuring glass electrode for pH measurement.

standard potential and to complete the cell and electrical circuit. This is called the *reference cell*. The algebraic sum of the potentials of the two half-cells is proportional to the concentration of hydrogen ions in the solution. A measuring glass electrode is shown in figure 14.2.

14.3.2 The glass electrode

The glass electrode operates on the principle that a potential is observed between two solutions that have different hydrogen ion concentrations when they are separated by a thin glass well. The electrode receives chloride ions from potassium chloride (KCL). The use of KCL has the benefit of being pH neutral. pH sensors typically use KCL solutions with concentrations lie between 3 molars to saturated. This potential is a function of the two concentrations. A buffer solution is contained in the permanently sealed glass electrode, which is surrounded by the solution whose pH is being measured. The buffer solution has a constant hydrogen ion concentration. The potential at the electrode therefore depends on the hydrogen ion concentration of the measured solution.

The liquid junction is provided by a small hole in the electrode over which a ground glass cap is placed. Thus the potassium chloride slowly escapes from the electrode into the measured liquid. The potential at the reference electrode is also constant. pH is a temperature-sensitive process parameter. Consequently, temperature error must be compensated. A simple resistance temperature detector (RTD) is incorporated to make this compensation. pH electrodes can be placed in the vessel or in the liquid pipeline. The glass calomel electrode can be used at liquid temperatures between 1 °C and 100 °C.

14.3.3 The reference electrode

The calomel electrode is the most commonly used reference electrode. The calomel (mercury and mercurous chloride) is contained in the inner tube and covers a platinum wire. A saturated solution of potassium chloride is in contact with the measured solution that surrounds the reference electrode. A calomel electrode is shown in figure 14.3.

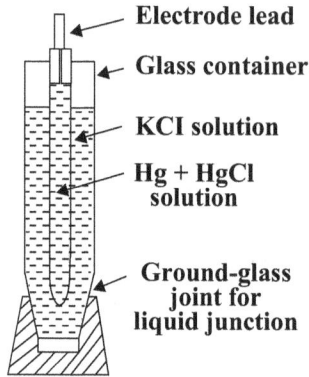

Figure 14.3. The calomel reference electrode.

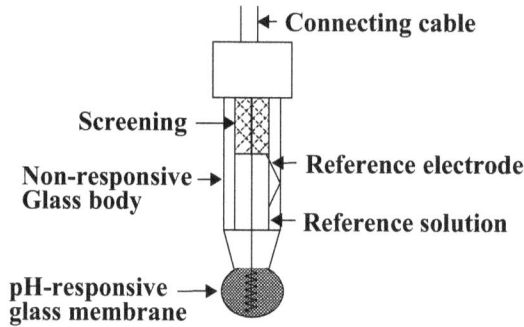

Figure 14.4. The combined pH probe.

14.3.4 The combined pH probe

The most common device used for pH measurement is the combined pH probe, which is shown in figure 14.4.

The combined glass probe consists of a glass probe containing two electrodes, i.e. a measuring electrode and a reference electrode, separated by a solid glass partition. The reference electrode is a screened electrode and is immersed in a buffer solution which provides a stable reference emf, usually 0 V. The tip of the measuring electrode is surrounded by a pH-sensitive glass membrane at the lowermost end of the probe, which permits the diffusion of ions according to the hydrogen ion concentration in the fluid outside the probe. The measuring electrode therefore generates an emf proportional to the pH. The characteristics of the glass electrode are dependent on the ambient temperature and subject to both zero drift and sensitivity drift. Thus, temperature compensation is needed, which is normally achieved by calibrating the system output before use by immersing the probe in solutions at reference pH values.

14.3.5 The practical range of pH measurement

The practical range of pH measurement is 1–12. Electrode contamination becomes a serious problem in highly alkaline liquids. The glass starts to dissolve in acid solutions containing fluoride, and this represents a further limitation on its use.

14.3.6 The voltage output of the pH probe

The net potential in the pH probe is given by

$$V_o = -2.30 \frac{RT}{F} \log \frac{C_H}{C_R}, \tag{14.2}$$

where
- R = the universal gas constant, 8314 J (kg mol K)$^{-1}$;
- T = the absolute temperature in K;
- F = Faraday's constant, 9.647×10^7 C (kg mol)$^{-1}$;
- C_H = the hydrogen ion concentration in the solution; and
- C_R = the concentration in the glass electrode = 1.0 for 1 N HCl.

Inserting all the values, the output voltage can be expressed as $V_o = -1.98 \times 10^{-4}$ T in pH units.

14.3.7 The pH amplifier

The pH probe has a source impedance of the order of 100 MΩ because the electrode is made of glass. A very high-input-impedance direct coupled amplifier would be a good choice. The input impedance of the amplifier should be of the order of 10^{10}–10^{12} Ω. An instrumentation amplifier should be made using an op-amp which has a high-input-impedance differential stage. The op-amp should preferably have a metal–oxide–semiconductor field-effect transistor (MOSFET) input differential stage and should be implemented using the bipolar complementary metal–oxide–semiconductor (BiCMOS) process. This will provide an input impedance of 10^{12} Ω. The Analog Devices AD 5449 op-amp is such a device.

14.4 The measurement of viscosity

Absolute viscosity is defined as

$$\tau = \mu \frac{dv}{dy}, \tag{14.3}$$

where
- τ = the shear stress between fluid layers in laminar flow in N m^{-2},
- μ = dynamic viscosity in N s m^{-2}, and
- dv/dy = the normal velocity gradient as indicated in figure 14.5.

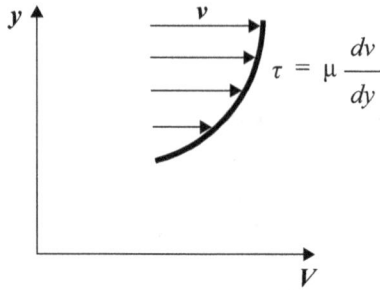

Figure 14.5. The relation between the viscosity gradient and fluid shear.

The kinematic viscosity is defined as

$$\mu_k = \frac{\mu}{\rho},$$

where μ_k is the kinematic viscosity, which is expressed in N s m kg^{-1} and ρ is the density of the fluid in kg m^{-3}. Some of the more common units are given below along with their conversion factors:

Dynamic viscosity:

$$1 \text{ N s m}^{-2} = 10 \text{ P} = 1000 \text{ centipoise (cP)}$$
$$= 1 \text{ kg m}^{-1} \text{s}^{-1}$$
$$1 \text{ P} \quad = 100 \text{ cP} = 1 \text{ dyn}-\text{s cm}^{-2}$$
$$= 0.1 \text{ N}-\text{s m}^{-2}$$
$$= 0.1 \text{ kg m}^{-1} \text{s}^{-1}.$$

Kinematic viscosity:

$$1 \text{ m}^2 \text{ s}^{-1} = 10^4 \text{ St (Stokes)}$$
$$1 \text{ St} \quad = 1 \text{ cm}^2 \text{ s}^{-1} = 100 \text{ centistokes (cSt)}$$
$$= 10^{-4} \text{ m}^2 \text{ s}^{-1}.$$

The two most common methods of viscosity measurement are:
1. The rotating concentric cylinder method, and
2. The capillary flow method.

14.4.1 Velocity distribution of a liquid placed between parallel plates

Let us consider two parallel plates as shown in figure 14.6. One of the plates is stationary and the other moves with a constant velocity V. The velocity profile for the fluid between the two plates is a straight line, and the velocity gradient is

$$\frac{dv}{dy} = \frac{V}{b}. \tag{14.4}$$

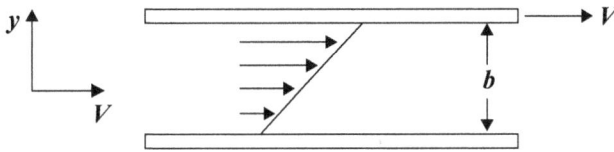

Figure 14.6. Velocity distributions between large parallel plates.

This principle could be used to measure the viscosity by measuring the force required to maintain the moving plate at the constant velocity V.

14.4.2 The rotating concentric cylinder method

The parallel plate system described in section 14.4.1 is difficult to construct; however, the parallel flat-plate situation can be realized using the rotating concentric cylinders shown in figure 14.7. The inner cylinder is stationary and connected to a torque-measuring device, while the outer cylinder is driven by a motor at a constant angular velocity of ω.

If the annular space b is very thin in comparison with the radius of the inner cylinder, then the rotating cylinder arrangement approximates the parallel plate situation, and the velocity profile of the liquid in the gap in between the cylinders may be assumed to be linear. Then,

$$\frac{dV}{dy} = \frac{R_2\omega}{b}, \tag{14.5}$$

where the distance y is measured in the radial direction and it is assumed that $b \ll R_1$. If the torque T is measured, the fluid shear stress can be expressed by

$$\tau = \frac{T}{2\pi R_1^2 L}, \tag{14.6}$$

where L is the length of the inner cylinder. The viscosity is determined by combining equations (14.3), (14.5), and (14.6):

$$\mu = \frac{T.b}{2\pi R_1^2 R_2 L \omega}. \tag{14.7}$$

If the bottom gap space a is small, then the bottom disk also contributes to the torque, modifying the calculation of the viscosity. The torque on the bottom disk is given by

$$T_d = \frac{\mu\pi\omega}{2a} R_1^4. \tag{14.8}$$

The total torques due to the bottom and the annular space are given by

$$T = \mu\pi\omega R_1^2 \left(\frac{R_1^2}{2a} + \frac{2LR_2}{b} \right). \tag{14.9}$$

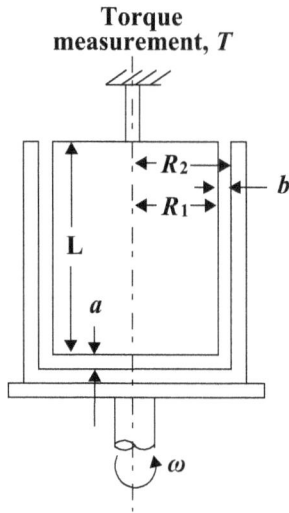

Figure 14.7. A rotating concentric cylinder system used to measure viscosity.

If the torque, angular velocity, and dimensions of the cylinders are measured, the viscosity may be calculated using equation (14.9).

14.4.3 The capillary flow method

In this method, the pressure drop is measured across a capillary tube in which the liquid is flowing in a laminar fashion. Consider the tube shown in figure 14.8. The Reynolds number is defined as

$$R_e = \frac{\rho v_m d}{\mu} \tag{14.10}$$

If laminar flow is present in the tube, the familiar parabolic velocity profile will be experienced, as shown in figure 14.8. If the fluid is incompressible and the flow is steady, the volume rate of flow Q is given by

$$Q = \frac{\pi R^4 (p_1 - p_2)}{8 \mu L} \text{ (m}^3 \text{ s}^{-1}). \tag{14.11}$$

A viscosity measurement can be made by measuring the volume rate of flow and pressure drop for the flow in such a tube. To ensure that the flow is laminar, a small diameter capillary tube is used. The small diameter reduces the Reynolds number calculated from equation (14.10). In equation (14.10) the product ρv_m may be calculated from

$$\rho v_m = \frac{\dot{m}}{\pi r^2},$$

where $\dot{m}$ is the mass flow rate.

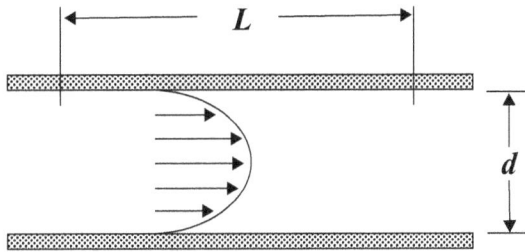

Figure 14.8. A laminar flow of liquid in a capillary tube.

When a viscosity measurement is made for a gas, the compressibility of the gas must be considered. Therefore, the expression for the mass flow of a gas under laminar flow conditions in a capillary is

$$\dot{m} = \frac{\pi R^4}{16 \mu R_0 T}(p_1^2 - p_2^2),$$

where R_0 is the gas constant for the particular gas.

It should be ensured that the flow in the capillary is fully developed, i.e., that the full parabolic velocity profile is established. This means that the pressure measurements should be taken far downstream from the entrance of the tube to ensure that required flow conditions are present.

It may be expected that the flow will have a parabolic velocity profile when

$$\frac{L}{d} > \frac{R_e}{8},$$

where L is the distance from the entrance of the tube.

14.4.4 Industrial viscosimeters

In industry, the instrument which is used for the measurement of viscosity is called the Saybolt viscosimeter. It uses the capillary tube principle to measure the viscosities of liquids. A schematic of the instrument is shown in figure 14.9. A cylinder is filled to the top with the liquid. The liquid is then allowed to drain from the bottom through the capillary tube. The time required for a given quantity of liquid to drain is recorded, and this time taken is calibrated for the viscosity of the liquid. Since the capillary tube is short, a fully developed laminar viscosity profile is not established, and it is necessary to apply a correction factor for the actual profile. If the velocity profile were fully developed, the kinematic viscosity would vary directly with the time taken for drainage; that is,

$$\mu_k = \frac{\mu}{\rho} = C_1 t.$$

Introducing the correction factor gives the following modified expression:

$$\mu_k = C_1 t + \frac{C_2}{t},$$

where C_1 and C_2 are constants which are to be found experimentally.

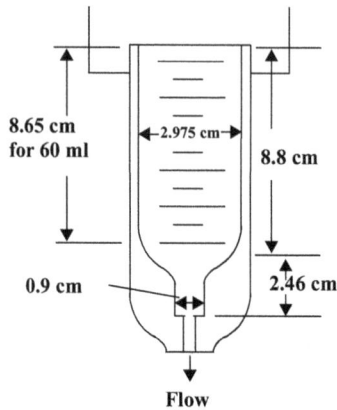

Figure 14.9. Schematic of the Saybolt viscosimeter.

Problem 14.1. The following design is used to measure pH: a pH electrode is connected through a shielded cable to a non-inverting amplifier as shown in figure 14.10. The input resistance of the non-inverting amplifier is given by

$$R = R_i\left(1 + A_0 \frac{R_F}{R_1}\right),$$

where A_0 = the open loop gain and R_i = the input resistance of the op-amp.

Find the output voltage V_o of the circuit when a 225 mV signal is generated at the electrode.

The following data are given:
- The resistance of the electrode = 10^8 Ω.
- The resistance (leakage) of the shielded cable = 2×10^8 Ω.
- $A_0 = 10^5$.
- $R_i = 10^6$ Ω.
- $R_F = 2$ KΩ.
- $R_1 = 1$ KΩ.

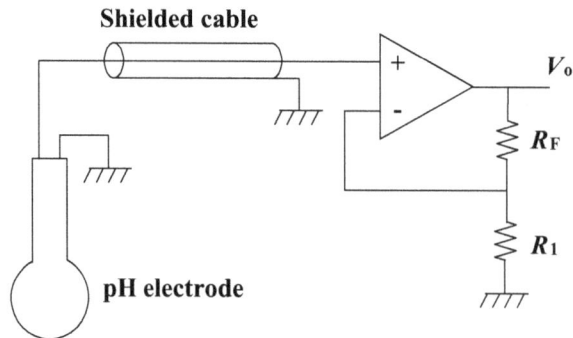

Figure 14.10.

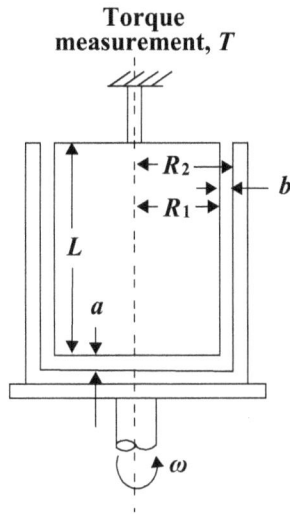

Figure 14.11. A rotating viscosity meter.

Problem 14.2. A rotating viscosity meter is shown in figure 14.11. It consists of two concentric cylinders of radii R_1 and R_2 with viscous liquid in between. When the outer cylinder is rotated by a motor at an angular velocity of ω, the suspension turns through an angle θ. Show that the viscosity of the liquid can be expressed as

$$\mu = \frac{K\theta b}{\omega L R_2 2\pi R_1^2},$$

where K, is the torsional constant. Neglect the viscous effect of the liquid at the bottom of the two cylinders.

IOP Publishing

Fundamentals of Industrial Instrumentation (Second Edition)

Alok Barua

Chapter 15

Dissolved oxygen sensors

Learning objectives:
- The working principle of the dissolved oxygen sensor.
- The electrodes of the dissolved oxygen probe.
- The one-layer model and the three-layer model of the electrode.
- The reactions at the anode and the cathode.
- Different electrodes and their construction.
- Instrumentation.
- Calibration of electrodes.

15.1 Introduction

Bioprocess operations make use of microbial, animal, and plant cells and cellular components such as enzymes to manufacture new products and to destroy harmful wastes. These processes require effective control techniques due to increased demands on productivity, product quality, and environmental responsibility. This is typically so in the case where the biomaterials used in the process are costly and require stringent control over product formation, as in animal cell culture.

The bioreactor is where the bioprocess operation takes place, and its design and controlled operation are very important for several aspects of production such as purity, quantity, efficiency, safety, etc. There are two kinds of bioprocesses, namely aerobic and anaerobic, depending on whether or not oxygen is required to carry out the bioprocess operation. Several types of bioreactors are designed and used in the laboratory in addition to those used in large-scale industrial applications. Some of these are continuous stirred tank, bubble column, airlift, see-saw, and packed bed reactors. As their names indicate, they are meant for aerobic bioreaction processes.

Bioreaction is a slow process that takes days, or at least hours, to complete. Nevertheless, some heat generation (or absorption) and a change in the pH value of the bioreactor fluid take place over time. The temperature and pH values of the bioreactor fluid are considered to be environmental variables and need to be controlled within a tight band for the microorganisms to survive.

The dissolved oxygen content of the bioreactor fluid is probably the most important single process variable for maximizing the product yield. There are some variables that cannot be measured in real time in any bioreactor. The other variables of the bioprocess can be predicted by a parameter estimation technique if the dissolved oxygen concentration is known. Thus, bioreaction processes require continuous measurement of the dissolved oxygen concentration as well as the temperature and pH value, since the last two parameters have a direct influence on bioreactions.

15.2 Dissolved oxygen sensing

Dissolved oxygen sensors or electrodes have been widely used in both research and industry. Compared with chemical analysis, the measurement of dissolved oxygen in water by a membrane-covered electrode offers several advantages:
- simplicity;
- less interference by other solutes in water;
- in-situ measurement with a lower time constant;
- continuous measurement;
- real-time control of the dissolved oxygen concentration in bioreactors or waste water treatment plants.

Dissolved oxygen (DO_2) sensors have been developed for different areas to meet the requirements of the specific applications. Examples are:
⇒ a steam-sterilizable DO_2 probe suitable for bioreactors;
⇒ oxygen microelectrodes for DO_2 measurement in human tissue;
⇒ a fast-response sensor for respiratory gas analysis;
⇒ the measurement of trace oxygen in boiler feed water.

Three broad areas in which the measurement of DO_2 concentration is essential are:
- biochemical engineering;
- microbiology;
- environmental engineering.

15.3 The operational principle of the polarographic electrode

When an electrode of a noble metal such as platinum or gold is made 0.6–0.8 V negative with respect to a reference electrode made of calomel or Ag/AgCl in a neutral potassium chloride solution, the dissolved oxygen is reduced at the surface of the cathode. This phenomenon can be observed in a current–voltage diagram, called a polarogram, of the electrode. As shown in figure 15.1(a), the current initially increases with an increase in the negative bias voltage. It then reaches a region where the current becomes constant. In this saturation region of the polarogram, the

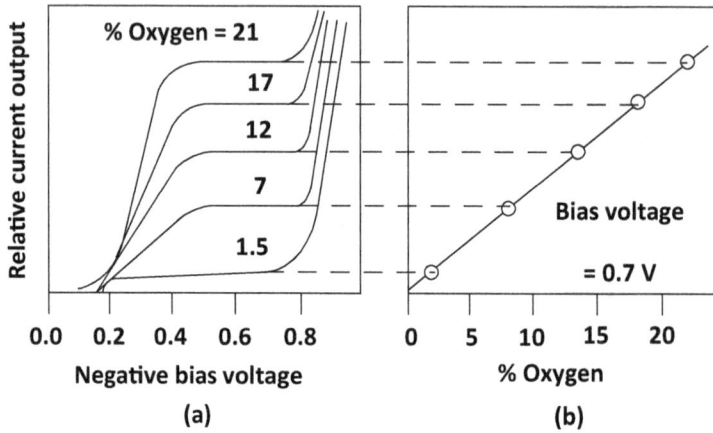

Figure 15.1. (a) A polarogram and (b) a calibration curve.

reaction of oxygen at the cathode is so fast that the rate of reaction is limited by the diffusion of oxygen to the cathode surface.

When the negative bias voltage is further increased, the current output of the electrode increases rapidly due to other reactions, mainly the electrolysis of water, which leads to the production of hydrogen. If a fixed voltage in the saturation region of the polarogram is applied to the cathode, then the current output of the electrode can be plotted for different dissolved oxygen concentrations. Figure 15.1(b) shows such a calibration curve. It should be noted that the current is proportional not to the actual concentration but to the activity or equivalent partial pressure of dissolved oxygen, which is often referred to as 'oxygen tension.' A fixed voltage of between 0.6 and 0.8 V is usually applied as the bias voltage (or polarization voltage) when using Ag/AgCl as the reference electrode. When the cathode, anode, and electrolyte are separated from the measured medium by a plastic membrane which is permeable to gas but not to most of the ions, and when most of the mass transfer resistance is confined within the membrane, the electrode system can measure the oxygen concentrations of various liquids. This is the basic operating principle of the membrane-covered polarographic DO_2 sensor.

A polarographic electrode is shown in figure 15.2. For polarographic electrodes, the reactions are as follows.

The reaction at the cathode is

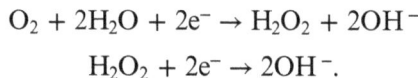

$$O_2 + 2H_2O + 2e^- \rightarrow H_2O_2 + 2OH^-$$

$$H_2O_2 + 2e^- \rightarrow 2OH^-.$$

The reaction at the anode is

$$Ag + Cl^- \rightarrow AgCl + e^-$$

The total reaction is

$$4Ag + O_2 + 2H_2O + 4Cl^- \rightarrow 4AgCl + 4OH^-.$$

The reaction makes the medium alkaline, and a small amount of hydrogen peroxide is also produced.

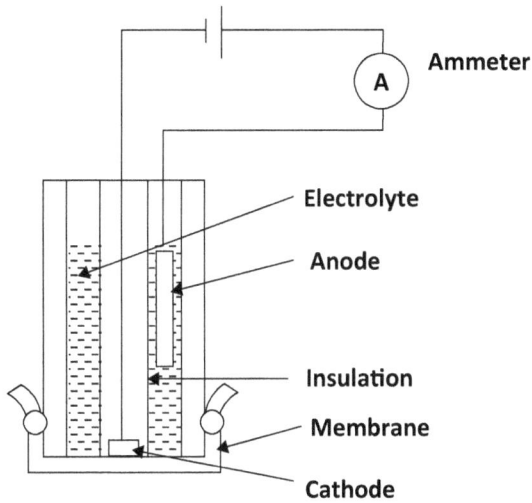

Figure 15.2. Polarographic electrode.

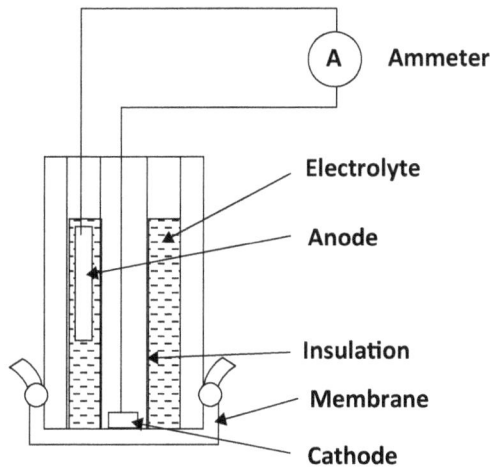

Figure 15.3. A galvanic electrode.

15.4 The operational principle of the galvanic electrode

The galvanic electrode is shown in figure 15.3. The galvanic electrode is different from the polarographic type, since it does not require external bias voltage for the reduction of oxygen at the cathode. When a basic metal such as zinc, lead, or cadmium is used as the anode and a noble metal such as silver or gold is used as the cathode, the voltage generated by the electrode pair is sufficient for the reduction of oxygen at the cathode surface.

The electrode reaction of the silver–lead galvanic electrode is as follows. The reaction at the cathode is

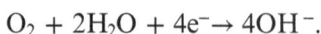

$$O_2 + 2H_2O + 4e^- \rightarrow 4OH^-.$$

The reaction at the anode is

$$Pb \rightarrow Pb^{2+} + 2e^-$$

The total reaction is

$$O_2 + 2\,Pb + 2H_2O \rightarrow 2Pb(OH)_2.$$

The electrolyte is not involved in the reaction but the anode is gradually oxidized. In this respect, the reaction is unlike that of the polarographic electrode. Therefore, the life of the probe depends on the exposed surface area of the anode. Regardless of whether the polarization voltage is applied internally (galvanic) or externally (polarographic), the principle of operation of the electrode remains the same. One important point should be noted: there are some gases which reduce in the presence of 0.6–1.0 V in the test medium, for example, halogens (Cl_2, Br_2, I_2) and oxides of nitrogen. This creates errors in the measurement system.

The basic principle of measurement for membrane-covered DO_2 sensors can be explained as follows: if the oxygen diffusion is controlled by the membrane covering the cathode, the current output of the probe is proportional to the the partial pressure in the liquid medium.

The following assumptions are made for the mathematical analysis of the pressure profile of oxygen in the liquid and the current output of the dissolved oxygen sensor:

1. The cathode is well polished and the membrane is tightly fitted over the cathode surface so that the thickness of the electrolyte layer between the membrane and the cathode is negligible.
2. The probe is immersed in liquid which is well stirred and agitated so that the partial pressure of oxygen at the membrane surface is same as that of the bulk liquid.
3. Oxygen diffusion occurs only in one direction, namely perpendicular to the cathode surface.

This is a 'one-layer' model that can be modified to include the effects of other layers.

At time zero, the partial pressure of oxygen in the liquid changes from zero to p_o. According to Fick's second law, the unsteady-state diffusion in the membrane is described by :

$$\frac{\partial p}{\partial t} = D_m \frac{\partial^2 p}{\partial x^2}, \tag{15.1}$$

where D_m is the oxygen diffusivity in the membrane and x is the distance from the cathode surface (figure 15.4). The initial and boundary conditions are:

$$p = 0 \text{ at } t = 0, \tag{15.2}$$

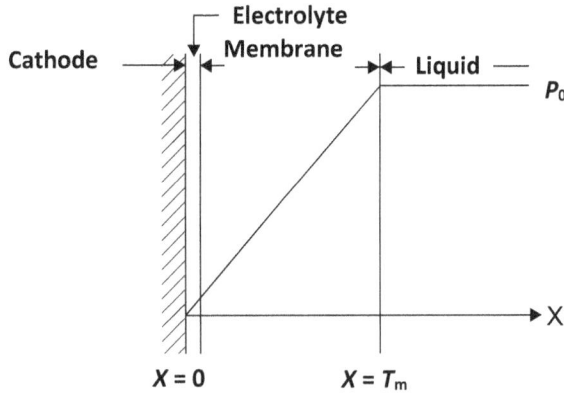

Figure 15.4. One-layer electrode model

$$p = 0 \text{ at } x = 0, \tag{15.3}$$

$$p = p_0 \text{ at } x = T_m, \tag{15.4}$$

where T_m is the membrane thickness. The boundary condition given by equation (15.2) assumes a very fast reaction at the cathode surface. The solution of equation (15.1) in normalized form with the boundary conditions of equation (15.2) is given by

$$\frac{p}{p_0} = \frac{x}{T_m} + \sum_{n=1}^{\infty} \frac{2}{n\pi}(-1)^n \sin \frac{n\pi x}{T_m} \exp(-n^2\pi^2 D_m t/T_m^2) \tag{15.5}$$

The current output of the electrode is proportional to the oxygen flux at the cathode surface.

$$I = NFAP_m\left(\frac{\partial p}{\partial x}\right)_{x=0}, \tag{15.6}$$

where

N = the number of electrons per mole of oxygen reduced,
F = Faraday's constant,
A = the surface area of the cathode, and
P_m = the oxygen permeability of the membrane.
The permeability, P_m, is related to the diffusivity, D_m, by the following expression:

$$P_m = D_m S_m, \tag{15.7}$$

where S_m is the oxygen solubility of the membrane. From equations (15.5) and (15.6), the current output (I_t) of the probe can be expressed as a function of time as follows:

$$I_t = NFA(P_m/T_m)p_0\left[1 + 2\sum_{n=1}^{\infty}(-1)^n \exp(-n^2\pi^2 D_m t/T_m^2)\right]. \tag{15.8}$$

The normalized partial pressure and the current under steady-state conditions can be obtained from equations (15.5) and (15.8), respectively:

$$\frac{p}{p_0} = \frac{x}{T_m}$$
(15.9)

and

$$I_S = \text{NFA}(P_m/T_m)p_0.$$
(15.10)

It can be observed that the pressure in the membrane is linear, as shown in figure 15.4, and the current is proportional to the partial pressure of oxygen in the bulk liquid. Measurements of the dissolved oxygen concentration performed using DO_2 probes are based on equation (15.10). From equation (15.8), we can see that the probe response depends on k, which is called the probe constant and is defined as follows:

$$k = \frac{\pi^2 D_m}{T_m^2}$$
(15.11)

The probe constant, k, is large when the membrane is thin and/or the oxygen diffusivity D_m is high. This means that the time constant is small and the probe response is fast. However, the assumption of membrane-controlled diffusion is no longer valid under these conditions. Thus, a compromise has to be made to achieve more realistic performance using DO_2 probes.

15.5 Limitations of the single-layer electrode model

The analysis is oversimplified by assumptions 1 and 2 which were made in section 15.4. In practical terms, a finite-thickness electrolyte layer is present between the cathode and the membrane because of the roughness of the cathode surface. Also, a stagnant film of liquid exists outside the membrane, even at very high liquid velocities. A more practical model of the electrode is shown in figure 15.5. Here, all three layers, namely the electrolyte, the membrane, and the liquid film are considered.

The effects of the different layers on the electrode behaviour can be computed using the 'one-layer' model, which is then extended to three layers. Under steady-state conditions, the oxygen flux, J, (figure 15.5) becomes equal for all layers. Therefore, we can write

$$\begin{aligned}
J &= K_0 p_0 \\
&= k_{LM}(p_0 - p_m) \\
&= k_m(p_m - p_e) \\
&= k_e p_e
\end{aligned}$$
(15.12)

where K_0 is the total mass transfer coefficient and the small k's represent the individual mass transfer coefficients corresponding to the liquid film (k_{LM}), the

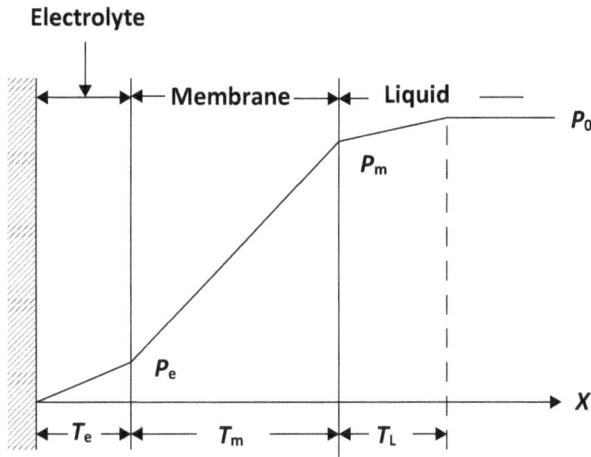

Figure 15.5. The three-layer electrode model.

membrane (k_m), and the electrolyte (k_e), respectively. The overall mass transfer resistance, $1/K_o$, is expressed as the sum of the individual resistances:

$$\frac{1}{K_0} = \frac{1}{k_{LM}} + \frac{1}{k_m} + \frac{1}{k_e}$$

(15.13)

Equation (15.13) can be rewritten using the oxygen permeability and the thickness of each layer as follows:

$$\frac{1}{K_0} = \frac{T_L}{P_L} + \frac{T_m}{P_m} + \frac{T_e}{P_e},$$

(15.14)

where T_L, T_e, P_L, and P_e are the liquid film thickness, the electrolyte thickness, the oxygen permeability of the liquid film, and that of the electrolyte layer, respectively. The liquid film is never stagnant around the DO_2 probe. Therefore, it is more accurate to use the convective mass transfer coefficient. Hence the condition for membrane-controlled diffusion becomes:

$$\frac{T_m}{P_m} >> \frac{T_L}{P_L} + \frac{T_e}{P_e}$$

(15.15)

This means that a relatively thick membrane with a low oxygen permeability is required, which increases the time constant of the probe. For given cathode dimensions, the resistance of the electrolyte is almost fixed. Also, since the electrolyte is contained inside the membrane, it does not affect the measurement. Therefore, the condition for accurate measurement of the dissolved oxygen concentration becomes:

$$\frac{T_m}{P_m} + \frac{T_e}{P_e} >> \frac{T_L}{P_L}.$$

(15.16)

When the individual resistances are taken into consideration, the steady-state output current can be written as

$$I_s = \mathrm{NFA}(P_m/\overline{T})p_0, \tag{15.17}$$

where $\overline{T}$ is defined as

$$\overline{T} = T_m + \frac{P_m}{P_L}T_L + \frac{P_m}{P_e}T_e. \tag{15.18}$$

In this case, the probe constant, k, is modified as follows:

$$k = \frac{\pi^2 D_m}{\overline{T}_t^2}, \tag{15.19}$$

where

$$\overline{T}_t = T_m + \sqrt{\frac{D_m}{D_L}}\,T_L + \sqrt{\frac{D_m}{D_e}}\,T_e. \tag{15.20}$$

Equations (15.17) and (15.19) show that the steady-state current decreases and the probe response time increases when there is a significant mass transfer resistance in the liquid film around the membrane. However, the decrease in the current can be addressed by using an additional signal conditioning circuit. Second, the effect of the liquid film resistance should be made negligible. This can be achieved by using membranes of low oxygen permeability and by thoroughly agitating the liquid around the probe.

15.6 Electrode design

The membrane-covered DO_2 probe basically consists of a cathode, an anode, and the electrolyte. In designing DO_2 probes, the following requirements are generally considered:

- There should be no drift in the measurement system.
- The current output of the probe should be sufficiently high, and its response to the dissolved oxygen concentration should be linear. However, if the current is low, it can be amplified by an additional low-noise amplifier.
- If the probe is placed in a flowing liquid, the measured output should be unaffected by the flow of the fluid.
- The instrument should have a fast response, although most DO_2 probes are sluggish in nature, which is inherent to the system.
- Ambient temperature compensation should be incorporated to avoid any change in the measured output due to a change in ambient temperature.
- The probe must be robust enough to withstand high pressures and repeated autoclaving due to the requirement for sterilization, which is necessary to avoid any contamination during bioreactions.

However, it is impossible to design a DO_2 sensor that has all the above features.

Therefore, some of the requirements are emphasized more than others, depending on the specific measurement requirement. Therefore, the probe must be custom built for some specific applications.

15.7 Details of some commercially available DO$_2$ sensors

Some of the commercially available DO$_2$ probes are discussed here. The constructional details and relative advantages and disadvantages are also discussed.

15.7.1 The Clark electrode

As shown in figure 15.6, this probe is characterized by a flat disk cathode and a reference electrode (Ag/AgCl) which is immersed in the electrolyte. Although the size of the cathode is almost similar, the membrane material and the electrolyte differ widely for different manufacturers. This design is most popular in commercial DO$_2$ probes for use in the laboratory or in the field. However, the probe output can become erroneous due to AgCl deposition on the anode surface, the deposition of silver on the cathode, the depletion of Cl$^-$ from the electrolyte, or a loose membrane. However, with routine maintenance and cleaning of the electrode, regular membrane replacement, electrolyte replenishment, and periodic calibration, such probes can be used for long periods of time. When a 25 μm Teflon membrane is used, 95% of the response times lie between 15 and 20 s. However, these electrodes exhibit hysteresis, which means that the calibration curves for increasing and decreasing oxygen concentrations do not overlap. This phenomenon is caused by the electrolyte, which acts as a reservoir of dissolved oxygen, and/or the accumulation and slow

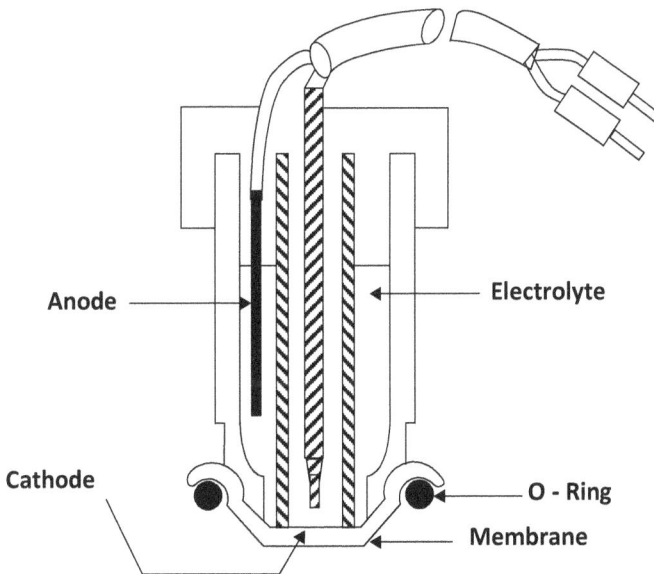

Figure 15.6. The Clark-type electrode.

decomposition of hydrogen peroxide in the vicinity of the cathode. Hysteresis is absent from the Mancy electrode discussed in the next subsection.

15.7.2 The Mancy electrode

A schematic of the galvanic probe originally designed by Mancy is shown in figure 15.7. Its major difference from the Clark electrode is the replacement of the electrolyte chamber by a thin film of electrolyte which is placed between the cathode and the membrane, which eliminates the hysteresis that is present in the Clark electrode.

Due to the relatively large diameter (0.6 cm) of the cathode, a microammeter could be directly connected to the probe. Its repeatability is better compared to that of earlier polarographic probes, but the useful probe life is somewhat less because the available surface area of the anode is relatively small. The anode surface is gradually oxidized over a period of time and can no longer be used. One more limitation is that the probe cannot be used for the measurement of flowing liquids. Both the Clark and Mancy probes need current amplification and an additional signal conditioning circuit.

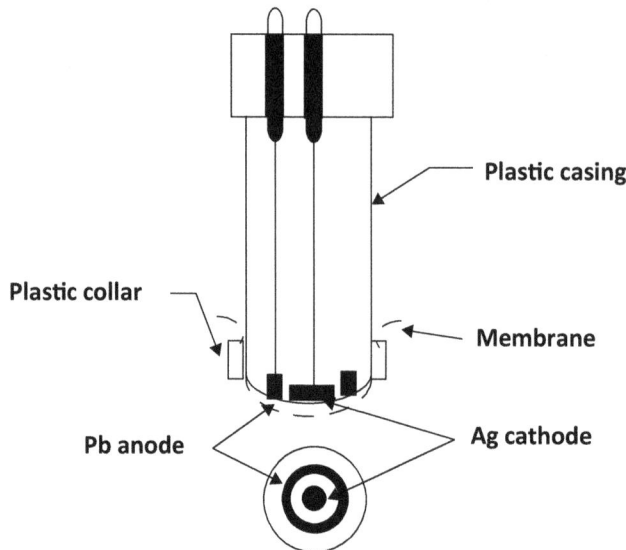

Figure 15.7. The Mancy electrode.

15.7.3 The Mackereth electrode

A schematic of the Mackereth electrode is shown in figure 15.8. It was noted that both the Clark electrode and the Mancy electrode do not have long-term stability. Moreover, their current outputs are low, in the range of μA. The Mackereth probe

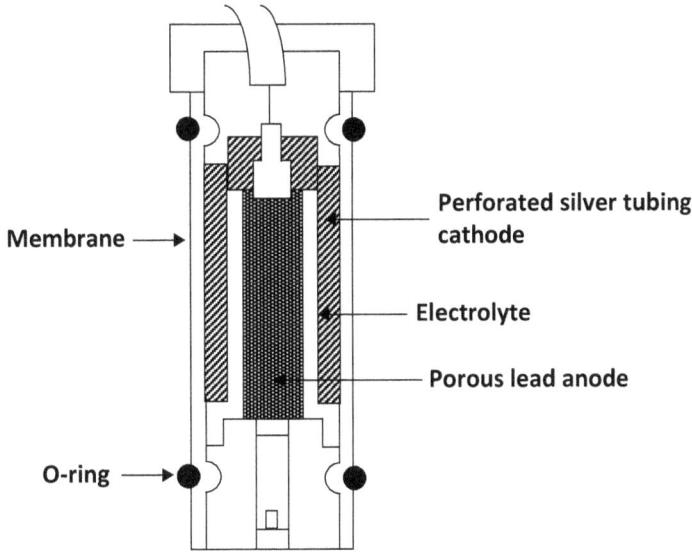

Figure 15.8. Mackereth electrode.

eliminate all these problems. Here, perforated silver tubing is used as the cathode and porous lead is used as the anode.

Its current output is much higher than those of other probes, so that current amplification is not necessary for this probe. Its repeatability is good over many months of continuous operation. However, its time constant is large. With a 25 μm Teflon membrane, this probe reaches 90% of its steady-state value in 1 min. Because of its long-term stability, this probe has been used for monitoring the dissolved oxygen concentration in continuous cultivation, which lasts for several weeks. The input impedance of the probe is low, which means the probe loads the system, especially when measurement is required in a small medium, since an amount of oxygen is consumed by the probe. The measured output becomes erroneous when the measured liquid becomes viscous at the end of any fermentation process.

15.7.4 The Borkowski–Johnson electrode

As shown in figure 15.9, in this DO_2 probe, the cathode is made from a silver spiral and a flattened lead wire forms the anode. A low-pH acetate buffer is used as the electrolyte to prevent interference by dissolved CO_2. The probe can withstand repeated steam sterilizations.

It is capable of operation for several months, and, most importantly, it has a linear response. With a 50 μm Teflon membrane, 90% of the steady-state value is reached in 1 min. As is the case for the Mackereth electrode, vigorous agitation of the liquid is required for reliable measurement. The probe is not suitable for viscous liquids, though measurement is possible up to a certain level of viscosity using a thicker membrane.

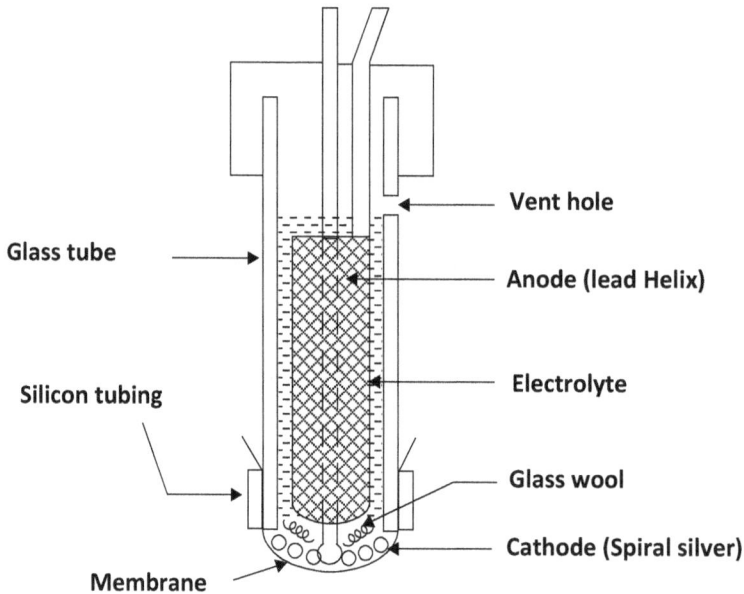

Figure 15.9. The Borkowski–Johnson electrode.

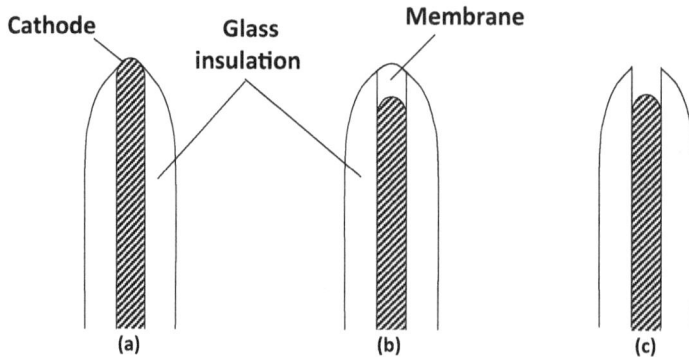

Figure 15.10. Different shapes of microcathodes (a) bare, (b) membrane covered, (c) recessed.

15.7.5 Microelectrodes

Microelectrodes are suitable for measurement when the system needs high resolution and a fast response time or a low time constant. If the cathode diameter is less than 1 μm, even a bare metal cathode as shown in figure 15.10(a) becomes insensitive to liquid flow and measures the local dissolved oxygen concentration. The probe performance is improved by covering the cathode with a membrane as shown in figure 15.10(b) or by extending the insulation as shown in figure 15.10(c) such that the diffusion gradient remains inside the recess. A thick platinum wire is first etched in an electrolyte solution to a fine point 0.2–1 μm in diameter and then insulated with a thin layer of glass. The membrane is applied by dip coating the electrode in polystyrene.

However, microprobes are fragile, they have poor repeatability, and their useful lives are short.

15.8 Electrode metals

For polarographic probes, platinum, gold, silver, and rhodium have been used as the cathode and Ag, Ag/AgCl, Ag/Ag$_2$O, and calomel have been used as the reference electrode. Gold is generally preferred to platinum as the cathode material because it is less susceptible to poisoning by noxious gases, notably hydrogen peroxide, for example. However, gold is not suitable for application in steam-sterilizable probes or microelectrodes, since gold and glass cannot be fused together. Gold-plated electrodes have been used in the manufacture of microelectrodes to avoid this problem. The reference electrode has to maintain a stable reference voltage to ensure that the polarographic probe has good performance. It also has to have a large surface area to avoid polarization. Ag/AgCl is normally used as the reference electrode. The deposition of silver ions on platinum, the oxidation of the catalytic surface, or the excessive deposition of AgCl on the reference electrode can cause a change of calibration or uncertainty in the measurement. The cathode surface should be cleaned with soft scouring powders and wet leather or using toothpaste. Excessive AgCl deposits can be removed by washing the electrode in 15% NH$_4$OH.

For galvanic probes, silver as the cathode and lead as the anode are most common choices, but silver–aluminum, platinum–aluminum, platinum–lead, gold–zinc, and gold–lead pairs have also been used. Although galvanic probes suffer less from contamination and survive autoclaving, their life is limited by the available surface area of the anode due to slow oxidation of anode surface. The probe life increases if less current is drawn from it. In other words, the probe life increases when it is used for monitoring low, rather than high, oxygen concentrations.

15.9 Electrolytes used in DO$_2$ probes

Since the electrode reaction occurs in the electrolyte solution, the composition, pH, and the volume of the electrolyte are directly related to calibration. For polarographic probes, the electrolyte is involved in the chemical reaction, so it is necessary to refill the electrolyte at regular intervals. In general, the pH of the electrolyte does not affect oxygen reduction on a clean metal surface; however, it influences the reading if the metal cathode is oxidized. The choice of electrode and electrolyte should properly matched so that there is no solubility problem between the two. The polarization voltage changes when the concentration of the electrolyte changes due to evaporation or the diffusion of water through the membrane. The loss of solvent through the membrane can be solved by having a large electrolyte reservoir which can effectively supply solvent to the electrolyte film where the solvent loss occurs. Another method is to use an electrolyte in the form of a gel or paste. An added advantage of the latter approach is that the residual current becomes smaller because the oxygen permeability is normally lower in more viscous media.

As mentioned in section 15.3, KCl is the most commonly used electrolyte that gives a constant calibration over 72 h. For galvanic probes, KOH is preferred to KCl because KCl creates a high residual current and does not maintain a clean anode surface.

15.10 The membrane

The ideal characteristics of a membrane for use in DO_2 probes are a relatively low oxygen permeability and a high oxygen diffusivity. The permeability has to be low to ensure that the membrane controls oxygen diffusion, while high diffusivity reduces the time constant of the probe. The membrane should be mechanically strong and chemically inert. Since the current output is directly related to the thickness and the oxygen permeability of the membrane, the probe sensitivity is directly affected by changes in the membrane properties. Other important factors are its CO_2 permeability and its water permeability. The water permeability of the membrane has to be low to prevent loss of water from the electrolyte solution, which causes an increase in electrolyte concentration and change in the calibration curve. Low CO_2 permeability of the membrane is also desirable for probes intended for use in aerobic cultures or in blood. Teflon, polyethylene, and polypropylene are the most popular membrane materials but silicon, polystyrene, and Mylar are also used. Polypropylene is better than Teflon in many respects. It has lower oxygen permeability, lower CO_2 permeability, and yet higher oxygen diffusivity. However, Teflon is more suitable than polystyrene and polyethylene for making steam-sterilizable probes due to its higher heat resistance. Moreover, Teflon also has low water permeability. Polystyrene is used in producing microprobes because it sticks well to the glass insulation and has relatively low oxygen permeability.

15.11 Signal conditioning circuits

Most DO_2 probes need current amplification. Op-amp-based circuits are used for this purpose. As mentioned in section 15.3, a bias voltage of 0.6–0.8 V (depending on the type of electrode) has to be externally applied for polarographic probes. For galvanic probes, a resistor is typically connected in series with the probe, and the voltage drop across the resistor is monitored using a voltmeter or a potentiometer.

15.12 General design considerations

One of the important considerations in DO_2 probe design is to minimize the zero current, which is defined as the current output of the probe at zero oxygen concentration. There are four major sources which contribute to this offset current, namely:
- electrochemically active impurities in the electrolyte,
- leakage resistance,
- incorrect polarization voltage,
- and back diffusion of oxygen.

Reducible or oxidized impurities in the electrolyte normally have a short-term effect because they are removed by the cathode during the early stages of measurement. Leakage resistance in between the anode and the cathode through the insulating material is not normally significant for large cathodes but cannot be ignored in microelectrodes. For example, to obtain an offset current of 1×10^{-11} A, the insulation resistance must be more than 7×10^{10} Ω. For this reason, glass is used as the insulating material for microelectrodes. This will lead to the necessity of using a high-input-impedance signal conditioning circuit. Moreover, one should wary of choosing an epoxy resin which may absorb water, thereby increasing the offset current by forming an electrical path between the anode and the cathode.

15.13 Calibrating DO_2 sensors

DO_2 probes/sensors can be calibrated in three ways: % saturation, partial pressure of oxygen, and actual concentration. In all calibrations, good temperature control ($\pm 0.1°$ C) of the test medium is required because the probe sensitivity changes substantially with temperature. It is advisable to use a resistance temperature detector (RTD) to measure the temperature during calibration.

15.13.1 Calibration based on % saturation

Gas-phase calibration is fast and easy. The output current from the probe is calibrated to 0% saturation in nitrogen while that in air or oxygen is set to 100%. This method is simple but not very accurate. The oxygen content can be removed either by reacting it with nitrogen or by adding 2% of sodium sulfite or sodium dithionite to 100% of the liquid by volume. These react with the dissolved oxygen to create anaerobic conditions.

15.13.2 Calibration based on partial pressure

DO_2 probes measure the partial pressure of dissolved oxygen but not the actual concentration. The reading based on % saturation can be directly calibrated in partial pressures if the barometric pressure is known. When water is equilibrated with air at temperature T, the partial pressure of oxygen, pO_2, can be expressed as follows:

$$pO_2 = [pB - p(H_2O)] \times 0.2095,$$

where
- pB = the temperature-corrected barometric pressure,
- $p(H_2O)$ = the vapor pressure of water at a given temperature, and
- 0.2095 = the fraction of oxygen in atmospheric air.

15.13.3 Calibration based on concentration

If the probe output is expressed as a partial pressure, it can be converted to a concentration if the solubility of oxygen in the liquid is known. The solubility is often expressed as the Bunsen coefficient, a, which is defined as ml O_2 absorbed by 1 ml of solvent at 0 °C and 1 atm of O_2:

$$a = \frac{V_g}{V_s} \frac{273.15}{T},$$

where V_g, V_s, and T are the volume of gas absorbed, the volume of the absorbing solvent and the absolute temperature, respectively.

For a sparingly soluble gas such as oxygen, pressures can be converted to concentrations using Henry's law:

$$C = p_0/H,$$

where H is the Henry's law constant. The conversion of a to H is as follows:

$$H = \frac{22414(760)}{1000a}.$$

In general, the probe calibration changes when it is used for long-term fermentation. This is generally caused by zero drift of the sensor. The sensitivity of the probe also decreases following long continuous use due to sensitivity drift. Moreover, calibration curve becomes nonlinear when the probe is placed in a liquid that has a high dissolved oxygen concentration. For most processes, measurement at 10 min intervals is sufficient, even with closed-loop control. Routine cleaning and sterilization of the probe even during continuous measurement removes all these problems.

IOP Publishing

Fundamentals of Industrial Instrumentation (Second Edition)

Alok Barua

Chapter 16

Gas chromatography

Learning objectives:
- The basics of chromatography.
- The chromatogram.
- What are packed bed columns and open tubular columns?
- What are the different partition forces that are present between solutes?
- Terms such as the partition coefficient, retention time, selectivity factor, etc.
- Quantitative analysis of chromatograms.
- Different types of equipment and different operational techniques used in gas and liquid chromatography.
- The different detectors used in chromatography, their purposes, and their classifications.

16.1 Introduction

Chromatography is a method used to separate and or analyze complex mixtures. Gas chromatography is a method of separating, identifying, and determining the components of complex mixtures of gases. In general, in chromatographic separation, the sample is dissolved in a mobile phase, which may be a gas, a liquid, or a supercritical fluid. The components to be separated are distributed between two mutually immiscible phases, a stationary phase and a mobile phase, which are brought into contact. The stationary phase forms the bed and the mobile phase percolates through it. The two phases are chosen so that the components of the sample distribute themselves between the mobile and stationary phases. Those components strongly retained by the stationary phase move slowly with the flow of the mobile phase. However, the components that are weakly held by the stationary phase travel rapidly. Therefore, due to these differences in flow rate, the sample components separate into discrete bands, or zones, that can be analyzed qualitatively and quantitatively.

With gas chromatography we can separate complex mixtures with great precision. Even very similar components, such as proteins that vary only by a single amino acid in composition, can be separated. Before the development of chromatography, the separation of these biological compounds was extremely tedious, time-consuming, and unreliable. Chromatography can be used to separate very delicate products, since the conditions under which it is performed are not typically severe. The data obtained by gas chromatography is useful for researchers or practicing engineers who want to know what material they have synthesized in the laboratory.

Therefore, gas chromatography is a critical aspect of many laboratories and industries; it allows scientists and engineers to separate, measure, and analyze organic molecules and gases. It is used in a wide range of applications including food analysis, quality assurance and control in manufacture, and even in forensic research.

16.2 Different methods of chromatography

- Gas chromatography makes use of a pressurized gas cylinder and a carrier gas, such as helium, to carry the solute through the column.
- Gas adsorption (gas–solid) chromatography involves a packed bed composed of an adsorbent that is used as the stationary phase. Common adsorbents are zeolite, silica gel, and activated alumina. This method is mainly used to separate mixtures of gases.
- Gas–liquid chromatography is a more common type of gas chromatography. It uses a column in which an inert porous solid is coated with a viscous liquid which works as the stationary phase. Diatomaceous earth is the most common solid used in this process. Solutes in the feed stream dissolve into the liquid phase and subsequently vaporize. The separation is thus based on relative volatilities.
- Capillary gas chromatography uses glass or fused silica capillary walls which are coated with an absorbent or other solvent. The column has only limited capacity because of the small amount of the stationary phase material. However, this method rapidly separates small amounts of gaseous mixtures.

16.3 The basics of chromatography

A schematic of a gas chromatograph is shown in figure 16.1. It consists of six parts: (1) a supply of carrier gas in a high-pressure cylinder, (2) a sample injection system, (3) the separation column, (4) the detector, (5) a recorder, and (6) a temperature-controlled compartment used to house the column.

16.3.1 The column

There are two types of columns, namely packed and open tubular or capillary columns. The packed columns are made of stainless steel, copper, or glass tubing that has a diameter of 1.6–9.5 mm and a length of 3 m. Capillary columns provide an open, unrestricted path for the carrier gas within the column. They are made from a long narrow tube that has a length of 50–150 m and a diameter a 0.25 mm; its inner

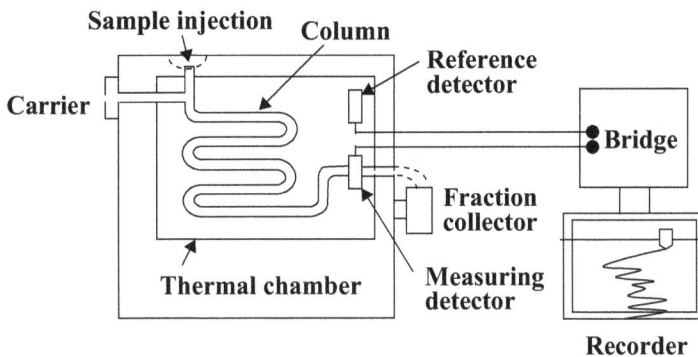

Figure 16.1. A schematic of a gas chromatograph.

wall is coated with the liquid stationary phase at a thickness of about 1 μ. Columns constructed of silicon-tetrahydride-treated Pyrex glass have the most desirable features. The sample capacity of a capillary column is principally determined by the thickness of the stationary phase on the column walls. The coating is usually made of silicone gum. Packed columns are usually formed into several coils and placed within a temperature-controlled compartment. The tubing of capillary columns can be coiled into an open spiral, a basket coil, or a flat pancake shape.

16.3.2 The sample injection system

The mobile phase comprises a solvent into which the sample is injected. The sample and the solvent flow through the column together; hence, the mobile phase is often called the 'carrier fluid.' The components of the sample to be separated have varying affinities for the material in the column that comprises the stationary phase. Depending on the materials which comprise the mobile phase, two general types of chromatographic process exist, namely:

(i) Gas chromatography: here, the mobile phase is a gas. Generally, the gas is inert in nature. The stationary phase is an adsorbent (solid) or liquid distributed over the surface of a porous, inert support.

(ii) Liquid chromatography: here, the mobile phase is generally a low-viscosity liquid which flows through the stationary phase bed. This bed may consist of an immiscible liquid coated onto a porous support, such as a thin film of the liquid phase bonded to the surface of an adsorbent (solid) or a solid of controlled pore size.

The sample must be introduced in a vaporous state in the smallest possible volume and in the shortest possible time without decomposition. Liquid samples that have volumes of 1–10 μl are usually injected by a microsyringe through a self-sealing silicone rubber septum. The most accurate and precise method for gas samples used a calibrated sample loop (0.5–10 ml) and a multiport rotary valve. The smaller the sample, the better the peak shapes. The stationary liquid phase separates the sample. In addition to possessing selectivity, the liquid phase should be chemically and thermally stable.

16.3.3 The temperature-controlled chamber

An oven is used to maintain precise temperature control around the column. Some form of ambient temperature compensation technique should be implemented.

16.3.4 The detector

The detector senses the presence of the individual components as they leave the column. Following amplification, the detector output is traced on a strip chart recorder, or it can be transmitted and recorded by a digital technique such as a data acquisition system. The different types of detectors generally used are: (i) the thermal conductivity detector, which is popular in gas chromatography and is used to analyze gases and organic compounds. It is used when sensitivity is not very important. (ii) The flame ionization detector, which is perhaps the most popular of all gas chromatography detectors. It is widely appreciated for its steadfast dependability and sensitivity when detecting organic vapors. (iii) The thermo ionic detector, which is not as widely used as the two previously described detectors. These detectors are used to gauge and measure the presence of compounds containing phosphorous or nitrogen. (iv) The flame photometric detector, which is similar to thermo ionic detectors and is used for specific applications. It detects components that contain phosphorous or sulfur atoms. The detectors used in chromatography are generally operated in one of two different ways: they respond either to the concentration of the solute or the mass flow rate. Those responding to the concentration yield a signal which is proportional to the solute concentration which traverses the detector. An elution peak is produced when the signal is plotted against time. For such detectors, the area under the peak is proportional to the mass of a component and inversely proportional to the flow rate of the mobile phase. Hence, it is important and necessary to keep the flow of the mobile phase constant for such detectors. In differential detectors that respond to mass flow rate, the peak area is directly proportional to the total mass and there is no dependency on the flow rate of the mobile phase.

Therefore, the detector is a critical component of a chromatograph; it forms the core of the equipment. The detector translates and provides information gathered by the chromatograph in a format usable by the user. The data can then be stored and used by recording devices and computers, allowing the user to make data-driven decisions.

16.3.5 The operational principle

(i) **Feed injection**

The feed (a mixture of various components) is injected into the mobile phase. The mobile phase flows through the system under gravity, capillary action, or the action of a pump. In the case of syringe–septum injection, a small (10 μl) sample is introduced into the pressurized column by a high-pressure syringe, though there are other feed or sample injection systems which are discussed in the later part of this subsection.

(ii) **Separation in the column**

Different components of the mixture (feed) have different rates of migration as the mixture is carried over the stationary phase by the mobile phase. These differential rates provide the separation between various components.

(iii) **Elusion from the column**

After the sample is displaced from the stationary phase, the different components in the sample are eluted from the column at different times. The smaller the affinity a molecule has for the stationary phase, the shorter the time spent in the column. Thus, the least retarded component elutes first and the most strongly retained component elutes last. Separation is achieved when one component is retarded sufficiently to prevent overlap with the zone of an adjacent solute as sample components elute from the column.

(iv) **Detection**

The different components emerging from the column are collected. The concentration and characteristics of each component are the evaluated using some analytical method such as infrared (IR) spectroscopy, nuclear magnetic resonance (NMR), etc.

(v) **The method of separation**

Consider a mixture consisting of three components x, y, and z. Initially, as the mixture is injected into the column, there is no separation between the three components (figure 16.2(a)). As the mobile phase drives the mixture down the column, a small separation becomes evident between the three components (figure 16.2(b)). Let us assume that the speeds at which the three components move are

$$v(z) > v(y) > v(x),$$

where $v(z) \rightarrow$ the speed of component z.

$v(y) \rightarrow$ the speed of component y.

$v(x) \rightarrow$ the speed of component x.

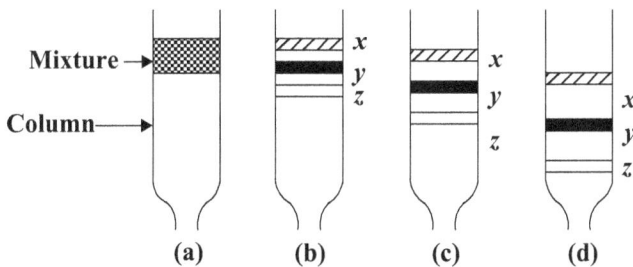

Figure 16.2. A schematic of the separation of a mixture of three gases.

In other words, x has the maximum affinity whereas z has the least (figure 16.2(c)). Thus z emerges from the column, first followed by y, followed by x (figure 16.2(d)).

16.3.6 The chromatic behaviour of solutes

(i) Partition coefficient (k): the partition coefficient or equilibrium coefficient is defined as the molar concentration of the analyte in the stationary phase divided by the molar concentration of the analyte in the mobile phase.

$$k = \frac{\text{molar concentration in stationary phase}}{\text{molar concentration in mobile phase}}.$$

If the output signal from the detector is plotted as function of time, a series of peaks is obtained. Such a plot is called a chromatogram, and a typical plot is shown in figure 16.3. Operating conditions such as pH and temperature affect the output of chromatography. The level of complexity of the sample is indicated by the number of peaks which appear. Qualitative information about the sample composition is obtained by the comparing peak positions with those of standards. Quantitative information regarding relative concentrations can be obtained from peak area comparisons. In this context, we shall introduce three important characteristics of the chromatogram.

(ii) Retention time

t_R (retention time): the time between sample injection and the time at which a peak reaches a detector at the end of the column is called the retention time. Each gas has a different retention time.

$t_M \rightarrow$ The time taken for the mobile phase to pass through the column is denoted by t_M.

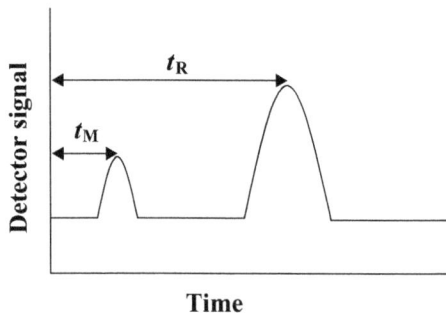

Figure 16.3. A chromatogram.

(iii) Retention factor (K'):

The retention factor describes the migration rate of a gas in a column. It is also called the capacity factor and it is expressed as follows:

$$K' = \frac{t_R - t_M}{t_R}$$

When a gas's retention factor is less than one, elution is very fast. Ideally, the retention factor for a gas should be between one and five.

(iv) Selectivity factor (α):

The selectivity factor describes the separation of two components (say A and B) in the column.

$$\alpha = \frac{K'_B}{K'_A}$$

and α is always greater than one. Therefore, in the above equation we have assumed that A elutes faster than B.

16.3.7 Quantitative analysis

Peak area integration:

In column chromatography, the signal generated by the detector is graphically recorded in the form of chromatographic peaks in the chromatogram (figure 16.3). The area under these peaks can then be integrated in a variety of ways and the resulting data relating to the composition of the samples can be studied. The ways in which the area under the peak can be calculated are as follows.

16.3.7.1 Height times width at half height

This method involves multiplying the actual peak height by the width at half height. A new baseline is drawn to measure the peak height and width at half height, since the normal (zero signal) baseline causes large deviations due to tailing.

16.3.7.2 Planimetry

In this method, the peak is traced by a planimeter, a mechanical device which measures the area by tracing the perimeter of the peak. Planimetry is less precise than height–width integration due to errors caused by placing the baseline and by tracing the peak outline.

16.3.7.3 Triangulation

In this method, tangents to the sides of the peak are drawn at the inflection points and the area of the triangle formed by these tangents and the baseline is determined. The height is measured from the baseline to the point where the tangents intersect.

16.4 Liquid chromatography

Only 20% of the known compounds can be analyzed by gas chromatography due to insufficient volatility or thermal stability. Liquid column chromatography does not have this limitation. The interchange or combination of solvents can provide special selectivity effects that are absent when the mobile phase is a gas. Traditional gas chromatography was achieved by gravity, hence analysis took place at a slow rate. In liquid chromatography, pressure is applied to the column, forcing the mobile phase through at a much higher rate. The pressure is applied using a pumping system.

The equipment commonly used for liquid chromatography consists of: (i) a solvent reservoir for the mobile phase, (ii) a solvent pump, (iii) a pre-column (except for bonded phases), (iv) a pressure gauge, (v) a sampling or injection device used to introduce the sample into the column, (vi) the separation column, and (vii) a detector.

16.4.1 The solvent delivery system

The solvent delivery system must supply a precise volume of solvent over a relatively broad flow range. One must also consider other features, namely compatibility with other components in the high-performance liquid chromatography system, compatibility with a wide choice of solvents, and a low level of noise caused by pulsations in the detector. The removal of dissolved air and other gases is also necessary. There are three main types of pumps used for the delivery, namely:

  (i) Reciprocating pumps;
  (ii) Syringe pumps;
  (iii) Constant-pressure pumps.

The choice of pump should be made after considering whether isolation or gradient elution is to be performed and determining the minimum detectability limit desired, the type of separation column, the detector to be employed for precise quantization, and the cost of the whole chromatograph system.

IOP Publishing

Fundamentals of Industrial Instrumentation (Second Edition)

Alok Barua

Chapter 17

Pollution measurement

Learning objectives:
- Sample collection.
- The principal pollutants in the atmosphere.
- Detection and measurement techniques.
- Gas detection based on optical filters.
- Color coding of the air quality index?

17.1 Introduction

The sources of pollution are many and varied. These include factories, automobiles, burning dumps, power stations, household furnaces, and so forth. Common harmful chemicals that are emitted into the atmosphere include hydrocarbons, the oxides of nitrogen, carbon monoxide, and sulfur dioxide. It is common for 10–70 mg m^{-3} of carbon monoxide to be present on busy streets with heavy vehicular traffic and more than 120 mg m^{-3} is considered dangerous. The various ill-effects of these pollutants on human health and the environment are well known. It is now recognized that many chemicals undergo photochemical decomposition and reaction in the atmosphere, creating new pollutants which may potentially be even more toxic than their precursors. For example, nitrogen oxides and hydrocarbons are the most important sources of the secondary pollutant called 'smog.'

Analytical instrumentation and photospectrometry play very important roles in making quantitative analyses of environmental pollution. The complete analysis process can be broadly divided into two methods, namely sample collection and sample analysis.

17.2 Sample collection

One of the most important steps in air pollution analysis is the collection of the sample. Once the sample has been collected, standard measurement techniques may be employed. The topic of sample collection can be divided to two steps: general considerations and the sampling train.

i) **General considerations**

The following points must be remembered when collecting pollution samples:

(a) **Sample size**

The volume of the air sample is dictated by the minimum pollutant concentration that must be measured, the sensitivity of the measurement, and the information desired. The sample size is generally chosen by trial and error. Samples of more than 10 m^3 may be required to determine ambient concentrations.

(b) **The rate of sampling**

The useful sampling rate varies with the sampling device and should be determined experimentally. Most sampling devices for gaseous constituents have a permissible flow rate of 0.003–0.03 m^3 min^{-1}.

(c) **The time and duration of sampling**

The time of day and duration of sampling are determined by the information that is desired. The sampling period only gives an indication of the average concentration during that period of a day or a week.

(d) **Sample storage**

Air samples should be stored for the minimum time duration before they are analyzed. They should be protected from heat and light. Moreover, the samples collected should not react with other constituents or with the material of the container. Gaseous samples are sometimes collected by adsorption onto a solid, in which case, the gases must not be lost by desorption before the analysis takes place.

ii) **The sampling train**

The requirements for intermittent air sampling are a vacuum source and a collector. An instrument is also required to measure the amount of air sampled.

(a) **Vacuum source**

Electric or hand-driven vacuum pump aspirators are generally used to draw the sample through the collection device. When vacuum devices are used to draw the sample through a filter in which pressure loss may build up during sampling, it is recommended that the flow should be maintained at a constant level so that there is no mechanical loading of the system.

(b) **Metering devices**

Flow measurement devices (see chapter 7) are of two general types. One measures the rate and the other measures the volume. The devices that measure flow rate are small and inexpensive but have the disadvantage of measuring only the instantaneous rate of flow.

Examples of this type include the rotameter and the Pitot tube. The devices that measure volume measure the total flow passing through them and are therefore more useful. Examples include the orifice plate and the Venturi meter. They are, however, bulkier and more expensive. All flow meters should be calibrated before use.

(c) **Collector**

The last component of the sampling train is the collector, which may be of various types, depending on the particular application.

17.3 Aerosol contaminants

One of the most commonly used means of collecting aerosol contaminants is filtration. Fibre filters (wood fibre, paper, glass fibre, asbestos), granular filters (glass or metal, porous ceramic, sand), and membrane filters (cellulose, esters) are used. A second type of collection device for both solid and liquid aerosols contaminants is the impinger. In this device, aerosols impinge on a surface exposed to an air stream. This device type includes both dry and wet impingers. More sophisticated collectors include electrostatic precipitators and thermal precipitators.

17.4 Gaseous contaminants

Gases and vapors may be collected by absorption in a liquid, adsorption on a solid surface, freezing or condensation, or filling an evacuated container. Absorption is one of the most widely used techniques for gaseous collection. The gas components of interest are dissolved in a suitable liquid or solution or on the surface of a hygroscopic solid.

17.5 Carbon monoxide detection

Carbon monoxide concentrations are measured by a technique known as non-dispersive infrared (IR) spectroscopy. The vast majority of covalently bonded molecules absorb IR radiation somewhere in the wavelength range of 3–15 µm. This represents the amount of energy required for atoms to start to vibrate relative to each other. Carbon monoxide is a very simple molecule that has only one mode of vibration. It absorbs IR radiation at wavelengths around 4.67 µm. The atoms take up the light energy and move slightly further apart than normal; they then re-liberate that energy in the form of heat when they are reunited.

Carbon monoxide detectors can be split into two types: those that depend on the measurement of energy or heat liberated after IR rays have passed through the sample and those that depend on electrochemical analysis.

i) **Detectors based on heat absorption**

Carbon monoxide analyzers pass an IR light beam alternately through a reference cell containing a non-absorbing gas such as N_2 and a sample cell containing air with carbon monoxide to be measured. The IR light at a wavelength of 4.67 µm is absorbed by the sample to an extent that corresponds to the carbon monoxide concentration. The IR beams that pass through the nitrogen-filled reference cell and the sample cell are received. However, the light beam which passes through the sample cell has a different intensity than the intensity of the beam that passed through the nitrogen-filled reference cell. This is due to light absorption by carbon monoxide in the sample cell. The difference in the measured intensities is calibrated in monoxide concentration.

ii) **Detectors based on electrochemical analysis**

When a gas is electrolyzed while maintaining a potential between electrodes immersed in an electrolyte, the current in the electrode circuit is proportional to the gas concentration because the electric potential at which electrolysis takes place depends on the carbon monoxide concentration. The electric potential is calibrated in carbon monoxide concentration.

17.6 NO_x measurement

When nitrogen monoxide (NO) or nitrogen dioxide (NO_2) in the sample reacts with ozone (which is produced by passing externally supplied O_2 over a UV lamp), part of the NO is oxidized to become NO_2. A portion of the generated NO_2 is in an excited state NO_2^* and radiates light when it returns to the ground state. This phenomenon is called chemiluminescence.

$$NO + O_3 \rightarrow NO_2^* + O_2$$

$$NO_2^* \rightarrow NO_2 + h.$$

The above reaction is very fast and only NO is involved; there is almost no effect on other gases. If the NO gas is at a low concentration, the amount of luminescence produced is in proportion to the concentration. NO_x analyzers separate the sampled gas into two parts. In one part, NO_2 is reduced to NO by the NO_x converter and used as the sample gas for measuring NO_x (NO + NO_2). In the other, the NO sample gas is used as it is. These sample gases are sampled and sent to the NO_x, NO, and reference gas lines by externally controlled solenoid valves every 0.5 s and are passed to the reaction chamber. In addition, the air that is separately sucked through the air filter is dried by silica gel and then introduced into the reaction chamber as ozone gas. In the reaction chamber, the sample reacts with ozone. The light emitted by the reaction is detected by a photodiode. The higher the intensity of light, the greater the photocurrent, which is calibrated as a concentration of NO_x.

One important thing to note is the affinity of water for NO_x gases. For this reason, it is important to remove practically all the water from the sample. A dryer is used for this purpose.

17.7 The sulfur dioxide analyzer

The analysis of sulfur dioxide is based on the principle of fluorescence spectroscopy. Sulfur dioxide has strong ultraviolet absorption at wavelengths between 200 and 240 nm. The absorption of photons at these wavelengths results in the emission of fluorescence. A zinc discharge lamp and an optical bandpass filter radiate UV light at 215 nm into the reaction chamber, where it interacts with the SO_2 molecules. The fluorescence is measured perpendicular to the beam using a photomultiplier tube. The amount of fluorescence is directly proportional to the concentration of sulfur dioxide.

One of the major difficulties with SO_2 is that it is extremely corrosive in nature and it has an affinity for water. The presence of any condensate quickly removes all traces of SO_2 from a measuring system. In addition, there is an absorption line for water very near the line for SO_2. This means that the presence of any water vapor will cause an erroneous reading. The IR sensor used to detect sulfur dioxide has the advantage of low cost but it is larger. However, IR sensors are among of the most preferred devices for measuring SO_2 in pollution control.

17.8 Ozone detection

Ozone detectors work on the principle of absorption spectrometry (photometric). The wavelength range for ozone is 220–330 nm and its absorption spectrum extends to 253.7 nm, which corresponds to the main emission wavelength of mercury vapor. The ozone absorption coefficient was therefore experimentally measured at 253.7 nm. To determine ozone concentration, a sample of ambient air is drawn through a flow cell which is exposed to ultraviolet light at a wavelength of 253.7 nm produced by a mercury vapor lamp. The intensity of ultraviolet light detected by the instrument changes depending on how much ozone in the sample absorbs it. A second source that has been scrubbed of ozone using a manganese dioxide (MnO_2) scrubber is then drawn through the absorption chamber, resulting in a higher intensity of light reaching the detector. The difference between the two intensities yields the concentration of ozone in the sample.

17.9 The detection of hydrocarbons

The flame ionization detector is the most commonly used detector for the analysis of organic compounds. Its sensitivity, linearity, and user repeatability make it the ideal detector for the measurement of hydrocarbon concentrations. The ionization of organic substances in the flame is carried out in two phases:

i. Cracking of organic compounds in the central zone of the flame and the formation of CH^*, CH_2^*, and CH_3^* radicals.
ii. Chemical ionization in contact with oxygen according to the reaction:

$$CH^* + O \rightarrow CHO^+ + e^-.$$

The molar response is defined as being proportional to the number of atoms of carbon of the molecule. Electrons (e^-) are extracted from the flame using a voltage (polarization) applied between the nozzle and the collecting electrode (chimney). The current obtained from the collecting electrode is converted into voltage by an I-to-V converter. It is subsequently digitized by an analog-to-digital converter (ADC) and displayed.

17.10 The air quality index

The air quality index (AQI) is an index or rating scale used for daily reports of the ambient air pollution recorded at monitoring sites. The AQI informs the public about the air quality and the associated health effects resulting from inhaling polluted air. The higher the AQI value, the greater the level of air pollution. Four major pollutants including suspended particulate matter (PM10) are considered in the calculation of the AQI. Public health departments enforce regulatory measures to improve the AQI.

17.11 Measurement and calculation of the air quality index

The concentration of a pollutant is measured by air quality monitoring equipment, which is located at a monitoring site. This pollutant concentration is converted to an index value using the following equation:

$$\text{index value} = \frac{\text{pollutant concentration}}{\text{pollutant goal concentration}} \times 100.$$

The 'pollutant goal concentration' used to calculate the index value is either the air National Environment Protection Measurement (NEPM) standard for that particular pollutant or the Environmental Pollution Protection (EPP) goal for air in the case of visibility.

The air NEPM standards and EPP goal for air used in calculating the index value are presented in table 17.1.

The EPP (Air) goal for visibility-reducing particles is 20 km, which means that one should be able to see clearly for at least 20 km. This goal is related to a light-scattering coefficient value measured using a nephelometer, which is an instrument for measuring the concentration of suspended particulates in a liquid or gas colloid. A scattering of 235 Mm^{-1} or less is equivalent to a visibility of more than 20 km (Mm^{-1} = per million meters).

Table 17.1.

Pollutant	Air NEPM standards	Averaging time (h)
Ozone	0.10 ppm	1
Nitrogen dioxide	0.12 ppm	1
Sulfur dioxide	0.20 ppm	1
Carbon monoxide	9 ppm	8
PM10	50 $\mu g \ m^{-3}$	24
Pollutant	**EPP (Air) goal (km)**	**Averaging time (h)**
Visibility	20	1

17.12 The meaning or interpretation of the air quality index reading

The higher the AQI value, the greater the level of air pollution and the greater the danger to health. If the AQI for suspended particles is 305 and this is the highest of the five pollutant scores, it is reported as an 'AQI of 305.' The AQI indicates the maximum safe level for a pollutant. Above this level, the pollutant has an undesirable impact on public health and the environment. If the AQI falls below 50, the air quality is considered 'good.' An AQI reading between 51–100 indicates 'moderate' air quality, and an AQI reading from 101–200 indicates 'poor' air quality. A reading of more than 201 indicates 'very poor' air quality and an AQI above 301 is regarded as 'critical' to human health. The color coding used for the AQI is shown in table 17.2.

Table 17.2. Color coding used for the AQI.

Index value	Description	Color
0–50	Good	Green
51–100	Marginal	Yellow
101–200	Unhealthy (poor)	Orange
201–300	Very unhealthy (very poor)	Red
301+	Critical	Purple

Chapter 18

Smart sensors

Learning objectives:
- An introduction to integrated, smart, and intelligent sensors.
- The working principle of intelligent sensors.
- What is meant by smart sensors?
- The difference between integrated and smart sensors.
- Some different fields of application of smart sensors.

18.1 Integrated, smart, and intelligent sensors

The recent development of many types of semiconductor sensors has led to a demand for either a lower unit cost or enhanced functionality to improve their application in the domain of instrumentation and measurement. These objectives can only be achieved through a higher level of device integration. In addition, more complex sensing problems are now being studied which require a higher level of inbuilt fast decision-making (signal/data processing) than that achievable using today's sensors. In order to more carefully define the different types of intelligent sensors, it is necessary to subdivide their processors into a signal conditioning circuit (including a data converter) and the main processing unit (e.g. a microprocessor with peripheral devices or a microcontroller). The signal conditioning circuit or preprocessor carries out low-level tasks such as amplification, filtering, analog-to-digital conversion, etc. A higher level of signal processing is carried out in the processing unit, which is digital. The output that is to be displayed or fed to a bus system should also be digital. Figures 18.1–18.3 illustrate the three levels of integration which make up sensor systems. There is no universal and widely accepted definition for smart sensors.

If a measuring device has at least one sensing element and a signal conditioning circuit and if they are integrated into a single chip, it can be called a smart sensor. It cannot be called an intelligent sensor with such a low level of intelligence or no intelligence. Instead, we will use the term 'integrated sensor' to describe this type of

doi:10.1088/978-0-7503-3755-7ch18

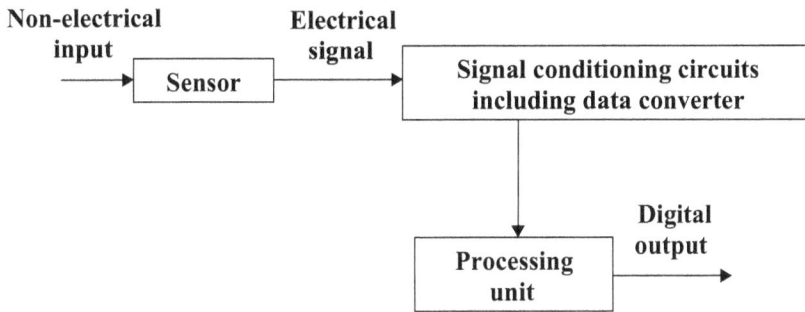

Figure 18.1. A sensor system with a signal conditioning circuit.

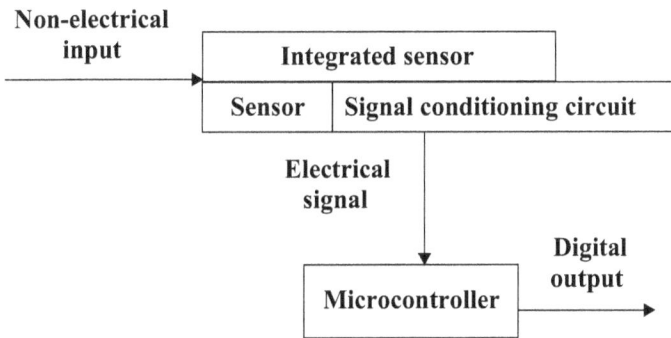

Figure 18.2. A schematic of an integrated sensor system.

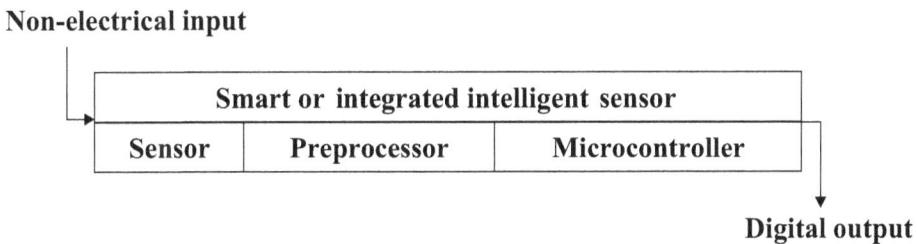

Figure 18.3. A smart sensor or integrated intelligent sensor.

low-level smart sensor for which most of the preprocessor is integrated, as shown in figure 18.2. The term 'smart' is reserved to denote the full or partial integration of the main processing unit, which also has intelligence. A schematic of a smart sensor is shown in figure 18.3. There is no confusion in this practical definition, because all smart sensors must be integrated and intelligent, while any sensor that has significant intelligence which has not been integrated can also be called an intelligent sensor. A microcontroller can be implemented for computational purposes as well as for decision-making using artificial intelligence (AI) or an expert system. If the entire sensor system is a system on a chip (SOC), then the microcontroller core can be fabricated within the chip.

The definition proposed by Brekenbridge and Husson considers AI and runs as follows:

'The sensor must have a data processing unit with automatic calibration and automatic compensation function, in which the sensor discards abnormal values or exceptional values. It must have an algorithm, which is re-configurable and must have a memory in the form of RAM where measurement data will be stored. Other desirable features are that the sensor is coupled to other sensors through a common bus, very similar to IEEE-488 and adapts to changes in environmental conditions, and has a discrimination function'.

This definition is not very simple (rather, it is long); however, it does include the essential functions required to define intelligence in a sensor. An intelligent sensor must possess one or more of the following three features:

⇒ it must perform a logical function;
⇒ it must communicate with one or more other devices;
⇒ it must make decisions using crisp or fuzzy data.

This definition describes the difference between an integrated sensor and an intelligent sensor.

18.2 The logical function of an intelligent sensor

In order for an intelligent sensor to perform a logical function, it clearly requires some type of processing unit. The processing unit is typically a microprocessor or microcontroller but could be some type of programmable logic device. The integration of the processing unit with the sensor may be possible through the use of application-specific integrated circuit (ASIC) technology or an SOC. An intelligent sensor may have a full-duplex form of communication with the user and so provide valuable information about errors or faults in the entire sensor system. Alternatively, an intelligent sensor may communicate with another device and so adapt or modify its own behaviour. This type of intelligence can provide an alarm or warning about abnormal operating conditions, or more cleverly provide a feedback control mechanism to perform autocorrection.

Intelligent sensors may offer a large degree of control via an embedded digital controller. An intelligent sensor may have some form of high-level adaptive control system that permits the control parameters to be automatically updated over time. Moreover, it may be the case that not all parameters are measurable. In this situation, a parameter estimation technique can be incorporated into the controller. The implementation of a sensor which can warn its user or adapt to environmental conditions requires some decision-making capability. The use of AI with a machine-learning-based algorithm can help with this issue. Traditionally, sensors use parametric data to make a decision. For example, the current passing through a diode can be used to provide overload protection to a device that exceeds its normal operating temperature.

However, the intelligent sensors of the future may use nonparametric methods such as artificial neural networks, AI, or expert systems which rely upon fuzzy data. The definition of an intelligent sensor used here includes any sensor system that contains a microprocessor or microcontroller unit. Consequently, there are a large number of instruments that can be classified as intelligent sensors. As analog output is becoming totally obsolete, it is expected that all intelligent and smart sensors will produce digital outputs, which is quite obvious from the use of microcontrollers.

18.3 Integration of the signal processing unit

The integration of the entire signal processing unit onto a single chip can offer many advantages. The strongest argument for the fabrication of an integrated sensor is to improve the signal-to-noise ratio (SNR). The electrical power output from a microsensor is often low and susceptible to stray capacitance, inductance, and noise. This is particularly true when, for example, a capacitive pickup is used in a microsensor. The lower the value of the stray capacitance, the higher the value of the unwanted noise. Moreover, the capacitance of a lateral resonant silicon structure may only be a few femtofarads. Thus, on-chip circuitry fabricated using metal-oxide-semiconductor field-effect transistor (MOSFET) technology is highly desirable in order to remove the effect of the high input capacitance of transmission cables and the subsequent instrumentation. It is obvious that this improves the response, sensitivity, and resolution of microsensors. There are other reasons for integrating the signal processing with the sensor, such as providing improved functionality at a lower cost, size, or weight.

Figure 18.4 shows a block diagram of the signal processing that takes place in a capacitive sensor and its associated electronic circuitry, which functions as a preprocessor. The output from such a system is digitized and then fed to the main processing unit, which is a microcontroller. The capacitive pressure microsensor has on-chip complementary metal-oxide–semiconductor (CMOS) switched-capacitor circuitry to perform accurate capacitance-to-voltage conversion, signal conditioning, and automatic compensation for the device's temperature using a reference

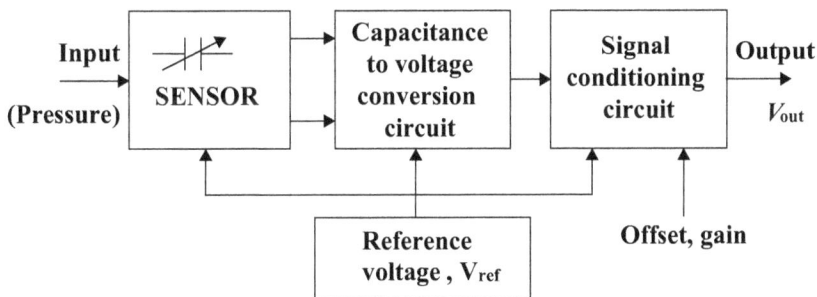

Figure 18.4. A schematic representation of the signal processing in a smart capacitive pressure sensor.

voltage. A similar situation arises for capacitive microphones. The temperature sensitivity of this smart sensor α_v is related to a reference voltage V_{ref} by

$$V_{ref} = V_0[1 + \alpha_v(T - T_0)],$$

and α_V can be programmed to have the desired number of bits of resolution.

18.4 Self-calibrating microsensors

Most sensors have at least two parameters that need to be set during the manufacturing process, namely the offset and the sensitivity or gain. During the manufacturing process, this is done while fabricating the sensor using silicon or another semiconductor device. Any adjustment of parameters requires semiconductor etching in a very large-scale integration (VLSI) process. In industry, which mostly produces smart sensors in silicon, the process of calibrating individual sensors is tedious, time-consuming, and expensive. It is undesirable but often essential. Moreover, the sensor may have to go through another calibration process during its lifetime to avoid the effect of any parametric drift if it is not possible to do so on the semiconductor (in this case, it has to be externally adjusted). Consequently, there is considerable need for a self-calibrating sensor which can carry out its own calibration, and this is particularly true when a high level of precision is required. The conventional calibration of a sensor may involve the laser trimming of integrated resistors. This means that the resistor film and patterning process must be compatible with VLSI and microsensor technology.

18.5 The self-testing of smart sensors

When sensors are produced by a manufacturing unit, it is necessary to check their specifications, which are prescribed before the development process. It is highly desirable for sensors to test their own functionality. Recent developments in the field of smart sensors are leading to sensors with some limited diagnostic capability. This is basically the ability of a sensor to determine whether it is functioning normally. A complete failure would usually be detected by the user if the output (either current or voltage) falls below its operating specification. In many cases, a sensor can fail to perform adequately even though it provides a reasonable output. In such cases, more sophisticated quality assurance is necessary. For instance, a noise characteristic (e.g. power spectral density) may be related to the physical property of the sensing element that is changing and affecting its performance and can thus be used to provide diagnostic information. A significant level of self-testing has been reported for a smart microaccelerometer whose basic mechanical performance can be routinely tested and thus diagnosed for faults, either catastrophic or parametric.

The increasing structural complexity of electronic components in smart sensors makes testing a challenging task, particularly under the constraints of quality and cost. The motivation for testing is to allow rapid detection after failure. Built-in self-test (BIST) is a design for a test technique in which the testing is accomplished through built-in hardware. For smart sensors with hundreds of transistors, the hardware required for testing can be integrated into the chip by dedicating a small percentage of the chip to the BIST circuits. In this way, BIST represents an important step toward regarding testing as one of the system functions. The principle of BIST is as follows: a BIST controller generates test patterns, controls the clock of the device under test (DUT), and collects and analyzes the responses. The self-test can be initiated using a single pin; the result ('accept' or 'reject') can be signaled via a second pin. Optionally, a serial bitstream of diagnostic data may be provided on a third pin available in the smart sensor chip.

18.6 Multisensing

Smart sensors often improve their performance through the use of other sensors to monitor undesirable dependent variables. As examples, the temperature sensitivity of a microsensor could be compensated by a diode which is fabricated on the same chip; alternatively, the temperature sensitivity of a strain gauge can be compensated by another strain gauge that is in the bridge circuit. A smart sensor can also eliminate spurious noise that influences the output reading and causes erroneous signals. For instance, an array of identical sensors can be employed and coupled to a microcontroller which calculates the average sensor output or perhaps discards any anomalous readings. The former approach has been adopted to make a smart pH sensor with ten identical sensing elements. Integrated sensors may also have a higher level of processing capability (i.e. intelligence). For example, the use of an array of dissimilar chemical microsensors and subsequent processing of the signal by an artificial neural network has led to the development of an intelligent artificial nose.

18.7 The outputs of smart sensors

The digital outputs produced by a smart sensor can be transmitted via a bus system, or the display device may be accommodated in the smart sensor itself. The simplest form of readout circuitry is often a current or voltage output. For example, the 4–20 mA current loop is a common standard and provides good immunity to noise in instrumentation and measurement systems. However, it is always convenient to convert the current to a voltage using an I-to-V (current-to-voltage) converter and subsequently condition the signal before it is displayed on a digital display device. A second approach is to use the principle of frequency modulation. For example, resonant microsensors produce an oscillatory output signal which can be either counted or converted to a voltage by a circuit integrated onto the sensor chip. The signal can then be digitized by an analog-to-digital converter (ADC) on the chip, and the digital output is either displayed or interfaced to a bus system. These functions are commonly integrated into smart sensors. One important consideration when deciding whether to fabricate an integrated sensor is the compatibility of the materials processing required, in particular, the temperature range over which the technology bases operate. CMOS technology is relatively inexpensive and readily available but is limited to a low temperature range. Silicon-on-insulator (SOI) technology can withstand a higher processing temperature but is a much more expensive process.

18.8 Applications of smart sensors and their future trends

Smart optical sensors are relatively easy to make because the silicon sensor technology is compatible with the integrated circuit (IC) processing. Charge-coupled device (CCD) image sensors with integrated image intensifiers are commonly used in digital photography. Such arrays of sensors are, however, expensive and prone to failure. Smart mechanical sensors, such as pressure sensors, were initially fabricated at relatively high cost and in low volumes. An example is the diaphragm pressure sensor fabricated on a silicon wafer. The automotive industry requirement to develop low-cost microaccelerometers for airbags has had a great impact on the development of smart sensors. Smart silicon Hall-effect devices have been fabricated which include a built-in offset for the null voltage and internal temperature compensation. However, the most impressive developments in the field of smart sensors will be in chemical and biological sensing. Traditionally, chemical sensors have suffered from various problems such as parametric drift and the degradation of sensitivity due to chemical reaction or aging. Interference caused by humidity and other chemical species also influences the output reading. The proper design and integration of intelligent chemical microsensors could overcome these difficulties and make a great impact on the smart sensor manufacturing industry. AI has also had a great impact on sensor selection. Many software packages are available for the automatic selection of a sensor from the huge gamut of sensors. AI can guide the user to select a particular sensor that is suitable for the user's application and is thus another method for smart sensor selection.

Chapter 19

Artificial intelligence and its application to sensor selection

Learning objectives:
- What is artificial intelligence?
- How do expert systems work?
- The function of an inference engine.
- Building a knowledge base.
- Expert-system-based sensor selection—a case study.

19.1 Introduction

Artificial intelligence (AI) refers to the simulation of human intelligence by machines that are programmed to think like humans and mimic their actions. However, from an engineering point of view, AI is about generating representation and procedures that automatically solve problems so far solved by humans.

In a broader sense, AI is the ability of a computer to perform tasks commonly associated with intelligent beings. The term is frequently applied to the endeavor of developing systems endowed with intellectual processes. It has been demonstrated that computers can be programmed to carry out very complex tasks—for example, discovering proofs for mathematical theorems or playing chess—with great proficiency. Still, despite continuing advances in computer data processing speeds and the availability of huge computer memories, there are, as yet, no programs that can match human flexibility across wider domains or in tasks requiring much everyday knowledge. However, there are some programs which have attained the performance levels of human experts and professionals in performing certain specific tasks, so that artificial intelligence in this limited sense is found in applications as diverse as medical diagnosis, speech and handwriting recognition.

AI is a cross-disciplinary subject involving cognitive science, computer science, engineering, and mathematics. From an engineering point of view, AI is about generating representations and procedures that automatically solve problems so far solved by humans. Practical AI is the engineering counterpart of cognitive science and is a blend of philosophy, linguistics, and psychology. An engineering approach to AI requires the development of programs, i.e. algorithms and databases that exhibit intelligent behaviour.

19.2 Elements of an AI system

A number of points have to be considered in the study of AI systems. The following concepts are fundamental to AI system development:
- Knowledge representation;
- Inference mechanisms;
- The ability to learn from experience and experts;
- The representation of uncertainty and incomplete reasoning;
- Search and matching principles (part of an inference engine);
- Non-monotonic reasoning;
- Problem decomposition;
- Dynamic knowledgebases that change over time;
- Types of reasoning;
- The programming languages used for implementation and the associated architectures.

19.3 Expert systems

In the context of developing an expert system for the selection of sensors, we recall the broader definition of an expert system. The expert system acts as AI-based software that covers a large number of sensors. It replaces the human expert in deciding which particular sensor is best suited to an engineer's goal.

Experts are difficult to come by, and the worldwide expert system design activity is to computerize and code the experts' knowledge in the field. However, for sensor selection, the issue is not so much the expert's opinion as to find one's path among the bewildering array of claims and counterclaims made by one kind of engineer over the other. The motivation is to glean all the information about a particular choice of sensor from literature and experience and to organize it into a knowledge base. This knowledge base is then structured to give an expert and unimpeded view of the different sensor choices.

The two main planks of any expert system design are the knowledge base and the inference engine. In addition, explanation generation modules and editing of the knowledge base are important. Expert systems are complex AI programs. They solve problems that are solved manually by human experts. To solve expert-level problems, they need access to a substantial domain of knowledge that must be made available as efficiently as possible.

The problems that expert systems deal with are very diverse. Examples abound in medicine, engineering design, fault detection, accounting, law, etc. MYCIN is an expert system which recommends appropriate therapies for patients with bacterial infections. It represents its diagnostic knowledge as a set of rules. Each rule is associated with a certainty factor, which is a measure of how well the antecedent of a rule supports its consequence. PROSPECTOR aids in mineral exploration. Each of its rules contains two confidence estimates. The first indicates the extent to which the presence of evidence described in the condition is part of the rule. The second confidence estimate measures how likely it is to validate any inferences.

It was observed in chapter 7 that the gamut of flow sensors/transducers is huge. One has to choose the correct flow sensor depending on the temperature, range of flow, open channel or closed channel, the pressure drop in the pipe, the nature of the fluid, etc. It is extremely difficult for a nonexpert user to choose the particular sensor/transducer that is perfect for a particular flow application. AI or expert-system-based software packages are available that can select sensors for flow measurement. Examples of these packages are FLOSEL, EXFLOW, etc. Recently, TRANSELEX was developed; it is a single expert system that selects sensors for temperature, pressure, and flow measurements. These are rule-based systems that can choose appropriate sensors for different process variables.

19.4 Languages used in AI programming

AI languages must have the capabilities of symbolic representation and symbolic manipulation. Programming in the AI languages Prolog and LISP is different from programming in an imperative language such as FORTRAN, PASCAL, C, or C++. Given the necessary facts and rules, AI languages use deductive reasoning to solve programming problems. Traditional languages such as C, C++, etc are procedural: the programmer must provide step-by-step procedures telling the computer how to solve a problem.

Prolog and LISP are languages that perform symbolic representation and manipulation. Any AI problem that can be solved in Prolog can also be accomplished in LISP. The syntax of Prolog is simpler than that of LISP. Moreover, Prolog has an in-built inference engine, which may be modified to some extent by introducing a 'cut' into a rule. Turbo Prolog is a true compiler, not just an interpreter. An interpreter must remain in the memory with the source code. For each execution of the program, the interpreter reads each line of code and then executes it. This is extremely slow and occupies precious random-access memory (RAM) space. The program cannot be executed unless the interpreter is present. However, an interpreter is always in control of the program being executed. A program can be stopped, examined, edited, and restarted at any stage. In Turbo Prolog, programs can be compiled into stand-alone .exe files that can run without a Prolog compiler. A full complement of standard predicates such as string operations, random file access, graphic display, etc. is available. Turbo Prolog has an integrated editor, which makes program editing, compiling, and debugging very easy. Expert system programs which are implemented in C are quickly growing in number, because C is versatile in both numeric computation and symbolic manipulation. However, if no numeric computation is involved, Prolog is the best choice for AI programs.

19.5 Knowledge bases

The crux of any expert system is its knowledge base. The larger the knowledge base, the better the expertise. It is always desirable that expert system packages used for sensor selection should have an extensive knowledge base. Each rule should be enriched with a large number of attributes, because the higher the number of attributes, the more specific the choice of sensor. Rules for sensor selection vary widely, as do the variables themselves, which range from flow to temperature, from conductivity to strain, and from viscosity to pollution measurement. The knowledge base should be always developed in consultation with domain experts. Some knowledge also should be gained from books and handbooks.

Knowledge representation in Prolog or Turbo Prolog is based on predicate logic. Such rule-based systems often operate by choosing the first rule that matches within each cycle. Thus, implicit in the ordering of the rules in the database is the concept that some rules are more applicable than others in leading to a solution. Conflict resolution is necessary when rules are not mutually exclusive, that is, several rules match at once. One of the advantages of the rule-based approach is the changing and growing world of the knowledge (fact) base. Every new sensor invented that corresponds to a particular process variable may give rise to a more all-embracing rule and make a few of the previous rules redundant.

There is also the scope for adding heuristic functions, that is, uncertainty factors can be assigned to the rules. But this is avoided because it would mostly reflect the developer's choices. Heuristics can also improve the fuzzy kinds of attributes attached to the facts in the antecedents of the rules.

It is inadvisable to use a more structured representation of the knowledge base for the simple reason that the number of rules for sensor selection is rather small and a more complete 'schema' that could represent the knowledge and consequent search devised by an elaborate inference engine cannot be justified in terms of time and space savings or making a software interface which is otherwise fairly transparent to the nonexpert user. The Prolog inference machine (or rule matcher) works by a backward chaining method. We could have organized the knowledge base in such a way that a 'breadth-first' technique was used. It was not felt necessary to build more finesse into the small expert system. Actually, rules framed using the [If...then...] approach simultaneously serve the purpose of explanation generation.

19.6 The inference engine

Turbo Prolog has a built-in structure for creating the database and a ready-to-run inference engine. In addition to these abilities, Turbo Prolog can add input and output commands, graphics, and sound, making it a fully fledged language that can take advantage of most of the features of a laptop, tablet, or any Android phone.

In Prolog, a production system consists of a set of rules, each consisting of a left-hand side that determines the applicability of the rule and a right-hand side that describes the action to be performed if the rule is applied. That is, Prolog is a production rule language in which programs are written as rules that prove relations among objects. A Prolog program consists of a set of such relations. Because of this, programming in Prolog is sometimes referred to as logic programming. In short, the knowledge is organized as rules, or the knowledge base is converted into a rule base.

Prolog uses a backward chaining search. In other words, given the sensor X, it tries to find out what can be X. A first clue is obtained from the main menu, where the user chooses a particular number corresponding to a variable. After this point, questions pertaining to other variables cannot be asked. This can be understood from the given rule that, while all other attributes are of the form 'positive (fact),' the first attribute concerns the variable that it measures.

19.7 EXSENSEL: a case study

EXSENSEL (EXpert SENsor SELection) is an expert system used to select sensors to measure process variables. At present, it works with 12 process variables and it has 94 rules in its knowledge base. The choice of process variable is menu driven. The variables are:
   1. Gas chromatography.
   2. Conductivity.
   3. Humidity.
   4. Level.
   5. Moisture.
   6. Nuclear radiation.
   7. Oxygen content.
   8. pH.
   9. Viscosity.
   10. Pollution.
   11. Reaction products.
   12. Strain.

This list excludes process variables such as dimensional metrology, odor, opacity, etc. for which it is difficult to build rules, since not all variables have a sensor-specific knowledge base. Moreover, the same sensor may be used in many ways in different measurements; hence, defining rules is difficult.

The rules of EXSENSEL contain no numeric measures of certainty factors, because it is possible to choose the correct sensor for each process variable. The main reason for this is that rich human expertise exists in the domain of process control instrumentation. Second, the purpose of EXSENSEL is to make the correct choice of sensor. It is not essential to consider all possible alternatives. Therefore, probability reasoning is not used in EXSENSEL. The configuration of a typical sensor selection expert system module is shown in figure 19.1.

19.8 A sample rule

EXSENSEL implements rule-based knowledge representation. The following discussion is generic in nature and applicable to any rule-based expert system. In rule-based systems, if the problem situation satisfies or matches the IF part of a rule, the

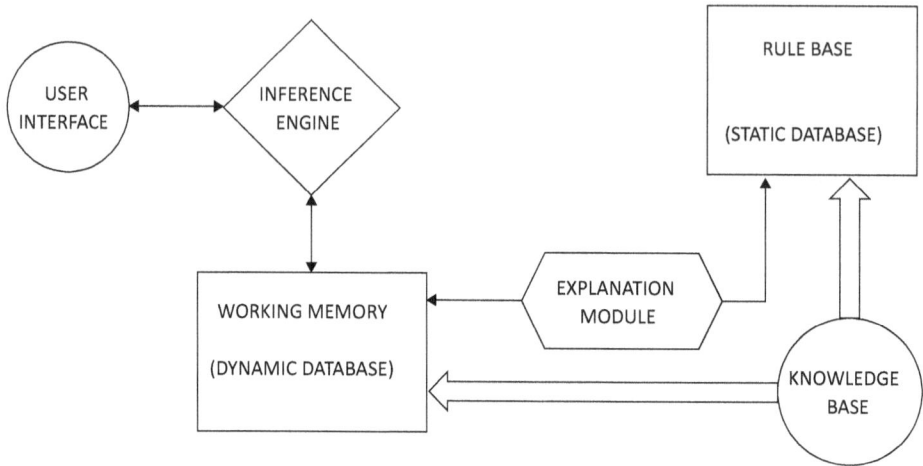

Figure 19.1. The configuration of a typical sensor selection expert system module.

action specified by THEN part of the rule is performed. A sample EXSENSEL rule for oxygen content follows.

If: the process variable is oxygen content, and good stability of the measurement system is required, and there is no water vapor in the sample gas, and the lowest range is 0%–2% oxygen, and the desirable resolution is 1/1500 of the range, and the desirable accuracy is ±2% of the full-scale reading, and the desirable linearity is ±1%,

then: use a thermomagnetic Rein 1 oxygen content sensor.

Explanation of the rule: This rule is only executed if all attributes are true. The first attribute of the rule is not queried, because only questions pertinent to oxygen content are asked, and it is decided in the menu that an oxygen content sensor is to be selected. Therefore, the system ask questions like:

'Do you want good stability of the measurement system?' If the answer is 'yes,' it asks the next question: 'Is the sample gas free from water?' If the answer is 'yes,' it asks the next question of the rule and so on.

If the first answer is 'no' or space bar only is pressed, then the second question related to thermomagnetic type Rein 1 analyzer will not be asked. Instead, the system asks the first question for the next sensor. If the first question for this sensor has already been answered by 'no,' the system goes straight to the third sensor without asking the same question again; since AI makes the system remember all the answers given, the same question is never asked again. When all the answers pertaining to a sensor are 'yes,' only then is the rule executed, that is, the suggestion is made.

19.9 An example knowledge base

Since the strength of an expert system lies in its knowledge base, an abridged knowledge base for viscosity sensors is given below:

sensor ('is a device using time to discharge a given volume of liquid through an orifice'):

measures ('viscosity),
positive ('that your fluid is Newtonian'),
positive ('low and moderate range')'
positive ('that high accuracy is not required')'
positive ('a laboratory or batch technique');
sensor ('is a device using timed fall of ball or rise of bubble'):-
measures ('viscosity),
positive ('that your fluid is Newtonian')'
positive ('to have a wide range')'
positive ('good accuracy'),
positive ('a laboratory or batch technique');
sensor ('is device using drag torque or a stationary element in a rotating cup'):-
measures ('viscosity),
positive ('that your fluid can be either Newtonian or non-Newtonian'),
positive ('to have wide range'),
positive ('to make recording possible'),
positive ('a laboratory or batch technique');
sensor ('is a device using timed fall of piston in cylinder'):-
measures ('viscosity),
positive ('that your fluid can be either Newtonian or non-Newtonian'),
positive ('to have range from 0.1 to 10,000 c.p.'),
positive ('that temperature can be very high'),
positive ('to have high repeatability even at 800 °C'),
positive ('that your vessel can be open or pressurized');
sensor ('is a device using torque to rotate an element in liquid'):-
measures ('viscosity),
positive ('that your fluid can be either Newtonian or non-Newtonian'),
positive ('to have wide range'),
positive ('that your vessel can be either or pressurized or in a vacuum');
sensor ('is a viscosity sensitive rotameter'):-
measures ('viscosity),
positive ('that your fluid can be either Newtonian or non-Newtonian'),
positive ('to have range from 3 to 30 000 c.p.'),
positive ('to be provided with a temperature compensator separately'),
positive ('that your vessel can be open, pressurized, or even vacuum');
sensor ('is a device using an ultrasonic probe'):-
measures ('viscosity),
positive ('that your fluid can be either Newtonian or non-Newtonian'),
positive ('range up to 50 000 c.p.');
sensor ('is consistency cup'):-
measures ('viscosity),
positive ('that your fluid can be either Newtonian or non-Newtonian'),
positive ('to use for solutions and suspensions');
sensor ('is rotational viscometer'):-
measures ('viscosity),

positive ('that your fluid can be either Newtonian or non-Newtonian');
sensor ('is blade type consistency transmitter'):-
measures (viscosity),
positive ('that your fluid can be either Newtonian or non-Newtonian'),
positive ('to have directly pneumatic output'),
positive ('range of consistency from 1.75% to 6.00%'),
positive ('a sensitivity of 0.02%'),
positive ('repeatability of 0.05%'),
positive ('your range of velocity of flow is 0.75 to 5 ft per second'),
positive ('that maximum line pressure would be 125 psi'),
positive ('a span adjustment ratio of 40:1');
sensor ('rotational type consistency transmitter'):-
measures (viscosity),
positive ('that your fluid can be either Newtonian or non-Newtonian'),
positive ('range of consistency from 0.75% to 6.00%'),
positive ('a sensitivity of 0.01%'),
positive ('that maximum line pressure would be 125 psi').

19.10 Amending programs

The rules are written in an extremely clear manner, so that whenever the user wants to add a rule for a particular sensor, they may do so simply by entering the Prolog file and just editing the file, writing the new rule as a positive new fact, such that:

- When the program is run, a question will be asked such as 'Do you want <new fact>?'
- The sequence of the question in the database is important. The earlier it is placed, the earlier it is asked in the query process. So, if a fact is to be added for many sensors, it should be placed before any uncommon questions. Thus, the system does not waste time searching blind paths, i.e. where there is a fact whose answer has already been received as 'no.'
- In the case in which all the rules for a given sensor form a subset of the rules of another sensor, the sensor with the greater number of rules is to be placed earlier.

19.11 General information for sensor selection packages

For any expert system program, general information regarding the mode of operation of the package is necessary before invoking the progam. EXSENSEL is no exception. The following paragraphs provide an example of such general information.

EXSENSEL is an expert system for the selection of sensors or transducers for the measurement of process variables other than pressure, temperature, and flow.

You have options to choose from 12 process variables. You can have a general description of the measurement of that variable if you want.

After that, the system will ask you a series of questions in order to select a sensor.

You should answer either 'y' or 'n.' Any set of characters beginning with 'y' will be taken as 'yes'. The same applies for 'n.' Any other answer will be taken as an error and hence the question will be asked again.

Please remember:

(i) The system chooses only one sensor at a time. Once meets your requirement, it will not look for other solutions based on sensors whose knowledge base is stored later in the program. So, in the first run, it is advisable to press the ENTER key for all questions. Then you know which options are available to measure the variable. Upon running the program for second time, you can exercise your choices.

(ii) You will then be given a brief summary of the selected sensor.

19.11.1 A sensor description

Here, as a case study, a description of an oxygen content sensor follows.

In a nonuniform magnetic field, if the atoms or molecules go to the weakest part of the field, they are called paramagnetic; if they move to the strongest part of the field, they are called diamagnetic.

Susceptibility can be mathematically expressed as follows:

Susceptibility per unit volume = the ratio of the magnetic intensity induced in a unit volume to the magnetic field intensity acting upon it. Its symbol is K. The ratio is positive for paramagnetic substances and negative for diamagnetic materials. The value of K for oxygen is remarkably high. This property of oxygen is utilized in various ways for the measurement of oxygen concentrations in gas mixtures.

19.12 Partial source code of EXSENSEL

EXSENSEL is a software package for sensor selection. It was developed for 12 process variables and it has 94 rules in its knowledge base. It was developed using the Turbo Prolog platform. Its knowledge base is expandable to accommodate many more sensors. The choice of process variables is menu driven. A general description for each process variable is available before invoking a particular set of rules. Once a sensor is selected, a brief summary of the selected sensor is available at the user's request. The knowledge base for one sensor and its rule in Prolog that measures viscosity are given below:

```
/* Expert system for the selection of a sensor */

database

xpositive (symbol,symbol)

xnegative (symbol,symbol)
```

```
predicates

sensor(symbol)

positive (symbol,symbol)

 ask(symbol,symbol,symbol)

store(symbol,symbol,symbol)

remember(symbol,symbol,symbol)

clear_facts

run

clauses

sensor(is_a_device_using_time_to_discharge_a_given_vo-
lume_of_liquid_through_an_orifice)if

measures (viscosity) if

positive (has, the_ fluid_is_Newtonian) and

positive (has, low_and_moderate_range) and

positive (has, high_accuracy_is_not_required) and

positive (has, a_laboratory_or_batch_technique).

/* this is a sample rule in Turbo Prolog. User can gen-
erate the other rules */

/* Inference Engine is described below */
```

```
ask(X,Y,yes):-

write(X," it ",Y,"\n"),

readln(Reply),

store(Reply,X,Y).

positive(X,Y) if xpositive(X,Y),!.

positive(X,Y) if

not(xnegative(X,Y)) and ask(X,Y,yes).

store(Reply,X,Y):-

frontchar(Reply,'y',_),!,

remember(X,Y,yes).

store(Reply,X,Y):-

frontchar(Reply,'n',_),!,

remember(X,Y,no),fail.

remember(X,Y,yes):-

asserta(xpositive(X,Y)).

remember(X,Y,no):-

asserta(xnegative(X,Y)).

run:-
```

```
sensor(X),!,

write("\nIt",X),

nl,nl,clear_facts.

run:-

write("\nUnable to determine what "),

write("your sensor is. \n\n"),clear_facts.

clear_facts:-

retract(xpositive(_,_)),fail.

clear_facts:-

retract(xnegative(_,_)),fail.

clear_facts:-

write("\n\nPlease press the space bar to exit\n"),

readchar(_).
```

Chapter 20

Objective test questions I

Q1. A first-order system is subjected to a unit step change in input. The time constant of the instrument is 1 s. Find the time at which the error reaches a maximum.
   (A) At 1 s
   (B) At 0 s
   (C) At 2 s
   (D) None of these.

Q2. A temperature sensor can measure in the range of 0 °C to 500 °C. The worst-case deviation from the best-fit straight line is found to be 5 °C. Find the maximum nonlinearity as a percentage of the full-scale reading.
   (A) 0.5%
   (B) 1.0%
   (C) 1.25%
   (D) 5%.

Q3. If the supply voltage of a Wheatstone bridge employing a single strain gauge in one of the arms decreases by 5%, then the bridge output sensitivity
   (A) reduces by 4%
   (B) increases by 4%
   (C) reduces by 2%
   (D) remains the same.

Q4. For a strain gauge, which of the following statement is NOT correct?
   (A) It should have high temperature coefficient of resistance.
   (B) It should possess a high gauge factor.
   (C) It should have a high nominal resistance.
   (D) It should possess a linear transfer characteristic.

Q5. As shown in figure 20.1, a force (F) applied to a steel block is to be measured using a single strain gauge (SG). The SG is to be mounted on the surface of column 1. Which of the following configurations (front view) gives the maximum change in the resistance of the SG?

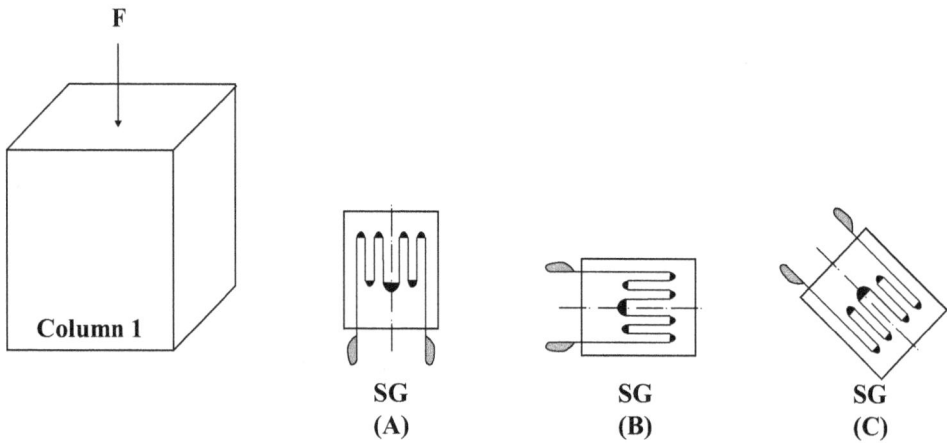

Figure 20.1.

(D) All the configurations give an equal change in resistance.

Q6. If Poisson's ratio is denoted by v, the gauge factor (λ) of a metallic strain gauge can be expressed as

(A) $\lambda = 1 + 2 \cdot v$

(B) $\lambda = \frac{1 + 2 \cdot v}{2}$

(C) $\lambda = 1 + v$

(D) $\lambda = 2 \cdot (1 + v)$.

Q7. Which of the following thermocouple pairs gives the lowest sensitivity?

(A) Chromel–Alumel

(B) Copper–constantan

(C) Platinum–rhodium platinum

(D) Iron–constantan.

Q8. The advantage of the four-wire method of resistance temperature detector (RTD) temperature measurement is that:

(A) It eliminates error due to contact resistance.

(B) The effects of contact resistance and lead wire resistance are eliminated.

(C) It eliminates error due to the connecting wire's resistance only.

(D) It eliminates the effect of contact resistance but the effect of the connecting wire's resistance is not eliminated.

Q9. In a rotameter, the inlet tube diameter is same as that of the float diameter. The tapering angle of the tube is 4 °. When the volumetric flow rate of the water is Q_1, the float height (from the inlet) is 60 mm. The height of the float is now reduced to 20 mm for the flow Q_2. The ratio Q_1/Q_2 of the flow rates is closest to

(A) 9.0

(B) 0.333

(C) 3.0

(D) 0.111.

Q10. A Pitot tube is used to measure the velocity of a gas in a duct. The velocity is proportional to
(A) the sum of the stagnation pressure and the static pressure
(B) the difference between the stagnation pressure and the static pressure
(C) the square root of the difference between the stagnation pressure and the static pressure
(D) the square root of the stagnation pressure.

Q11. A differential pressure transmitter is used to measure the flow rate in a pipe. The sensitivity of the transmitter is reduced by 4% due to ageing effects. If all other aspects of the flow measuring system are constant, the flow measurement sensitivity
(A) reduces by 4%
(B) increases by 4%
(C) reduces by 2%
(D) remains unaffected.

Q12. Which of the following flowmeters is unsuitable for the hydrocarbon industry:
(A) the Venturi meter
(B) the electromagnetic flowmeter
(C) the orifice meter
(D) the positive displacement flowmeter.

Q13. An example of a reversible transducer is
(A) a potentiometer
(B) a linear variable differential transformer (LVDT)
(C) a piezoelectric crystal
(D) a thermistor.

Q14. Which of the following is a synthetic piezoelectric material?
(A) Rochelle salt
(B) quartz crystal
(C) lithium niobate
(D) gallium arsenide.

Q15. A pressure of the order of 10^5 pounds per square inch (PSI) can be measured by
(A) a bellows gauge
(B) an ionization gauge
(C) a Pirani gauge
(D) a Bourdon tube.

Q16. Match the following

	List 1		List 2
A.	Dall tube	1.	Vacuum pressure measurement
B.	Pirani gauge	2.	Flow measurement
C.	Gyroscope	3.	Temperature measurement
D.	Radiation pyrometer	4.	Angular velocity measurement

(A) A-2, B-1, C-4, D-3

(B) A-2, B-3, C-1, D-4

(C) A-1, B-2, C-4, D-3

(D) A-2, B-3, C-4, D-1.

Q17. Laser light that has a wavelength of 635 nm passes through 2 cm of tissue and 4 cm of glass. The refractive indices of tissue and glass are 1.33 and 1.6, respectively. The velocities of laser light in glass and in tissue are in the ratio of

(A) 1.33:1.6

(B) 1.6:1.33

(C) 0.8:1.33

(D) 1.6:0.67.

Q18. In a bioreactor, the only parameter that is measureable on site is

(A) the dissolved oxygen concentration

(B) the product concentration

(C) the biomass concentration

(D) the dissolved carbon monoxide concentration.

Q19. The pH value of distilled water is:

(A) 0

(B) 14

(C) 5

(D) 7.

Q20. Which of the following transducers is a zeroth-order instrument?

(A) a potentiometer

(B) a U-tube manometer

(C) a mercury-in-glass thermometer

(D) an elbow meter.

Answers to objective test questions I

Question Number	Answer
1	**B**
2	**B**
3	**D**
4	**A**
5	**A**
6	**A**
7	**C**
8	**B**
9	**C**
10	**C**
11	**A**
12	**B**
13	**C**
14	**C**
15	**D**
16	**A**
17	**A**
18	**A**
19	**D**
20	**A**

Chapter 21

Objective test questions II

Q1. The characteristic equation of a system is
$\frac{d^2y}{dt^2} + \frac{dy}{dt} + 25y = 0$. The damping ratio of the system is:
  (A) 1
  (B) 0.2
  (C) 0.1
  (D) 0.

Q2. The rank of matrix $A = \begin{bmatrix} 1 & 2 & 3 \\ 3 & 1 & 1 \\ 4 & 3 & 4 \end{bmatrix}$ is:
  (A) 0
  (B) 1
  (C) 2
  (D) 3.

Q3. The maximum power output of a three-phase delta-connected induction motor is 300 kW. If the windings are now changed to a star connection, the maximum power output should be:
  (A) 100 kW
  (B) 300 kW
  (C) 100 $\sqrt{3}$ kW
  (D) 300 $\sqrt{3}$ kW.

Q4. Which of the following can act as a null detector for AC bridge measurements?
  (A) The d'Arsonval galvanometer
  (B) the ballistic galvanometer
  (C) the vibration galvanometer
  (D) the fluxmeter.

Q5. A system with impulse response $h(t)$ is excited by a signal $u(t)$. The output $y(t)$ can be written as:
  (A) $\int_0^1 u(t)h(t)dt$
  (B) $\int_0^1 u(t-x)h(x)dx$

(C) $\int_0^1 u(t-x)h(t)\mathrm{d}x$

(D) $\int_0^1 u(t)h(t-x)\mathrm{d}t$.

Q6. When a current of $(3\sqrt{2}\ \sin 314\ t + 4)$ A is passed through a true root mean square (RMS) ammeter, the meter reading is:

(A) 3 A

(B) $3\sqrt{2}$ A

(C) 5 A

(D) 4 A.

Q7. The maximum power that can be transferred to the load R_L shown in figure 21.1 is:

(A) 2 W

(B) 5 W

(C) 10 W

(D) 100 W.

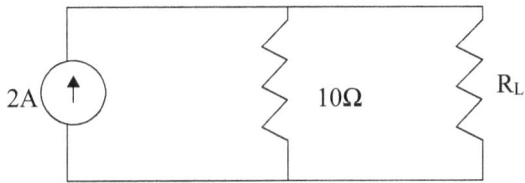

Figure 21.1.

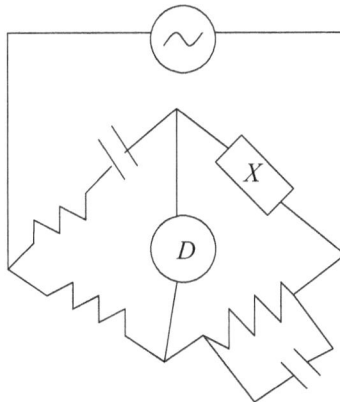

Figure 21.2.

Q8. To balance the bridge shown in figure 21.2, the element X should be a:
  (A) a pure inductor
  (B) a pure capacitor
  (C) a pure resistor
  (D) an R–L series element.

Q9. In a practical constant-current source, the load impedance is:
  (A) equal to the source impedance
  (B) much greater than the source impedance
  (C) much less than the source impedance
  (D) independent of the source impedance.

Q10. The instrument and metering current transformer are qualified if the error is at least:
  (A) less than 5%
  (B) less than 1%
  (C) less than 0.01%
  (D) less than 10%.

Q11. The standard output voltage from a potential transformer in a substation is:
  (A) 220 V AC
  (B) 220 V DC
  (C) 110 V AC
  (D) 110 V DC.

Q12. In the circuit shown in figure 21.3, the voltmeter, which has an internal resistance of 1200 ohms, reads:
  (A) 24 V
  (B) 21.81 V
  (C) 34.28 V
  (D) 40 V.

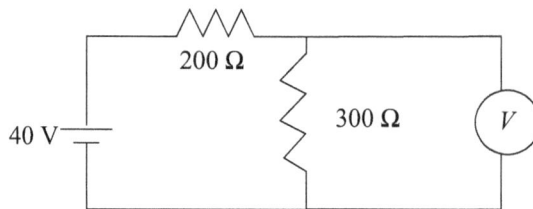

Figure 21.3.

Q13. The time constant in seconds for the circuit shown in figure 21.4 is:
  (A) 1
  (B) 2
  (C) 1/2
  (D) 1/4.

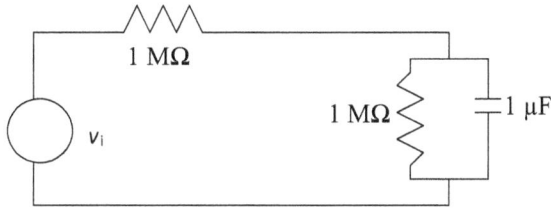

Figure 21.4.

Q14. The resistances $100 \, \Omega \pm 6 \, \Omega$ and $200 \, \Omega \pm 8 \, \Omega$ are connected in series. The uncertainties are expressed in terms of standard deviation. The equivalent resistance is:
(A) $300 \, \Omega \pm 2 \, \Omega$
(B) $300 \, \Omega \pm 10 \, \Omega$
(C) $300 \, \Omega \pm 14 \, \Omega$
(D) $300 \, \Omega \pm 48 \, \Omega$.

Q15. The time base of an analog oscilloscope is generated by:
(A) a sinusoidal oscillator
(B) a square-wave generator
(C) a sawtooth generator
(D) a crystal oscillator.

Q16. Which of the following bridges is used for the measurement of very low resistances?
(A) The Wheatstone bridge
(B) Kelvin's double bridge
(C) The Schering bridge
(D) The Anderson bridge.

Q17. The speed of an alternator in a typical thermal power plant in the 50 Hz system is:
(A) 3080 rpm
(B) 2000 rpm
(C) 3000 rpm
(D) 2500 rpm.

Q18. The unbalanced voltage in the bridge circuit shown in figure 21.5 is:
(A) $IRX/2$
(B) IRX
(C) $2IRX$
(D) IX.

Q19. The circuit shown in figure 21.6 is:
(A) a differential amplifier
(B) an integrator
(C) a differentiator
(D) a rectifier.

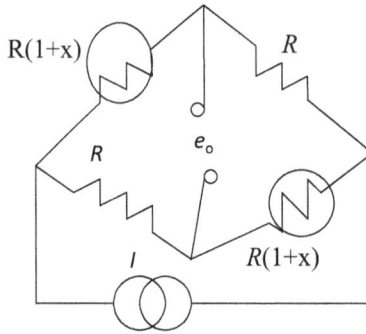

Figure 21.5.

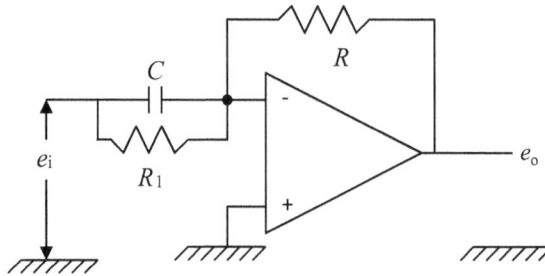

Figure 21.6.

Q20. One can design an oscillator with a notch filter by:
   (A) putting it into negative feedback
   (B) putting it into positive feedback
   (C) by cascading it at the output
   (D) by connecting at the input.

Q21. The transfer function of a second-order filter circuit is given by:

$$G(s) = K/(s^2 + \sqrt{2}\,s + 1).$$

The filter is
   (A) a Butterworth filter
   (B) a Chebyshev filter
   (C) a Bessel filter
   (D) an all-pass filter.

Q22. For the circuit shown in figure 21.7, the voltage gain is:
   (A) −1
   (B) 0
   (C) +1
   (D) +2.

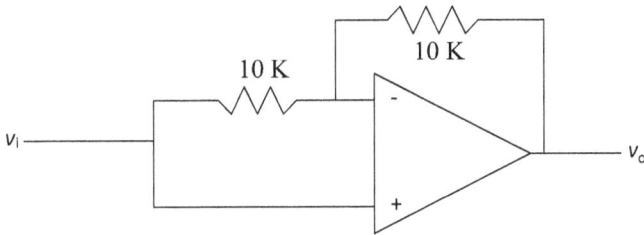

Figure 21.7.

Q23. The boolean expression $A\bar{B} + \bar{A}\ B$ is equal to:
(A) $\bar{A}\bar{B} + AB$
(B) $\bar{A}\ B + \bar{B}\ A$
(C) $A + \bar{A} + B + \bar{B}$
(D) $\bar{A}\ \bar{B}$.

Q24. A six-bit successive-approximation analog-to-digital converter (ADC) reaches the stable state in:
(A) 2^6 steps
(B) 2^6-1 steps
(C) 2 steps
(D) 6 steps.

Q25. The address latch enable (ALE) signal in the 8085 microprocessor is used:
(A) to multiplex data bus and address bus
(B) to start the arithmetic logic unit
(C) to enable data flow
(D) to enable data transfer from memory.

Q26. The priority of RST 6.5 in 8085 microprocessor is:
(A) higher than that of RST 5.5
(B) higher than that of RST 7.5
(C) equal to the priorities of RST 5.5 and RST 7.5
(D) equal to the priority of RST 6.

Q27. Which of the following is the fastest ADC?
(A) the counter control type
(B) the successive-approximation type
(C) the dual slope type
(D) the flash type.

Q28. ASCII codes are transmitted as:
(A) 8-bit data
(B) 7-bit data preceded by a parity bit
(C) 7-bit data preceded by a sign bit
(D) 4-bit data.

Q29. For a l2-bit ADC with a 10 V full-scale range, the quantization error is:
(A) 2.44 mV
(B) 4.88 mV

(C) 255 mV

(D) 512 mV.

Q30. In the digital circuit of figure 21.8, if A = 0 and B = 1:

(A) D = 0, C = 0

(B) D = 1, C = 0

(C) D = 0, C = 1

(D) D = 1, C = 1.

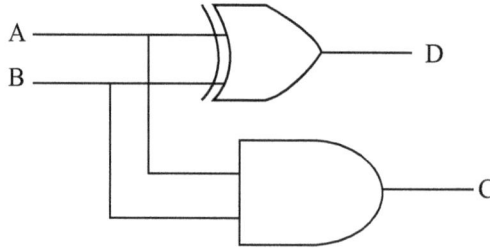

Figure 21.8.

Q31. If A, B, C, and D make up a four-bit input word, P is the output, and P'= 0, the digital circuit in figure 21.9 is:

(A) an adder

(B) a subtracter

(C) a two's complement

(D) an odd-parity checker.

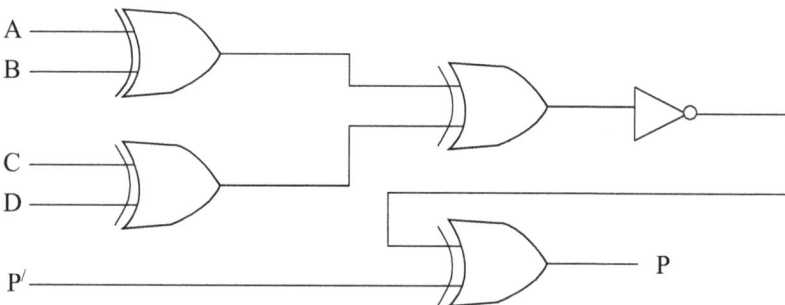

Figure 21.9.

Q32. The master–slave configuration in a J-K flip-flop is used:

(A) to isolate the input from the output

(B) to avoid race conditions

(C) to get two outputs from two inputs

(D) to provide four inputs and two outputs.

Q33. The cutoff frequency of an antialiasing filter that precedes an ADC that has a 100 μs conversion time should be at least:
(A) 10 MHz
(B) 10 kHz
(C) 5 kHz
(D) 20 kHz.

Q34. The phase-locked loop (PLL) is used for:
(A) locking into a particular frequency
(B) maintaining a constant phase difference
(C) compensating the phase of an amplifier
(D) tuning amplifier gain.

Q35. The IEEE-488 protocol is used for:
(A) serial communication
(B) parallel communication with printers
(C) parallel data transmission and controlling multiple instruments
(D) current-loop transmission.

Q36. Two-way data communication is possible:
(A) for all types of serial communication links
(B) for simplex serial communication links
(C) for half-duplex serial communication links
(D) for full-duplex serial communication links.

Q37. In serial communication, the parity bit is used for:
(A) error checking
(B) data bit separation
(C) time multiplexing
(D) handshaking.

Q38. The circuit diagram of an inverting voltage amplifier is shown in figure 21.10. The function of the resistance R_3 is to:
(A) set the gain
(B) reduce the drift
(C) eliminate the effect of the input offset current
(D) eliminate the effect of the input offset voltage.

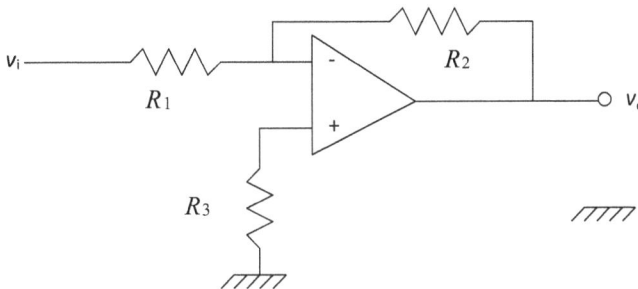

Figure 21.10.

21-8

Q39. The input resistance of the circuit shown in figure 21.11 is:
(A) +10 k
(B) +20 k
(C) −10 k
(D) infinity.

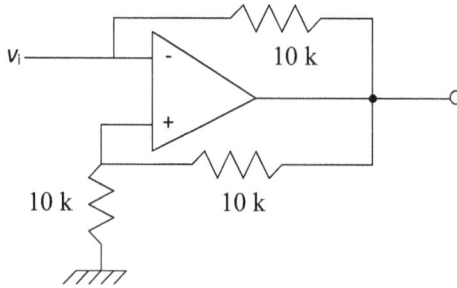

Figure 21.11.

Q40. The input–output voltage relationship of the circuit shown in figure 21.12 is:
(A) $e_i = e_0$
(B) $e_i = e_0$, for $e_i > 0.0$ v
(C) $e_i = e_0$, for $e_i > 0.7$ v
(D) $e_i = -e_0$, for $e_i > 0.0$ v.

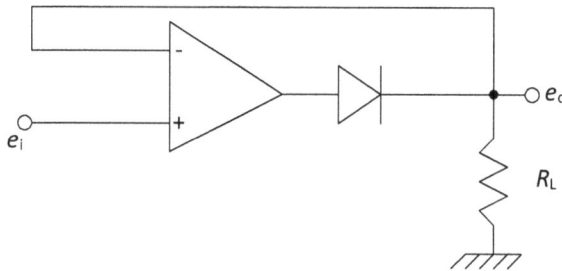

Figure 21.12.

Q41. A signal has a maximum frequency component of 500Hz. The minimum carrier channel band required is:
(A) 250 Hz
(B) 1 kHz
(C) 5 kHz
(D) 2.5 kHz.

Q42. A notch filter at 50 Hz is used to:
(A) remove power-frequency interference from a signal
(B) extract the power frequency
(C) pass a 50 Hz signal
(D) remove the harmonics of a 50 Hz signal.

Q43. If the Fourier transform of $h(t)$ is $H(jw)$, the Fourier transform of $h(t-2)$ is:
(A) $2 H (jw)$
(B) $e^{-2jw} H (jw)$
(C) $e^{+j2w} H (jw)$
(D) $e^{+j2}H(jw)$.

Q44. The lengths of two discrete time sequences $x_1 [n]$ and $x_2 [n]$ are five and seven, respectively. The maximum length of the linear convoluted sequence $y[n]$ is:
(A) 5
(B) 6
(C) 7
(D) 11.

Q45. A example of a flowmeter that is independent of fluid density is:
(A) the rotameter
(B) the electromagnetic flowmeter
(C) the Venturi meter
(D) the orifice meter.

Q46. Which of the following flowmeters works on the constant-pressure-drop principle?
(A) the Pitot tube
(B) the Venturi meter
(C) the rotameter
(D) the orifice meter.

Q47. For a thermocouple pair (A,B), the extension wires (C,D):
(A) should be identical pair elements.
(B) should have identical temperature–emf relationships
(C) can be any two dissimilar metals
(D) should have very small temperature–emf sensitivity.

Q48. A metal strain gauge with a gauge factor of 2.0 and a nominal resistance of 120 ohms undergoes a change in resistance of 2.4 milliohms. The strain sensed is:
(A) 10^{-6}
(B) 10^{-5}
(C) 10^{-4}
(D) 10^{-3}.

Q49. Optical encoders are used for the measurement of:
(A) stress
(B) low pressure
(C) angular displacement
(D) temperature.

Q50. A Hall probe 2 mm thick carries a current of 200 mA. When it intercepts a field of strength 0.5 T, it produces a voltage of 400 mV. The Hall coefficient is then:
(A) 10^{-3}
(B) 8×10^{-3}
(C) 0.5×10^{-2}
(D) 4×10^{-2}.

Q51. A pressure of the order of 10^{-10} torr can be measured by:
(A) a bellows gauge
(B) an ionization gauge
(C) a Pirani Gauge
(D) a Bourdon gauge.

Q52. A pressure gauge measures a pressure as a gauge pressure of 5 kPa. If the atmospheric pressure is 100 kPa, the absolute measured pressure is:
(A) 95 kPa
(B) 100 kPa
(C) 105 kPa
(D) 20 kPa.

Q53. The sensitivity of a semiconductor strain gauge is given by $S = K_1 \in + K_2 \in^2$, where K_1 and K_2 are constants and $\in$ is strain. If the operational range is $\pm \in_{max}$, the percentage nonlinearity is:
(A) $K_2 /K_1 \pm \in_{max} \times 100\%$
(B) $K_2/(2K_1) \pm \in_{max} \times 100\%$
(C) $K_2 \in^2 \times 100\%$
(D) $K_1 /K_2 \times 100\%$.

Q54. Which of the following thermocouples can operate in the temperature range of 1500 °C–1700 °C?
(A) iron–constantan
(B) copper–constantan
(C) Chromel–constantan
(D) platinum rhodium–platinum.

Q55. The purpose of a wave trap in power-line carrier communication is:
(A) to block the power-frequency component to the bus bar end
(B) to pass the carrier frequency to the bus bar end
(C) to pass the power frequency to the bus bar end and block the carrier frequency
(D) to trap the switching transients.

Q56. A dynamic system has two roots, $+j$ and $-j$. The system is:
(A) overdamped
(B) critically damped
(C) underdamped
(D) undamped.

Q57. In any radioactive decay, the disintegration rate of the nuclei is:
(A) constant at all times
(B) inversely proportional to the number of nuclei at any time

(C) directly proportional to the number of nuclei at any time

(D) inversely proportional to the half-life.

Q58. The attenuation of a narrow monochromatic x-ray beam in a metal plate of thickness d is given by the equation:

(A) $I = I_0 \, e^{\mu d}$

(B) $I = I_0 \, e^{-\mu d}$

(C) $I = I_0 \, e^{-\mu/d}$

(D) $I = I_0 \, e^{-\mu/2d}$.

Q59. Which of the following flowmeters is an integrating type?

(A) the orifice meter

(B) the rotameter

(C) the hot-wire anemometer

(D) the positive displacement flowmeter.

Q60. Which of the following flowmeters is a non-obstructing type?

(A) the Venturi meter

(B) the turbine flowmeter

(C) the Doppler shift flowmeter

(D) the hot-wire anemometer.

Q61. The signal-processing circuit associated with an LVDT is a:

(A) charge amplifier

(B) phase-sensitive detector

(C) push–pull amplifier

(D) logarithmic amplifier.

Q62. The pressure drop across an orifice plate for a particular flow rate is 5 kg m^{-2}. If the flow rate is doubled, the corresponding pressure drop is:

(A) 2.5 kg m^{-2}

(B) 5 kg m^{-2}

(C) 10 kg m^{-2}

(D) 20 kg m^{-2}.

Q63. Which of the following transducers is a reversible type?

(A) the LVDT

(B) the thermistor

(C) the piezoelectric transducer

(D) the strain gauge.

Q64. Which of the following gases has paramagnetic properties?

(A) H_2S

(B) N_2

(C) O_2

(D) SO_2.

Q65. Which of the following gases can be used as a carrier in gas chromatography?

(A) oxygen

(B) nitrogen

(C) carbon dioxide

(D) methane.

Q66. Which of the following photo detectors has the highest sensitivity?
  (A) the light-dependent resistor (LDR)
  (B) the p–n photodiode
  (C) the p–intrinsic–n (PIN) photodiode
  (D) the avalanche photodiode.
Q67. Which of the following is a commonly used piezoelectric material?
  (A) iron–constantan
  (B) a complementary metal–oxide–semiconductor device
  (C) lead zirconium titanate
  (D) gallium arsenide.
Q68. The primary and secondary windings of an LVDT are wound on a cylinder made of:
  (A) high-permeabilty steel
  (B) medium-permeability steel
  (C) brass
  (D) strong paper board.
Q69. To measure shaft torque, one can use two gauges oriented:
  (A) with both gauges parallel to the axis of the shaft
  (B) at 45° with respect to the shaft axis
  (C) at 60° with respect to the shaft axis
  (D) at 90° with respect to the shaft axis.
Q70. The technique normally used for CO_2 concentration measurement in a gas is:
  (A) x-ray analysis
  (B) the zirconia probe method
  (C) ultraviolet (UV) spectroscopy
  (D) infrared (IR) spectroscopy.
Q71. In gas chromatography, the retention time of constituent gases depends on:
  (A) the fractionating column length only
  (B) the carrier gas pressure only
  (C) the fractionating agent only
  (D) all of the above.
Q72. A push–pull arrangement is preferred in a capacitance transducer because it:
  (A) reduces nonlinearity
  (B) increases the signal-to-noise ratio
  (C) eliminates the effect of cable capacitance
  (D) increases the bandwidth.
Q73. The gauge factor for a semiconductor strain gauge is of the order of:
  (A) 0.1
  (B) 2.0
  (C) 4.0
  (D) 100.

Q74. The required loop gain G and the total phase shift φ of a phase shift oscillator should be:
(A) $G = 0.5$, $\varphi = 180°$
(B) $G = 1.0$, $\varphi = 360°$
(C) $G = 2.0$, $\varphi = 360°$
(D) $G = 1.0$, $\varphi = 90°$.

Q75. The standard range for electrical signal transmission in industry is:
(A) 0–20 mV
(B) 4–20 mA
(C) 0–5 V
(D) 1–5 V.

Q76. The characteristic of a control valve is shown in figure 21.13. The valve is:
(A) a quick-opening valve
(B) a linear type
(C) an equal percentage type
(D) an ON–OFF type.

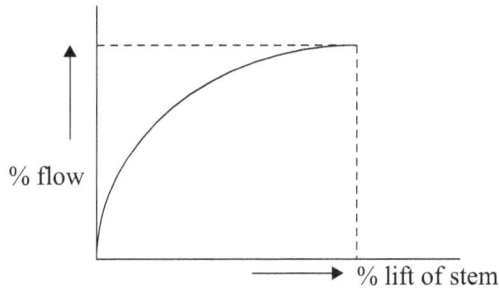

Figure 21.13.

Q77. The steady-state error of a closed-loop system (shown in figure 21.14) for a unit step input is:
(A) zero
(B) 1/2
(C) 2/3
(D) 2.

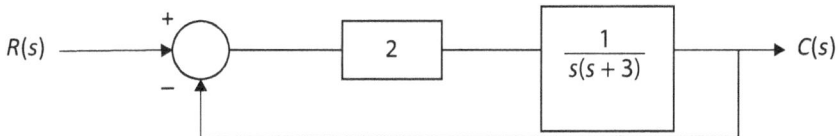

Figure 21.14.

Q78. Identify the transfer function which has all-pass filter characteristics:

(A) $\frac{k}{s+1}$

(B) $\frac{k(s+1)}{(s+2)}$

(C) $\frac{s-1}{s+2}$

(D) $\frac{s-1}{s+1}$.

Q79. Refer to the closed-loop system shown in figure 21.15. For a unit step input, the steady-state value of the output c(t) should be:

(A) 2

(B) 1/2

(C) 1

(D) 2/3.

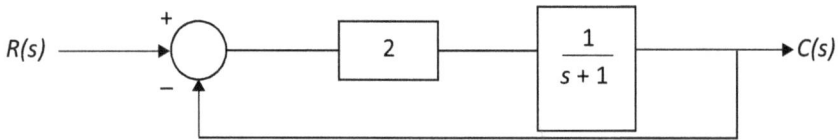

Figure 21.15.

Q80. A system with the transfer function $1/(1+\tau s)$ subjected to a step input takes 10 s to reach 50% of its final steady-state value. The value of τ is:

(A) 6.9 s

(B) 10 s

(C) 14.4 s

(D) 20 s.

Q81. The transfer function of a PD controller is given by (here, the Ks are real constants):

(A) $Ko + K_1/s + K_2s$

(B) $Ko + K_2s$

(C) $K_1s + K_2/s$

(D) $Ko + K_1/s$.

Q82. A system whose characteristic equation is $s^3 + s^2 + 4s + 4 = 0$

(A) has one pole in the right half-plane

(B) has two poles in the right half-plane

(C) is oscillatory

(D) is asymptotically stable.

Q83. The average value of the voltage V_o in the circuit of figure 21.16, which has an ideal diode, is:

(A) 50 V

(B) 40 V

(C) 20 V

(D) 10 V.

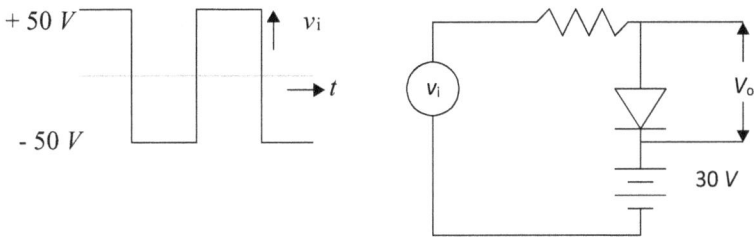

Figure 21.16.

Q84. The psychrometer is used for the measurement of:
(A) relative humidity
(B) density
(C) dynamic pressure
(D) flow rate.

Q85. Which of the following temperature transducers has the highest sensitivity?
(A) the RTD
(B) the thermistor
(C) the thermocouple
(D) the radiation pyrometer.

Q86. Which of the following bridges is suitable for the measurement of mutual inductance?
(A) Kelvin's double bridge
(B) the Owen bridge
(C) the Anderson bridge
(D) the Campbell bridge.

Q87. Which of the following instruments can measure alternating current only?
(A) the permanent magnet moving coil (PMMC) type
(B) the Hall-effect sensor
(C) the dynamometer type
(D) the induction type.

Q88. In a cathode ray oscilloscope, the ramp input voltage is applied to the:
(A) focusing electrodes
(B) cathode
(C) horizontal deflection plates
(D) vertical deflection plates.

Q89. Which of the following instruments can measure changes in the thermal conductivity of a gas?
(A) the katharometer
(B) the Pitot tube
(C) the Golay detector
(D) the load cell.

Q90. When the input terminals of a pH meter are shorted, it reads a pH value of:

(A) 0

(B) 14

(C) 7

(D) 5.

Q91. The transfer function of a system has p poles and z zeros. The system is called 'strictly proper' if:

(A) $p<z$

(B) $p = z$

(C) $p>z$

(D) $p \geq z$.

Q92. Of the following transfer functions, the one that refers to an unstable system is:

(A) $\dfrac{z - 0.5}{z + 0.5}$

(B) $\dfrac{z + 0.5}{z - 0.5}$

(C) $\dfrac{z}{z + 2}$

(D) $\dfrac{z}{z - 1}$.

Q93. Which of the following actuators has highest output power-to-weight ratio?

(A) the DC servomotor

(B) the AC servomotor

(C) the pneumatic actuator

(D) the hydraulic actuator.

Q94. The transfer function of a first-order time-delay system is given by $G(s) = \dfrac{e^{-s\tau_d}}{1 + sT}$. Its response for a unit step input $u(t)$ is given by:

(A) $(1 - e^{-t/T})\, u\,(t)$

(B) $(1 - e^{-t/T})\, u\,(t - \tau_d)$

(C) $(1 - e^{-t}\tau_d)\, u\,(t - T)$

(D) $e^{-t}\, \tau_d\, u\,(t - \tau_d)$.

Q95. A cascade controller is used when:

(A) the process gain is too small

(B) the process gain is too large

(C) the process has two widely different time constants

(D) oscillation of the output is not permitted.

Q96. The function of the reset action in process control is to:

(A) reduce the steady-state error

(B) reduce the response time

(C) reduce the oscillation in the response

(D) increase the overall gain.

Q97. The symbol in figure 21.17 represents a:

(A) a p-metal–oxide–semiconductor (MOS) device

(B) an n-MOS device

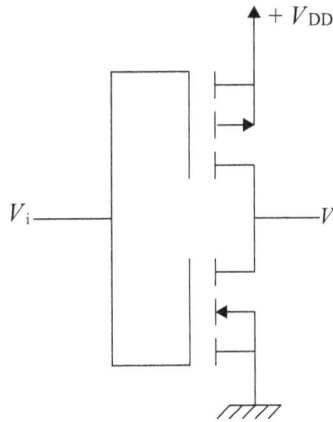

Figure 21.17.

(C) a complementary MOS (CMOS) device

(D) a junction field-effect transistor (JFET).

Q98. The equations of a system are $\frac{dx_1}{dt} = x_2$,

$\frac{dx_2}{dt} = x_1 x_2 + x_2 + u$. It is a nonlinear system since it does not obey:

    (A) the superposition theorem

    (B) Thevenin's theorem

    (C) Norton's theorem

    (D) the mean value theorem.

Q99. A three-phase four-pole induction motor is running at a speed of 1350 rpm. Its slip is:

    (A) 0.9

    (B) 1.1

    (C) 0.3

    (D) 0.1.

Q100. In optical fibre:

    (A) light travels in a straight line

    (B) the core refractive index is higher than that of the cladding

    (C) the refractive indices of the core and cladding are equal

    (D) the optical signal does not experience any loss.

Answers to the objective test questions II:

Q1	C	Q2	C	Q3	A	Q4	C	Q5	B
Q6	C	Q7	C	Q8	B	Q9	C	Q10	B
Q11	C	Q12	B	Q13	C	Q14	B	Q15	C
Q16	B	Q17	C	Q18	A	Q19	C	Q20	A
Q21	A	Q22	C	Q23	A	Q24	D	Q25	A
Q26	A	Q27	D	Q28	B	Q29	A	Q30	B
Q31	D	Q32	B	Q33	C	Q34	A	Q35	C
Q36	D	Q37	A	Q38	C	Q39	C	Q40	B
Q41	B	Q42	A	Q43	B	Q44	D	Q45	B
Q46	C	Q47	B	Q48	B	Q49	C	Q50	B
Q51	B	Q52	C	Q53	B	Q54	D	Q55	C
Q56	D	Q57	C	Q58	B	Q59	D	Q60	C
Q61	B	Q62	D	Q63	C	Q64	C	Q65	B
Q66	D	Q67	C	Q68	D	Q69	B	Q70	D
Q71	D	Q72	A	Q73	D	Q74	B	Q75	B
Q76	A	Q77	A	Q78	D	Q79	D	Q80	C
Q81	B	Q82	C	Q83	B	Q84	A	Q85	B
Q86	D	Q87	D	Q88	C	Q89	A	Q90	C
Q91	C	Q92	C	Q93	D	Q94	B	Q95	C
Q96	A	Q97	C	Q98	A	Q99	D	Q100	B

Chapter 22

Solutions to problems

Chapter 2

Problem 2.1. A first-order temperature transducer is used to measure the temperature of an oil bath. If the temperature exceeds 100 °C, the heat supplied to the oil bath should be stopped within 5 s of reaching 100 °C. Determine the maximum allowable time constant of the sensor if a measurement error of 5% is allowed.

Solution: measurement error $= e^{\frac{-t}{\tau}}$

For the given problem, the measurement error is 0.05.

For $t \leqslant 5$ s, $\tau \leqslant 1.66$ s (the answer).

Problem 2.2. A first-order instrument has a time constant of 0.5 s. It is measuring a process parameter that is sinusoidal in nature and has a frequency of 3 Hz. Determine the dynamic error of the system.

Solution:

$$d_e = \frac{y/k}{x} - 1$$

Here, $\frac{y/k}{x} = \frac{1}{\sqrt{1 + (\omega\tau)^2}}$

$\omega = 2\pi 3$

$\tau = 0.5$

$$\frac{y/k}{x} = \frac{1}{\sqrt{1 + (2\pi 3 \times 0.5)^2}} = \frac{1}{\sqrt{89.7364}} = \frac{1}{9.472} = 0.105$$

∴ dynamic error $= 0.105 - 1 = -0.89$; 89 % (the answer).

Problem 2.3. A first-order temperature sensor is suddenly immersed in a liquid that has a temperature of 100 °C. If, after 3 s, the sensor shows a temperature of 80 °C, calculate the instrument time constant. Also calculate the error in the temperature reading after 2 s.

Solution:

$$y = 100\,(1 - e^{-t/\tau})$$

Here, $t = 3$ s, $y_{t(3\,\text{sec})} = 80°$ C

Therefore, $80 = 100(1 - e^{-3/\tau})$

$$0.8 = 1 - e^{-3/\tau}$$

$\therefore e^{-3/\tau} = 1 - 0.8 = 0.2$ or, $e^{3/\tau} = 5$

$$\frac{3}{\tau} = 2.303 \times \log 5$$

$\therefore \tau = \dfrac{3}{2.303 \times \log 5} = 1.86$ s

Now, $y_t = 2$ sec $= 100(1 - e^{\frac{-2}{1.86}})$

$$= 100(1 - e^{-1.0732})$$

$$= 100(1 - 0.341)$$

$$= 65.8°C$$

Error $= -34.2$ °C (the answer).

Problem 2.4. A pressure transducer is to be selected to measure the pressure of a vessel. The pressure variation can be considered to be a sinusoidal signal whose frequency lies between 1 and 4 Hz. Several sensors are available, each with a known time constant. Select the appropriate sensor if a dynamic error of $\pm 1\%$ is acceptable.

Solution: the magnitude of the dynamic error

$$|d_e| \leqslant 0.01$$

The magnitude ratio is $\dfrac{y/k}{x}$ and $0.99 \leqslant \dfrac{y/k}{x} \leqslant 1.01$.

For a first-order system, $\dfrac{y/k}{x}$ never exceeds one.

So, $0.99 = \dfrac{y/k}{x} = \dfrac{1}{\sqrt{1 + (\omega\tau)^2}}$

Now, $\omega = 2\pi 4$ and $\omega^2 = 631$

$\therefore 0.981\,(1 + \omega^2\,\tau^2) = 1$

or $1 + \omega^2\,\tau^2 = 1.01936$

So, $\tau = 5.6$ms.

For a 1 Hz signal frequency, $\tau = 0.0226$ s.

The values of τ lie in between 5.6 msec and 0.0226 sec (the answer).

Problem 2.5. A periodic signal is to be measured by a first-order instrument that has a time constant of 3 s. If a dynamic error of $\pm 5\%$ can be tolerated, find the highest-frequency input signal that can be measured by the instrument.

Solution: the dynamic error, $d_e = \dfrac{y/k}{x} - 1$.

Since the instrument is first order,

$$0.95 \leqslant \frac{1}{\sqrt{1 + (\omega\tau)^2}}.$$

The given time constant, $\tau = 3$ s,

$$\therefore \sqrt{1 + (\omega\tau)^2} \leqslant \frac{1}{0.95}.$$

Or, $1 + (\omega\tau)^2 \leqslant 1.108$
  Or, $(\omega\tau)^2 \leqslant 0.108$
  Or, $\omega^2 \leqslant \frac{0.108}{9}$
  Or, $\omega^2 \leqslant 0.012$
$\therefore \omega \leqslant 0.1095$ (the answer).

Problem 2.6. An accelerometer that is second order in nature is to be selected to measure a sinusoidal signal whose frequency is less than 100 Hz. If a dynamic error of ±6% is allowed, choose a sensor for a damping ratio of 0.6.
  Solution: signal frequency, $f \leqslant 100$ Hz.
  To meet the ± 5% error constraint, one has to fulfill the requirement of

$$0.95 \leqslant \left| \frac{y/k}{x} \right| \leqslant 1.05, \tag{2.7}$$

where $\frac{y/k}{x}$ is the dynamic error of the second-order system.
  Equation (2.7) must be valid for the frequency range of $0 \leqslant \omega \leqslant 628$ rad s^{-1}.
  A number of instruments with different natural frequencies (ω_n) can perform this task. From equation (2.7), we can write

$$1.05 \geqslant \frac{1}{\left[\left\{ 1 - \left(\frac{\omega}{\omega_n}\right)^2 \right\}^2 + \left\{ 2\xi\left(\frac{\omega}{\omega_n}\right) \right\}^2 \right]^{\frac{1}{2}}}$$

$$0.95 \leqslant \frac{1}{\left[\left\{ 1 - \left(\frac{\omega}{\omega_n}\right)^2 \right\}^2 + \left\{ 2\xi\left(\frac{\omega}{\omega_n}\right) \right\}^2 \right]^{\frac{1}{2}}}$$

where ξ is the damping ratio of the system.
  $\omega_n \leqslant 730$ rad s^{-1} (the answer).

Problem 2.7. A first-order temperature alarm unit with a time constant of 2 min is subjected to a sudden 100 °C rise because of a fire. If an increase of 50 °C is required to activate the alarm, what is the delay in signaling the temperature change?

Solution:

We recall the response equation of a first-order instrument to a step input:

$$y(t) = Kx_s\left(1 - e^{-\frac{t}{\tau}}\right).$$

Here, the time constant $\tau = 2$ min. From the given data,

$$50 = 100\left(1 - e^{-\frac{t}{\tau}}\right).$$

Solving the above equation, we get the answer $t = 1.386$ min.

Chapter 3

Problem 3.1. A steel bar of rectangular cross-section (2 cm × 1 cm) is subjected to a tensile force of 20 kN. A strain gauge is placed on the steel bar as shown in figure 3.13. Find the change of resistance of the strain gauge if it has a gauge factor of two and a resistance of 120 Ω in the absence of an axial load. The Young's modulus of elasticity of steel is equal to 2×10^8 kN m^{-2}.

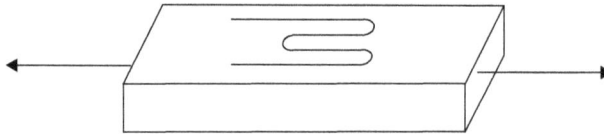

Figure 3.13.

Solution:

Given the gauge factor (λ) = 2, the Young's modulus of elasticity (E) = 2 × 10^8 kN m^{-2}, the force (P) = 20 kN, and the cross-sectional area of the bar (A) = 2 × 1 (cm)^2:

The stress on the bar (σ_a) = $\frac{P}{A}$ = $\frac{20}{0.02 \times 0.01}$ 10^5 kN m^{-2}.

Therefore, the axial strain (ε_a) = $\frac{\sigma_a}{E}$ = $\frac{10^5}{2 \times 10^8}$ = 5 × 10^{-4}.

Again $\frac{\Delta R}{R} = \varepsilon_a.\lambda$, where R is the resistance of the strain gauge and ΔR is the change of resistance. Therefore, $\Delta R = 120 \times 5 \times 10^{-4} \times 2 = 0.12$ Ω (the answer).

Problem 3.2. A strain gauge whose resistance is 120 Ω and whose gauge factor is 2.0 is under zero strain. A 200 kΩ fixed resistance is connected in parallel with it. How much strain does the combination of the resistance and the strain gauge represent?

Solution:

$$R_{\text{equivalent}} = \frac{200 \times 10^3 \times 120}{200 \times 10^3 + 120} = 119.9280 \ \Omega.$$

Therefore, $\frac{\Delta R}{R} = \frac{R_{\text{equivalent}} - R_{\text{gauge}}}{R_{\text{gauge}}} = \frac{119.9280 - 120}{120} = -5.99 \times 10^{-4}$.

We know the gauge factor is defined as $\frac{\frac{\Delta R}{R}}{\frac{\Delta l}{l}}$.

$\therefore$ Strain $= \frac{\Delta l}{l} = \frac{\frac{\Delta R}{R}}{\text{Gauge factor}} = \frac{-5.99 \times 10^{-4}}{2} = -299.5 \times 10^{-6} = -299.5\,\mu\text{m m}^{-1}$ (the answer).

Problem 3.3. Two strain gauges with resistance values of 120 Ω and gauge factor of $\lambda = 2$ are installed on a cantilever beam as shown in figure 3.14. The beam is fixed at one end and a load P is applied at the free end. One gauge is mounted above the beam and other below the beam. Both the gauges are aligned in the axial direction The gauges are connected to a Wheatstone bridge that uses two more fixed resistances of 120 Ω each and the bridge excitation is 2.0 V DC.

(a) Show the bridge arrangement used for ambient temperature change compensation.

(b) The unbalanced voltage of the bridge is fed to a voltmeter that has a resolution of 2 μV. Find the minimum microstrain it can detect.

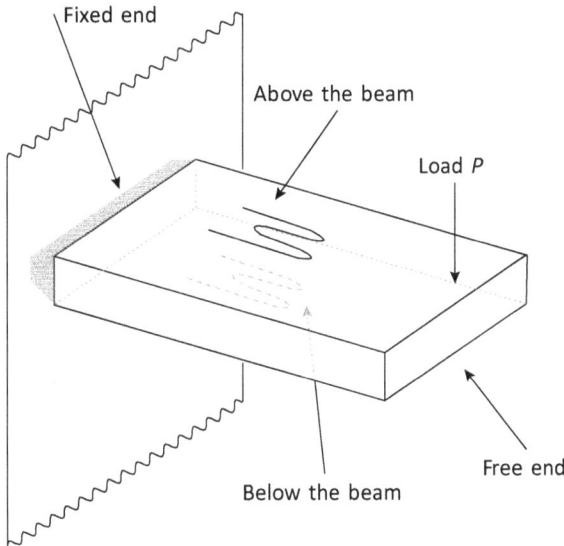

Figure 3.14.

Solution:

(a) The bridge arrangement for ambient temperature compensation is shown in figure 3.15

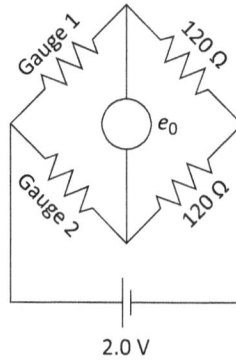

Figure 3.15.

(b) The bridge unbalanced voltage

$$e_0 = \frac{\Delta}{R} \cdot \frac{E_{ex}}{2} = \frac{\lambda \cdot \varepsilon}{2} \, E_{ex},$$

where ε is the axial strain and E_{ex} is the bridge excitation.

If $e_0 = 2\,\mu V$, then $\varepsilon = \frac{10^{-6} \times 2 \times 2}{2 \times 2} = 1\,\mu$ strain (the answer).

Chapter 4

Problem 4.1. Based on figure 4.18, answer the following questions:

(a) Determine the sensitivity of the bridge for small x (< 0.1).

(b) Also determine the linearity of the output voltage with respect to x when x is large.

(c) What happens to the linearity and sensitivity when the circuit of figure 4.19 is used instead of using the Wheatstone bridge signal conditioning circuit?

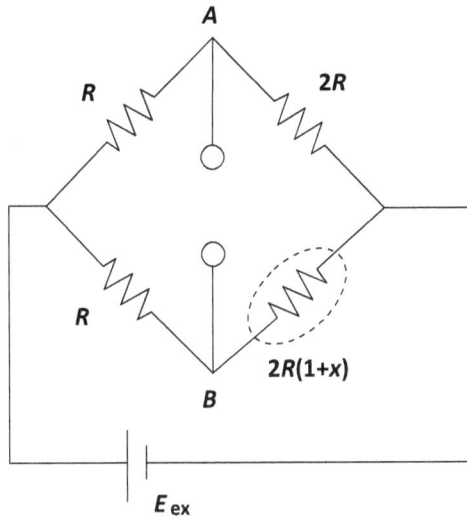

Figure 4.18.

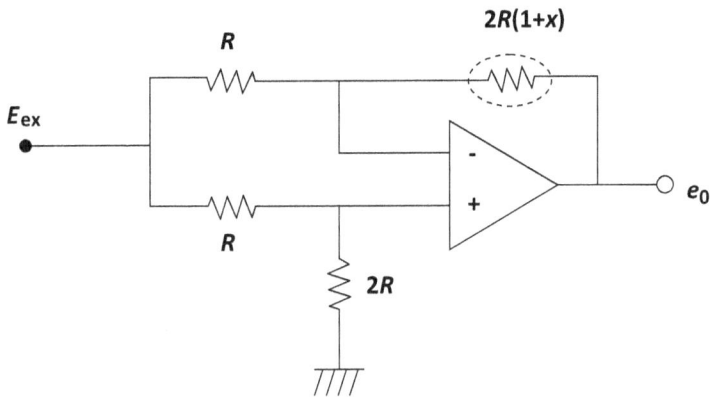

Figure 4.19.

Solution:
  (a)

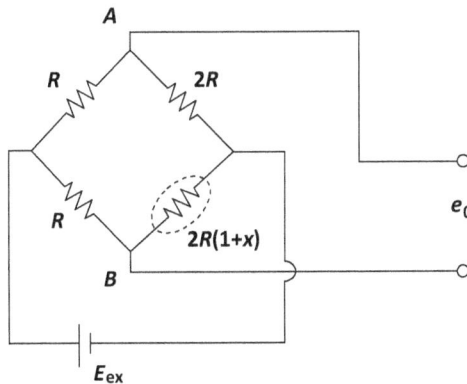

$$e_A = \frac{2}{3} E_{ex}$$

$$e_B = \frac{2(1 + x)}{(3 + 2x)} E_{ex}$$

$$\therefore \ e_0 = \left[\frac{2}{3} - \frac{2(1 + x)}{(3 + 2x)} \right] E_{ex}$$

or, $\dfrac{e_0}{E_{ex}} = \dfrac{1}{3} \dfrac{[6 + 4x - 6 - 6x]}{(3 + 2x)}$

$$= \frac{-2x}{3(3 + 2x)}$$

$$= \frac{-\frac{2}{3}x}{3\left(1 + \frac{2}{3}x\right)}.$$

When x is small,

$$\left| \frac{e_0}{E_{ex}} \right| = \frac{2}{9}x$$

or sensitivity $= \dfrac{\left| \dfrac{e_0}{E_{ex}} \right|}{x} = \dfrac{2}{9}$.

When x is large,

(b)

$$\frac{e_0}{E_x} = \frac{-\frac{2}{9}x}{\left(1 + \frac{2}{3}x\right)} = \frac{Kx}{\left(1 + \frac{2}{3}x\right)} \quad \text{where } K = -\frac{2}{9}.$$

By putting $x = 1$ (which is a large value compared to $x < 0.1$), we get

$$\frac{e_0}{E_{ex}} = \frac{3}{5}K.$$

The straight line passing through $(0, 0)$ and the point $(2/3 \ K, 1)$ in a e_0/E_{ex} versus x curve is given by

$$\frac{e_0}{E_{ex}} = \frac{3}{5}Kx$$

or $e_0 = \frac{3}{5}K_1 x$ (where $K_1 = KE_{ex} = $ constant).

But the actual value of the output voltage is

$$e'_0 = \frac{K_1 x}{\left(1 + \frac{2}{3}x\right)},$$

∴ the error voltage

$$\Delta e_0 = K_1 x \left[\frac{3}{5} - \frac{3}{(3+2x)} \right]$$

$$= \frac{3K_1 x}{5} \left[\frac{3+2x-5}{3+2x} \right]$$

$$= \frac{3K_1 x}{5} \frac{(2x-2)}{(3+2x)}.$$

To see where the error is at a maximum with respect to x, we differentiate Δe_0.

$$\therefore \frac{\partial \Delta e_0}{\partial x} = \frac{\partial}{\partial x} \left[\frac{3K_1}{5} \frac{\left(2x^2 - 2x\right)}{(3+2x)} \right]$$

$$= \frac{\partial}{\partial x} \left[\frac{3K_1}{5} \frac{\left(2x^2 + 3x - 5x - 7.5 + 7.5\right)}{(3+2x)} \right]$$

$$= \frac{3K_1}{5} \frac{\partial}{\partial x} \left[x - 2.5 + \frac{7.5}{(3+2x)} \right]$$

or $\frac{\partial \Delta e_0}{\partial x} = \frac{3K_1}{5} [1 - \frac{15}{(3+2x)^2}]$.

∴ for the maximum value of Δe_0, we get
$(3+2x)^2 = 15$
or, $x = \frac{\sqrt{15}-3}{2} = 0.436$

$$\therefore \Delta e_{0max} = \frac{3K_1}{5} \left[\frac{2 \times 0.436 - 2}{3 + 2 \times 0.436} \right]$$

$$= -0.127 \times \frac{3K_1}{5}$$

$$\therefore \text{linearity} = \frac{\text{max shift}}{\text{full-scale deflection}} \times 100.$$

Here, we have assumed the full-scale deflection to be $\frac{3K_1}{5}$.

$$\therefore \text{linearity} = \frac{-0.127 \times 3K_1/5}{3K_1/5} \times 100 = -12.7\% \text{ (the answer)}.$$

(c) We now use an op-amp instead of a Wheatstone bridge circuit.

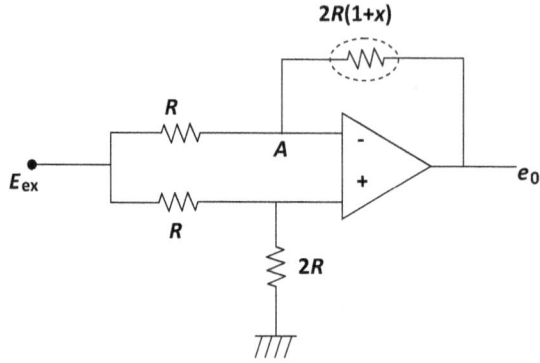

The voltage at node A

$$e_A = \frac{2}{3}E_{ex}.$$

Now applying Kirchoff's current law (KCL) at node A, we get

$$\frac{E_{ex} - e_A}{R} = \frac{e_A - e_0}{2R(1 + x)}$$

or $E_{ex} - \frac{2}{3}E_{ex} = \frac{\frac{2}{3}E_{ex} - e_0}{2(1 + x)}$

or $\frac{1}{3}E_{ex} = \frac{1}{3}\frac{E_{ex}}{(1 + x)} - \frac{1}{2}\frac{e_0}{(1 + x)}$

or $\frac{1}{3}E_{ex}[1 - \frac{1}{(1 + x)}] = -\frac{1}{2}\frac{e_0}{(1 + x)}$

or $[\frac{1}{3}E_{ex}]x = -\frac{1}{2}e_0$

or $\frac{e_0}{E_{ex}} = -\frac{2}{3}x$

or $|\frac{e_0}{E_{ex}}| = \frac{2}{3}x.$

Therefore,

sensitivity $= \frac{|\frac{e_0}{E_{ex}}|}{x} = \frac{2}{3}$..

We see that the sensitivity has improved compared to the previous case. We also see that the final output is also linear with respect to x. Thus, the above circuit is an improvement over the Wheatstone bridge.

Problem 4.2. An RTD is required to measure temperature correctly in the temperature range of 75 °C to 125 °C with a resolution of 0.5 °C. Design a signal conditioning circuit for the above.

The following data are given:
- α of the RTD = 0.002 °C^{-1},
- the resistance of the RTD = 350 Ω at 25 °C, and
- P_D = 30 mW °C^{-1}.

Solution:

The signal conditioning circuit should produce 0 V at 75 °C and 5 V at 125 °C, i.e. we define our output range to be 0 V to 5 V. Thus, we use a Wheatstone bridge together with an instrumentation amplifier. The Wheatstone bridge is necessary to give an output of 0 V at 75 °C.

The output of the Wheatstone bridge will be very small and hence an instrumentation amplifier is used. The following circuit is proposed:

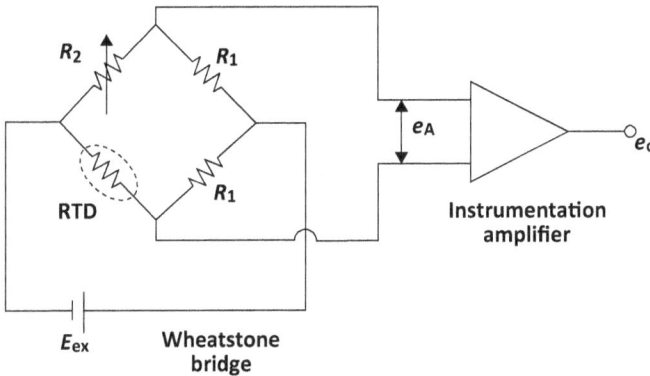

A variable resistance R_2 is necessary so that the bridge remains balanced at 75 °C.

We take E_{ex} = 10 V.

Let

R_{To} = the resistance of the RTD at 25 °C and

R_T = the resistance of the RTD.

$R_T = R_{To}(1 + \alpha \Delta T)$.

$$\therefore R_T|_{75°} = 350(1 + 0.002 \times 50) = 385 \ \Omega.$$

To keep the bridge balanced at 75 °C, R_2 is adjusted to 385 Ω.

We now need to achieve a resolution of 0.5 °C, therefore the output should change by 0.05 V for every 0.5 °C change in temperature.

Error may be introduced due to the self-heating of the RTD. To ensure that this is not so, we choose the temperature change due to self-heating to be 10% of the given resolution, therefore ΔT (for self-heating) = 0.05 °C.

We know that $\Delta T = \frac{P}{P_D}$, where P = power dissipation in the RTD (in W). P_D = the dissipation constant of the RTD (W °C^{-1}). Therefore, the maximum allowable power dissipation in the RTD, $P = \Delta T \cdot P_D = 0.0015$ watt, which is equal to the $I^2 R_T$ loss, or $I^2 R_T = 0.0015$. The resistance at 75 °C = 385 Ω, or $I = 0.001\,97$ A ≈ 0.002 A.

From the Wheatstone bridge circuit, we see that

$I(R_T + R_1) = 10$ V

or, $R_T + R_1 = 5000$

or, $R_1 = 4615$ Ω.

We choose a standard value of resistance, namely 4700 Ω, i.e. $R_1 = 4700$ Ω.

As the temperature increases, the value of the resistance increases but still does not introduce much error.

Again, from the Wheatstone Bridge we see that

$$e_A = \left[\frac{385 + x}{5085 + x} - \frac{385}{5085} \right] \times 10,$$

where x = the change in the resistance of the RTD.

As x is very small compared to 5085 Ω, we can write

$$e_A = \left[\frac{385}{5085} + \frac{x}{5085} - \frac{385}{5085} \right] \times 10$$

$$= \frac{10x}{5085}.$$

Now

$e_o = K e_A,$

where K = the gain of the instrumentation amplifier. Now, $e_o = 0.05$ V for a 0.5 °C change in temperature. $x = 0.35$ Ω for a 0.5 °C change in temperature, therefore $K = 72.64$. Thus, the gain of the instrumentation amplifier should be adjusted to 72.64.

Problem 4.3. Find the temperature versus output voltage relation for a thermistor using the following linearization techniques:

(a) Using a resistance in parallel to a thermistor.

(b) Using a thermistor in a Wheatstone bridge.

Solution:

We know that

$$R = R_0 \exp\left\{ \beta\left(\frac{1}{T} - \frac{1}{T_0} \right) \right\}.$$

Let $T = T_0 + \Delta T$

or $R = R_0\, e^{\beta\left(\frac{1}{T_0 + \Delta T} - \frac{1}{T_0} \right)}$

$$= R_0\, e^{-\beta \Delta T / T_0^2}$$

$$= R_0\, e^{-x},$$

where

$$x = \beta \Delta T \, / \, T_0^2.$$

(a) Using parallel resistance.

To give a linear output, the parallel resistance used must have the same value as that of the thermistor at temperature T_0.

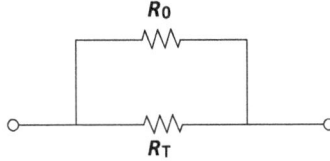

Thus, the equivalent resistance (R_{eq}) is

$$R_{eq} = \frac{R_T R_0}{R_T + R_0} = \frac{R_0^2 e^{-x}}{R_0(1 + e^{-x})} = R_0 \frac{1}{(1 + e^x)}$$

$$\therefore R_{eq} = R_0(1 + e^x)^{-1}.$$

Now, $e^x = 1 + x + x^2/2 + \cdots$.

$\therefore (1 + e^x) = 2 + x + x^2/2 + x^3/6 + \cdots$.

Using the long division method, we find the value of $(1 + e^x)^{-1}$:

$$2 + x + {x^2}\!/\!{2} + {x^3}\!/\!{6} + \cdots \overline{\big)\, 1} \qquad \left({1}\!/\!{2} - {x}\!/\!{4} + {x^3}\!/\!{48} \cdots \right.$$

$$\underline{1 + {x}\!/\!{2} + {x^2}\!/\!{4} + \cdots}$$

$$- {x}\!/\!{2} - {x^2}\!/\!{4} - {x^3}\!/\!{12}$$

$$\underline{- {x}\!/\!{2} - {x^2}\!/\!{4} - {x^3}\!/\!{8}}$$

$${x^3}\!/\!{24} \cdots$$

Therefore,

$$R_{eq} = R_0(1/2 - x/4 + x^3/48 \cdots).$$

Now, $x = \frac{\beta \Delta T}{T_0^2}$, which is a very small value. Therefore, $\frac{x^3}{48}$ and the higher-order terms become negligible.

Therefore, we see a more or less linear relation between the resistance R_{eq} and the temperature change ΔT.

(b) In this scheme, all the resistances are made equal to the thermistor resistance R_0 at temperature T_0.

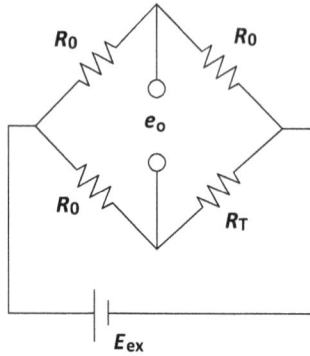

From the circuit, we see

$$\frac{e_o}{E_{ex}} = \left[\frac{1}{2} - \frac{R_0(e^{-x})}{R_0(1 + e^{-x})} \right]$$

or, $\dfrac{e_o}{E_{ex}} = \dfrac{1 + e^{-x} - 2e^{-x}}{1 + e^{-x}}$

$$= -\left[\frac{e^{-x} - 1}{e^{-x} + 1} \right] = \left[\frac{e^x - 1}{e^x + 1} \right]$$

$$e^x - 1 = x + \frac{x^2}{2} + \frac{x^3}{6} + \cdots$$

$$e^x + 1 = 2 + x + \frac{x^2}{2} + \frac{x^3}{6} \cdots .$$

By long division, we find that,

$$2 + x + \frac{x^2}{2} + \frac{x^3}{6} + \ldots) \, x + \frac{x^2}{2} + \frac{x^3}{6} + \ldots \quad \left(\frac{x}{2} - \frac{x^3}{24} + \ldots \right.$$

$$\underline{x + \frac{x^2}{2} + \frac{x^3}{4}}$$

$$\frac{x^3}{12}$$

$$\therefore \frac{e_o}{E_{ex}} = \frac{x}{2} - \frac{x^3}{24} + \cdots .$$

Thus, we see that when a Wheatstone bridge is used, the relation between the output voltage and the temperature change is mostly linear.

Problem 4.4. A K-type thermocouple is used to measure a certain temperature range. The signal conditioning circuit is kept at a distance from the place of measurement. Design a signal conditioning circuit, taking into account the noise as well as the cold junction compensation. Use a thermistor for cold junction compensation.

Solution:

Since the signal conditioning circuit is placed at a distance from the place of measurement, we require compensating cables. Here, we are using a type K or Chromel-Alumel thermocouple. Thus, we can use copper–constantan as the compensating cables.

A thermocouple is an active transducer that produces only a very few milivolts as the output voltage. Hence it is highly susceptible to noisy environments. Noise can also be introduced into the signal conditioning circuit from the power line. To remove this, we may use a capacitor between the power supply and the circuit.

Noise is also introduced by:

(a) capacitive coupling
(b) magnetic fields

To remove this, we may use a shielded pair of twisted cables with the shield grounded.

It is then passed through a low-pass filter and then fed to an amplifier. Thus we have the design shown in figure P4.4(a).

An instrumentation amplifier with a high common mode rejection ratio needs to be used so that any common-mode signals (noise) are attenuated.

For cold junction compensation, we use a thermistor.

As the thermistor is a nonlinear device, we need to use linearization techniques. The output is then passed through an amplifier to make the sensitivities of figure P4.4(a) and figure P4.4(b) identical, so as to take care of any ambient temperature fluctuation.

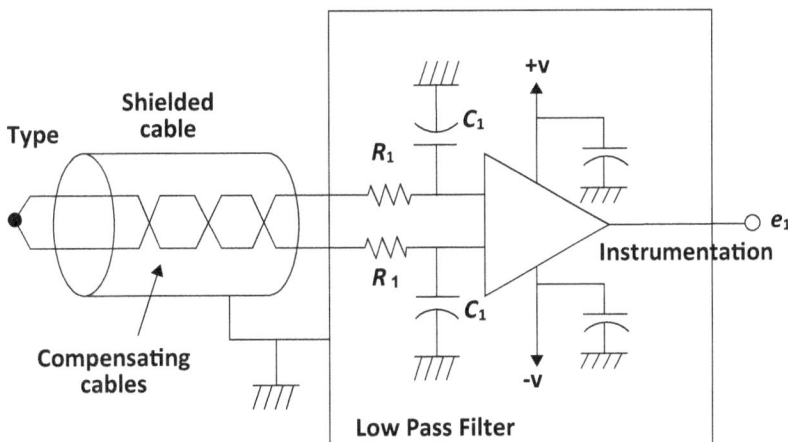

Figure P4.4(a). A thermocouple with measuring circuit.

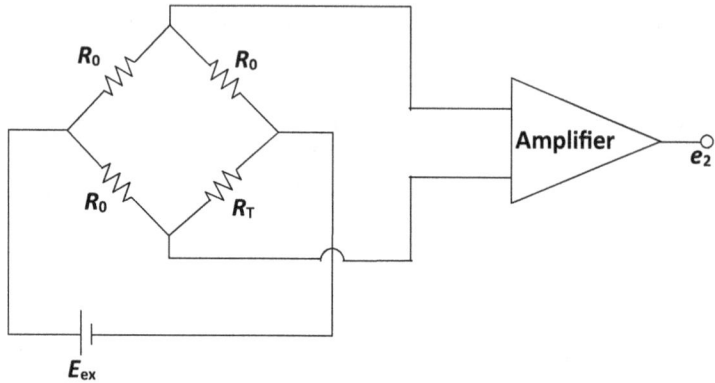

Figure P4.4(b). A cold junction compensation circuit.

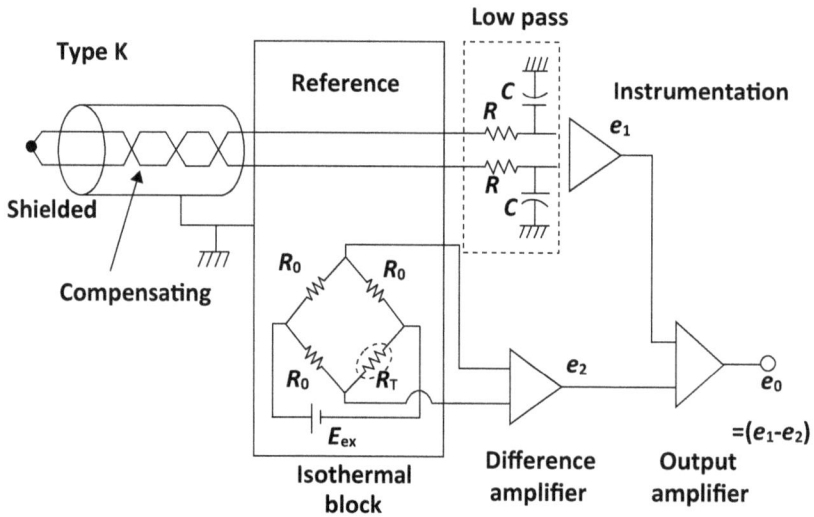

Figure 4.4(c). The complete signal conditioning circuit for a thermocouple.

The cold junction compensation circuit should be placed close to the reference junction temperature so that any fluctuation in the temperature of the reference junction can be picked up and taken care of. Ideally, the cold junction compensation circuit and the reference junction should be placed within an isothermal block. The complete circuit is shown in figure P4.4(c).

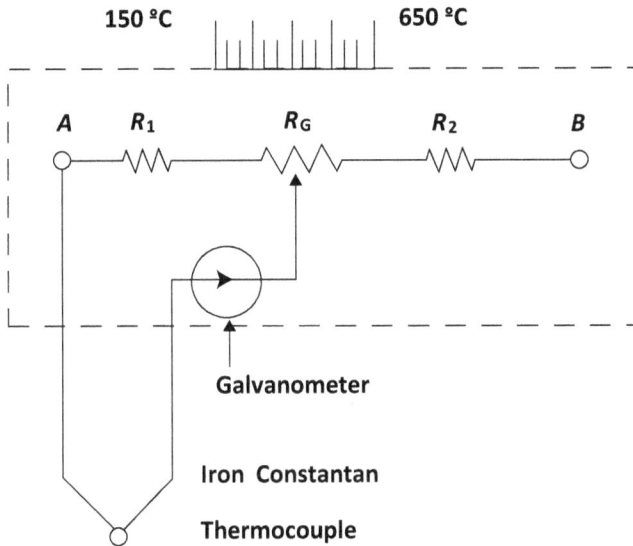

Figure 4.20.

Problem 4.5. A simple potentiometer circuit, as shown in figure 4.20, is used to measure the emf of an iron–constantan thermocouple. A fixed voltage of 1.215 V is applied across points A and B. A current of 3 mA flows through the resistors. The range of the temperature variation is from 150 °C to 650 °C. Find the values of R_1, R_2, and R_G for an ambient temperature of 25 °C.

The following data are provided:
- the emf at 25 °C = 1.022 mV,
- the emf at 150 °C = 14.682 mV, and
- the emf at 650 °C = 53.512 mV.

Solution:
The current = 3 mA
or $i = \frac{E_{standard}}{R_1 + R_2 + R_G}$
or $R_1 + R_2 + R_G = 405\ \Omega$.
The measured voltage at 150 °C is the emf at 150 °C − the emf at 25 °C = 13.66 mV.
Similarly, the measured voltage at 650 °C is the emf at 650 °C − the emf at 25 °C = 52.49 mV.
At 150 °C, from the circuit, we can see:
$R_1 \times$ current = the measured voltage at 150 °C,
or $R_1 = \frac{13.66}{3} = 4.55\ \Omega$.
Similarly, at 650 °C,
$R_1 + R_G = 17.496\ \Omega$
or $R_G = 12.95\ \Omega$,

$\therefore R_2 = 405 - (R_1 + R_G) = 387.5 \, \Omega$. Thus, the answer is:

$R_1 = 4.55 \, \Omega$

$R_2 = 387.5 \, \Omega$

$R_G = 12.95 \, \Omega$.

Problem 4.6. An RTD has a resistance of 600 Ω at 25 °C and a temperature coefficient of 0.005 °C^{-1}. The RTD is used in a Wheatstone bridge circuit with $R_1 = R_2 = 600 \, \Omega$ (figure 3.3). The variable resistance R_3 nulls the bridge. If the bridge excitation is 12 V and the RTD is in a bath at 0 °C, find the values of R_3 required to null the bridge when:

    (a) The self-heating effect of the RTD is not considered.

    (b) The self-heating effect of the RTD is considered and it is known that the dissipation constant of the RTD is 20 mW °C^{-1}.

Solution:

    (a) Without the self-heating effect of the RTD:

$$R_3 = \frac{\text{RTD resistance 25 °C}}{1 + \alpha_0(25 - 0)} = \frac{600}{1 + 0.005 \times 25} = 533.33 \, \Omega,$$

where α_0 is the temperature coefficient of resistance of the RTD.

    (b) With the self-heating effect, the dissipated power due to self-heating

$$= \left(\frac{12}{600 + 533.33}\right)^2 \times 533.33 = 59.79 \text{ mW}. \quad \text{Therefore, the temperature}$$

rise $= \frac{59.79}{20} = 2.98$ °C.

So, the RTD resistance due to a rise in temperature of 2.98 °C

$= 533.33(1 + 0.005 \times 2.98)$

$= 533.33 \times 1.0149$

$= 541.27 \, \Omega$ (the answer).

Problem 4.7. An iron–constantan thermocouple is to be used to measure temperatures between 0 °C and 200 °C. With the reference junction at 0 °C, the emf outputs at temperatures of 100 °C and 200 °C are 5.268 and 10.777 mV, respectively. Find the nonlinearity at 100 °C as a percentage of the full-scale reading.

    Solution: the emf at 200 °C = 10.777 mV.

    If the device were linear, then the emf at 100 °C would equal 10.777/2 mV = 5.3885 mV

    Therefore, the percentage nonlinearity $= \frac{5.268 - 5.3885}{10.777} \times 100 = -\frac{0.1205}{10.777} = -1.118\%$ (the answer).

Problem 4.8. The resistance R_T of a resistive transducer is modeled as $R_T = R(1-Kx)$, where K is the constant of transformation and x is the input quantity being sensed. The resistor is connected to an op-amp as shown in figure 4.21. The value of R is 100 Ω, K is 0.004, and x is 75. Find the output voltage V_o. (assume the op-amp as an ideal device)

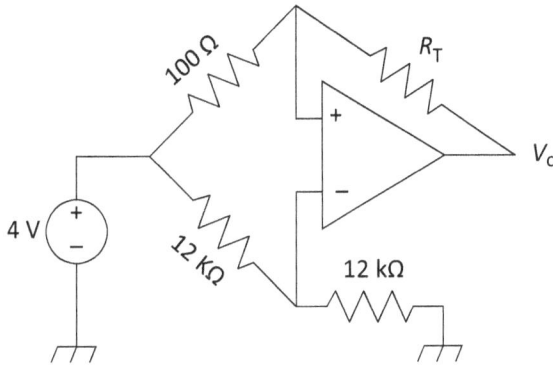

Figure 4.21.

Solution:
Let us redraw the circuit of figure 4.21

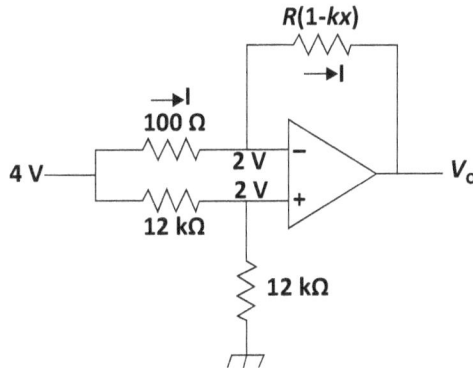

The voltage at the non-inverting terminal of the op-amp is $\frac{4}{12+12} \times 12 = 2$ V. Since the inverting and non-inverting input terminals of the op-amp are connected by a high resistance, the voltage at the inverting input terminal of the op-amp is also 2 V. Therefore, the current I can be expressed as

Current $I = \frac{2}{100}$ (in Amperes); $2 - Vo = IR(1 - Kx)$

$$\therefore Vo = -\frac{(-Kx)4}{2} = 0.004 \times 75 \times 2 = 600 \text{ mV (the answer).}$$

Problem 4.9. A two-wire platinum RTD with a resistance R_0 of 10 Ω at 0 °C is shown in figure 4.22. This is an experimental setup used to measure temperatures between 0 °C and 100 °C. All other arms of the bridge are resistances of fixed value

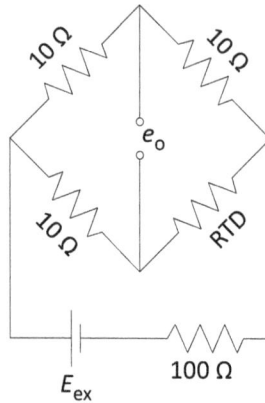

Figure 4.22.

$R = 10\ \Omega$. To avoid the self-heating effect, the power dissipation of the RTD should be less than 1 mW. Calculate the maximum bridge excitation voltage E_{ex} that can be applied in the circuit.

Solution:
$R_0 = 10\ \Omega$.

The maximum permissible current through the RTD, $I = \sqrt{\dfrac{P}{R_0}} = \sqrt{\dfrac{10^{-3}}{10}} = 10^{-2}$ A.

$\therefore$ Maximum permissible bridge excitation $= I \times 2R_0 + 2\,I \times 100$
$$= 10^{-2} \times 2 \times 10 + 2 \times 10^{-2} \times 100$$
$$= 2.2\ \text{V (the answer)}.$$

Problem 4.10. An RTD is represented using a linear model over the range of 0 °C to 100 °C by the equation

$$R_T = R_0(1 + 0.003T),$$

where T is the temperature in °C. $R_0 = 100\ \Omega$ with a tolerance of $\pm 3\ \Omega$. The actual model of the RTD is

$$R_T = R_0(1 + 0.003T + 6 \times 10^{-7}T^2).$$

What is the worst-case error magnitude in the measurement system?

Solution:
The maximum negative error occurs at $T = 0$ °C and when $R_0\,|_{\text{actual}} = 97\ \Omega$; it is equal to $-3\ \Omega$. The maximum positive error occurs at $T = 100$ °C and when $R_0 = 103\ \Omega$; it is equal to $(103 - 100)(1 + 0.003\ T) + 6 \times 10^{-7} \times (100)^2 \times 103$ $= 3.9 + 0.618 = 4.518\ \Omega$ (the answer).

Chapter 5

Problem 5.1.   In figure 5.26, let x_i be a periodic motion with a significant frequency content at up to 500 Hz. The excitation frequency is 5000 Hz. The output signal obtained is then passed through a low-pass filter and then to an oscilloscope with an input impedance of 10^6 Ω. It is desired that ripple due to higher frequencies should not be more than 5% of the unfiltered value.

    a) Find the frequency range after the modulation process.

    b) Design a low-pass filter for the above application. Also calculate the value of
       R in figure 5.26.

    c) Calculate the time lag introduced by the low-pass filter.

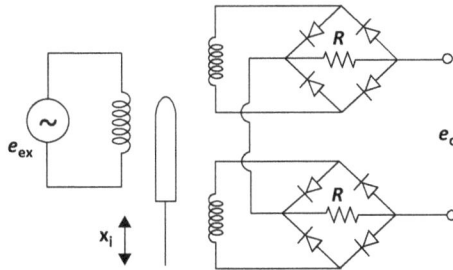

Figure 5.26.

Solution:

  (a) The diode bridge shown in the figure acts as a phase-sensitive demodulator. The frequency of the motion is 500 Hz = (f_m). The carrier frequency (f_c) = 5000 Hz.

     After the modulation process, in addition to the motion frequency, we have frequencies $2f_c + f_m$ = 10500 Hz and $2f_c - f_m$ = 9500 Hz.

     Thus, this will be the operational frequency range.

  (b) In order to design the low-pass filter, we first need to choose the order of the filter that will give us our desired result. First, we consider a first-order filter.

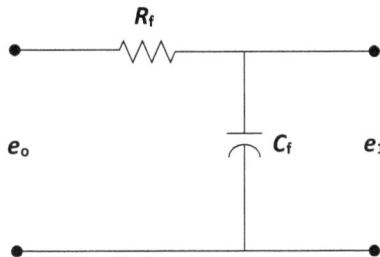

Now, $\frac{e_1}{e_o}(j\omega) = \frac{1}{j\omega R_f C_f + 1}$.

It is a requirement that ripples due to higher frequencies must be less than 5% of the unfiltered value. For higher frequencies, we consider 9500 Hz and above, i.e.
$0.05 = \frac{1}{\sqrt{\omega^2 \tau_f^2 + 1}}$, where $\tau_f = R_f C_f$ and $\omega = 2\pi \times 9500$ or $\tau_f = 0.0003348$ s.

For this τ_f, we have the amplitude ratio

$$\left|\frac{e_1}{e_o}\right| = \frac{1}{\sqrt{\left(2\pi f_m \tau_f\right)^2 + 1}} = 0.69.$$

Thus, we see that the higher-frequency portion of x_i would be distorted considerably. So, a first-order filter will not serve our purpose. We now consider a second-order filter, as shown below:

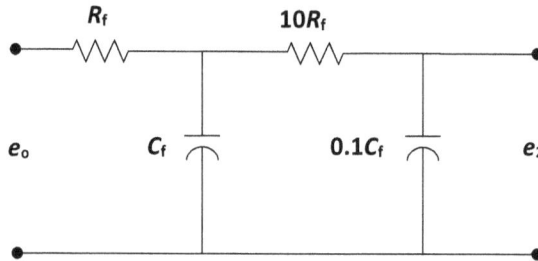

$$\therefore \frac{e_2}{e_o}(j\omega) = \frac{1}{\sqrt{\left(j\omega R_f C_f\right)^2 + 1}}.$$

Now, $0.05 = \frac{1}{(\omega \tau_f)^2 + 1}$ or $\tau_f = 0.0000731$ s. The amplitude ratio for the above case is

$$\left|\frac{e_2}{e_o}\right| = \frac{1}{\left(2\pi f_m \tau_f\right)^2 + 1} = 0.95.$$

Thus, we see that the amplitude ratio from $\omega = 0$ to $\omega = 3140$ rad s^{-1}, i.e. from 1.0 to 0.95, is nearly flat. Thus the waveforms of the transients will be faithfully reproduced.

In order to determine the values of R_f and C_f, we use the fact that each stage in the process must have an impedance ten times greater than that of the previous stage. The oscilloscope has a input impedance = 10^6 Ω.

$\therefore 10 R_f = 10^5 \Omega$ or $R_f = 10^4 \Omega$.

Now, $C_f R_f = 0.0000731$ or $C_f = 73 \times 10^{-6}/10^4 = 0.0073$ μF.

Thus, we have the following filter.

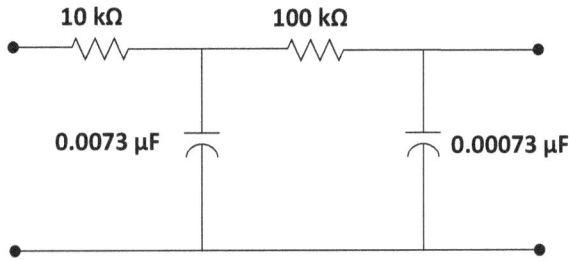

(c) To determine the phase angle,

$$\frac{e_2}{e_o}(j\omega) = -\tan^{-1}\frac{2\omega\tau_f}{1 - \omega^2\tau_f^2}$$

$$= -25.8°.$$

This phase angle introduces a delay of $(25.8)/(57.3 \times 3140) = 144$ μs (the answer).

Problem 5.2. In figure 5.27, a voltmeter with a finite input impedance R_m is connected. Find the expression for frequency at which the phase shift is zero.

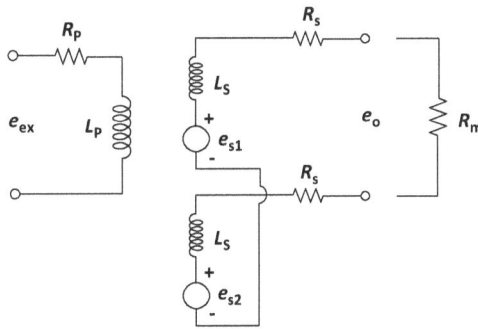

Figure 5.27.

Solution:
Analyzing the above circuit, we observe that

$$i_p R_p + L_p D_{i_p} - (M_1 - M_2)D_{i_s} - e_{ex} = 0$$

$$(M_1 - M_2)D_{i_p} + (2R_s + R_m)i_s + 2L_s D_{i_s} = 0$$

or $\dfrac{e_o}{e_{ex}}(D) = \dfrac{R_m(M_2 - M_1)D}{[(M_1 - M_2)^2 + 2L_pL_s]D^2 + [L_p(2R_s + R_m) + 2L_sR_p]D + (2R_s + R_m)R_p}$

or

$\dfrac{e_o}{e_{ex}}(s) = \dfrac{R_m(M_2 - M_1)s}{\left[(M_1 - M_2)^2 + 2L_pL_s\right]s^2 + \left[L_p(2R_s + R_m) + 2L_sR_p\right]s + (2R_s + R_m)R_p}$

or $\dfrac{e_o}{e_{ex}}(jw) = \dfrac{j\omega R_m(M_2 - M_1)}{j\omega[L_p(2R_s + R_m) + 2L_sR_p] + [(2R_s + R_m)R_p - \omega^2\{(M_1 - M_2)^2 + 2L_pL_s\}]}.$

Thus, we see that a zero phase shift is possible when

$$(2R_s + R_m)R_p - \omega^2\left[(M_1 - M_2)^2 + 2L_pL_s\right] = 0$$

or $\omega = \sqrt{\dfrac{(2R_s + R_m)R_p}{(M_1 - M_2)^2 + 2L_pL_s}}.$

At that frequency,

$$\dfrac{e_o}{e_{ex}}(j\omega) = \dfrac{R_m(M_2 - M_1)}{L_p(2R_s + R_m) + 2L_sR_p} \quad \text{(the answer)}.$$

Problem 5.3. Design a linear variable differential transformer (LVDT) based on the following details:

- the supply frequency = 10 kHz,
- the length of the core (L_a) = 20 cm,
- the maximum distance from the core to the null position (x_{max}) = 6 cm,
- the error introduced due to nonlinearity (ε) = 10%,
- the maximum emf induced in the secondary coils (e_o) = 5 V, and
- the current in the primary required to obtain the maximum emf in the secondary = 20 mA.

Assume that the number of turns in the secondary (N_S) is four times the number of turns in the primary (N_P).

Solution:

The net induced emf $e_o(j\omega)$ of the secondary coils is given by

$$e_o(j\omega) = j\omega I_p\left[\dfrac{4\pi N_pN_s\mu_0px}{3s ln(r_0/r_i)}\left(1 - \dfrac{x^2}{2p^2}\right)\right]$$

or

$$|e_o| = \omega|I_p|\left[\dfrac{4\pi N_pN_s\mu_0px}{3s ln(r_0/r_i)}\left(1 - \dfrac{x^2}{2p^2}\right)\right] \tag{P5.1}$$

We know from the text that the ratio of r_i/L_a is about 0.05, therefore the inner radius of the LVDT assembly is $0.05 \times 20 = 1$ cm. The ratio of r_o/r_i varies between two and eight. Let us take $r_o/r_i = 4$; then, the outer radius of the LVDT assembly is 4 cm.

The length of the primary winding

$$p = x_{max}/\sqrt{2\varepsilon}$$
$$= 6/\sqrt{2 \times 0.1}$$
$$= 13.4 = 14 \text{ cm (say)}.$$

The length of the secondary winding

$$S = p + x_{max} = 20 \text{ cm}.$$

In equation (P5.1), when $|e_o| = 5$V, $|I_p| = 20$ mA

Putting the values of p, s, r_o/r_i, and ω into equation (P5.1) and noting that the maximum emf is induced at the maximum distance from the null position, we find

$$5 = \left[\frac{6.28 \times 10^4 \times 20 \times 10^{-3} \times 4\pi \times (N_pN_s)4\pi \times 10^{-7} \times 0.14 \times 0.06}{3 \times 0.20 \times \ln(4)}\right] \times \left[1 - \frac{1}{2}\left(\frac{0.06}{0.14}\right)^2\right].$$

μ_o, the permeability of free space, $= 4\pi \times 10^{-7}$ H m^{-1}

$$= \frac{6.28 \times 12070 \times 10^{-8}}{4.2} \times (N_pN_s)$$

or $N_pN_s = 27427$

or $4N_p^2 = 27427$ (as $N_s = 4 N_p \rightarrow$ given)

or $N_p^2 = 6856$

or $N_p \approx 83$.

We take $N_p = 85$. Then, $N_s = 340$. Thus, the complete design parameters for the LVDT are:

- $L_a = 20$ cm
- $r_i = 1$ cm.
- $r_o = 4$ cm
- $p = 14$ cm.
- $s = 20$ cm.
- $N_p = 85$ turns.
- $N_s = 340$ turns (the answer).

Chapter 6
Problem 6.1.

In figure 6.17, four strain gauges are placed over a diaphragm.

For the above arrangement, draw the signal conditioning circuit.

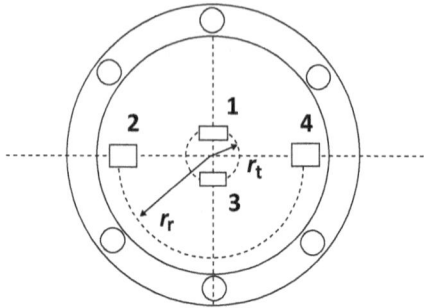

Figure 6.17.

1. Find the sensitivity of the circuit in mV Pa^{-1}.
2. Calculate its natural frequency in vacuum.
3. Find the maximum allowable pressure for a nonlinearity of 2%.
4. Find the full-scale output.

The following data are given:
- $r_t = 0.015$ m, $r_r = 0.064$ m.
- The diameter of diaphragm (D) = 0.15 m.
- The thickness of the diaphragm (t) = 1.28 mm.
- Poisson's ratio (ν) = 0.26.
- The excitation voltage (E_{ex}) of the Wheatstone bridge = 5 V.
- The gauge factor = 2.
- The gauge resistance = 120 Ω
- The density of the diaphragm material (ρ) = 6×10^3 kg m^{-3}.
- The modulus of elasticity of the diaphragm material (E) = 2.1×10^{11} Pa.

Solution:
Strain gauges 1 and 3 are placed close to the center and are therefore used to read the tangential strain, which is maximized at the center. Gauges 2 and 4, on the other hand, are placed close to the periphery and are oriented to read radial strain, which reaches its maximum value there. The bridge circuit is shown below.

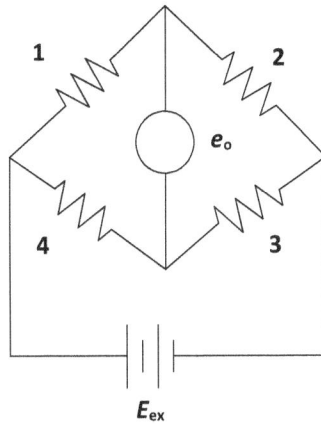

The radial and tangential stresses at any point are given by equations (6.1) and (6.2), respectively. Recalling those equations, we write

$$S_r = \frac{3PR^2\nu}{8t^2}\left[\left(\frac{1}{\nu} + 1\right) - \left(\frac{3}{\nu} + 1\right)\left(\frac{r}{R}\right)^2\right]$$

$$S_t = \frac{3PR^2\nu}{8t^2}\left[\left(\frac{1}{\nu} + 1\right) - \left(\frac{1}{\nu} + 3\right)\left(\frac{r}{R}\right)^2\right].$$

Let us write $\frac{3PR^2\nu}{8t^2} = A$ (say), and we are given $R = 0.075$ m. Then, at $r = r_t$, $S_t = 4.572A$ and $S_r = 4.344A$. Again, at $r = r_r$, $S_r = -4.285A$ and $S_t = -0.139A$.

The diaphragm is in a state of biaxial stress, and both the radial and tangential stresses contribute to the radial and tangential strains. We know that the tangential strain $(\in_t) = \frac{S_t - \nu S_r}{E}$. At $r = r_t$, $\in_t = \frac{3.4426A}{E}$. Again, the radial strain $(\in_r) = \frac{S_r - \nu S_t}{E}$.

At $r = r_r$, $\in_r = -\frac{4.249A}{E}$. Let us take $\frac{A}{E} = B$ (say). Therefore, $\in_t = 3.4426B =$ and $\in_r = -4.249B$.

The gauge factor is equal to two; therefore, upon the application of pressure, the resistance changes for strain gauges 1 and 3 are equal, and those of gauges 2 and 4 are also equal. Initially, all their values are $R = 120$ ohms. The modified values of resistance of the gauges are as follows:

$R_1' = R_3' = R_o(1 + \lambda \, \varepsilon_t)$ and $R_2' = R_4' = R_o(1 + \lambda \, \varepsilon_r)$.

The unbalanced voltage e_o is given by

$$e_o = \left[\frac{R_2'}{R_1' + R_2'} - \frac{R_3'}{R_4' + R_3'}\right]E_{ex}.$$

Substituting the values of R'_1, R'_2, R'_3 and R'_4, we get

$$\frac{e_0}{e_{ex}} = -\frac{1}{2R_o + 2R_o\varepsilon_t + 2R_o\varepsilon_r}[2R_o\varepsilon_t - 2R_o\varepsilon_r]$$

$$= \frac{-1}{1 + (\varepsilon_t + \varepsilon_r)}[\varepsilon_t - \varepsilon_r]$$

$$= \frac{-1}{1 - 0.8064B}[7.6916B]$$

Now substituting all the values we get

$$B = \frac{3PR^2\nu}{8t^2} \times \frac{1}{E}$$

$$= \frac{3 \times (0.075)^2 \times 0.26P}{8 \times (1.28)^2 \times 10^{-6} \times 2.1 \times 10^{11}}.$$

$$= 1.6 \times 10^{-9}\,P$$

Therefore, $\dfrac{e_0}{e_{ex}} = -\dfrac{7.6916 \times 1.6 \times 10^{-9}P}{1 - 0.8064 \times 7.6916 \times 1.6 \times 10^{-9}P}$

$$\text{and } \frac{e_0}{P} \approx -7.6916 \times 1.6 \times 10^{-9}e_{ex},$$

$\therefore$ therefore the sensitivity in (mV/Pa) is $S = \frac{e_0}{P} = -7.6916 \times 1.6 \times$ $5 \times 10^{-6} = -0.615 \times 10^{-4}$ (the answer).

(2) We know that the natural frequency is given by equation (6.7). Recalling the equation,

$$f = \frac{10.21}{\pi R^2}\sqrt{\frac{Et^2}{12(1 - \nu^2)\rho}}\,Hz$$

therefore the frequency is

$$f = \frac{10.21\,t}{\pi R^2}\sqrt{\frac{E}{12(1 - \nu^2)\rho}}\,Hz$$

$$= 1308\,Hz \text{ (the answer).}$$

(3) The maximum allowable nonlinearity is 2%. We know that the relation between pressure and the deflection of the central point of the diaphragm is given by equation (6.3):

$$p = \frac{16Et^4}{3R^4(1 - \nu^2)}\left[\frac{y_c}{t} + 0.488\left(\frac{y_c}{t}\right)^3\right].$$

The nonlinearity is introduced by the second term.

$$\therefore\ 0.488\left(\frac{y_c}{t}\right)^3 \leqslant \frac{2}{100}$$

or $\left(\frac{y_c}{t}\right)^3 \leqslant \frac{2}{48.8}$

or $\frac{y_c}{t} = \sqrt[3]{\frac{2}{48.8}} = 0.345.$

Therefore, from the above expression, we get the maximum allowable pressure as follows:

$$P = \frac{16 \times 2.1 \times 10^{11} \times (1.28)^4 \times 10^{-2}}{3 \times (0.075)^4 \times (1 - 0.26^2)}[0.345 + 0.02]$$

$= 371.96$ kPa (the answer).

(4) We know that the sensitivity is

$S = -0.615 \times 10^{-4}$ mV Pa^{-1}.

The maximum allowable pressure is 371.96 kPa. Therefore, the full-scale output is given by

$$|e_o| = 0.615 \times 10^{-4} \times 371.96 \times 10^3$$
$$= 0.615 \times 37.196 \text{ mV}$$
$$= 22.87 \text{ mV (the answer)}.$$

Problem 6.2. A McLeod gauge has a bulb of volume 110 cm^3. Its capillary diameter is 1.2 mm. Initially, the reading was found to be 3 cm. Later, it was found that the observed reading was wrong. Find the error in the measured pressure if the true reading is 2.5 cm.

Solution:

The volume of the bulb is V_B:

$$V_B = 11 \times 10^4 \text{ mm}^3.$$

The volume of the capillary for the initial reading is observed to be

$$V_{C1} = \frac{2(1.2)^2}{4} \times 30.0$$
$$= 33.93 \text{ mm}^3.$$

The pressure $p_1 = \frac{33.93 \times 30}{11 \times 10^4 - 33.93}$

$= 0.009256$ torr

$= 1.2332$ Pa.

However, the exact capillary volume is

$$V_{C2} = \frac{2 \times (1.2)^2}{4} \times 25$$
$$= 28.27 \text{ mm}^3.$$

The corresponding pressure

$$p_2 = \frac{28.27 \times 25}{11 \times 10^4 - 28.27} = 0.00642 \text{ torr}.$$
$$= 0.857 \text{ Pa}.$$

This is the true pressure.

Therefore, the error in measurement was

$$\frac{0.857 - 1.2332}{0.857} \times 100\%$$

=−43.9% (the answer).

Problem 6.3. A pressure sensor has the following specifications:
- its sensitivity at the design temperature = 10 V MPa^{-1},
- the zero drift = 0.01 V °C^{-1}, and
- the sensitivity drift = 0.01 (V MPa^{-1}) °C^{-1}.

What will the true value of the pressure be when the sensor is used at an ambient temperature of 20 °C above the design temperature and the device output is 7.4 V?
Solution:
zero drift = 20 °C × 0.01 V °C^{-1} = 0.2 V
sensitivity = 10 + 0.01 × 20 = 10.2 V
Therefore, the output = 10.2 (pressure) +0.2 = 7.4 V.
Hence, 10.2 (pressure) = 7.2 V, therefore pressure = 0.71 MPa (the answer).

Chapter 7
Problem 7.1. Calculate the flow rate of water flowing through a pipe of diameter 0.1 m using an orifice with a hole diameter of 0.05 m. The differential head read over a mercury column manometer for vena contracta taps is 0.25 m. The flowing water temperature is 35 °C. The manometer temperature is 24 °C. The standard temperature at which the data are derived is 15 °C.
The following data are given:
- $\rho_{35°C} = 997$ kg m^{-3}.
- $\rho_{24°C} = 994$ kg m^{-3}.
- $\rho_{15°C} = 999.8$ kg m^{-3}.
- $\mu_{35°C} = 0.06$ poise. (1 poise = 0.1 N·s m^{-2}).

Solution:
The pipe diameter ratio is given by $\beta = \frac{d_{\text{orifice}}}{D_{\text{pipe}}} = 0.5$. We know the volume flow rate (Q) is expressed as $Q = K. A. \sqrt{\rho(\rho_m - \rho_f)} \sqrt{\frac{2gh}{\rho_c}}$, where
- Q = the volumetric flow rate
- K = the discharge coefficient
- A = the cross-sectional area of the orifice
- ρ_m = the density of the manometer liquid (here, it is mercury)
- ρ_f = the density of water in the pipe = $\rho_{35°C}$
- ρ_c = the density of water at the manometer temperature = $\rho_{24°C}$
- h = the difference of height in the mercury columns

22-30

The cross-sectional area of the orifice $A = \frac{\pi d^2}{4} = 0.001963$, and we were given

$$\rho_m = 13600 \text{ kg m}^{-3}.$$

$$\text{Therefore,} \quad Q = 0.015383\ K. \tag{P7.1}$$

The Reynolds number

$$R_e = \frac{\rho V d}{\mu}$$

$$\text{or} \quad R_e = \frac{\rho_{24} Q . d}{\frac{\pi d^2}{4}\mu_{35}} = 4.218\ 10^6\ Q. \tag{P7.2}$$

If we choose $R_e = 10^4$, then $K = 0.6631$ from the table in appendix I and from equation (P7.1), $Q = 0.0102 \text{ m}^3 \text{ s}^{-1}$. From (P7.2.2), $Q = 0.00237 \text{ m}^3 \text{ s}^{-1}$.

These are widely different, so we choose $R_e = 3 \times 10^4$ and $K = 0.6451$. From equation (P7.1), $Q = 0.00992 \text{m}^3\text{s}^{-1}$, and from equation (P7.2), $Q = 0.00711 \text{ m}^3 \text{ s}^{-1}$. Hence, we conclude that the flow rate is $Q = 9.923\ 10^{-3} \text{ m}^3 \text{ s}^{-1}$ (the answer).

Problem 7.2. Hydrogen flows in a pipe of diameter 0.05 m with mass flow rate of $9.6 \times 10^4 \text{ kg m}^{-2}$ and a temperature of 20 °C. If, at the given temperature and pressure, the flow rate is 0.493 m³ min⁻¹ for a manometer height of 55 cm of mercury, calculate the orifice diameter.
- The specific heat $\gamma = 1.4$
- $\mu = 0.003$ poise.
- $R = 28.81$ m °C⁻¹.
- The rational expansion factor: $Y = 1 - [0.41 + 0.35\beta^4](\Delta p/\gamma p_1)$.
- β = the pipe diameter ratio.

Solution:
We know that the volume of gas is given by

$$V = C_f \frac{RT}{p},$$

where C_f = Compressibility factor. First, let us consider the fluid to be incompressible. Then, $V_s = V_1 = C_{f1}\frac{RT}{p_1}$.

Here, $C_{f1} = 1$
$T_1 = 293$ K

$$\therefore V_s = V_1 = \frac{28.81 \times 293}{9.6 \times 10000} = 0.0879 \text{ m}^3 \text{ kg}^{-1}.$$

The flow rate $q = 49.3 \ 10^{-2} \ \text{m}^3 \ \text{min}^{-1}$

$$= 82.2 10^{-3} \ \text{m}^3 \ \text{s}^{-1}.$$

We also know that

$$q = K \ \beta^2 \ A_1 \ Y V_s \sqrt{\frac{\Delta P}{V_1}} \sqrt{2g},$$

where Y = (compressible flow rate)/(incompressible flow rate).

Now, $A_1 = \frac{\pi D^2}{4}$

so $K \ \beta^2 \ Y = 0.036$

and $Y = 1 - [0.41 + 0.35\beta^4](\frac{55 \times 13.6}{1.4 \times 9.6 \times 1000})$

$$= 1 - 0.0228 \ \text{(neglecting the } \beta^4 \text{ term)}$$

$= 0.977$

$$\text{or} \quad K \ \beta^2 = 0.0368. \tag{7.2.1}$$

The Reynolds number is given by

$$R_e = \frac{\rho v d}{\mu} = \frac{4\rho q}{\pi \mu \beta D} = \frac{7937}{\beta}.$$

We first choose a value for R_e of 10^4. Then, $\beta = 0.7937$. From the table in appendix I we find that $K > 0.7$, which does not satisfy equation (7.2.1). Hence, we choose $R_e = 3 \times 10^4$, where $\beta = 0.26$. From the table in appendix I, $K = 0.61$, which it is very close to the value given by equation (7.2.1). Therefore, we choose $\beta = 0.26$, and the diameter of the orifice is $\beta \times D$, which is equal to 0.13 m (the answer).

Problem 7.3. A Venturi meter is to be used to measure the flow rate of water in a pipe of diameter $D = 0.20$ m. The maximum flow rate is 2136 m^3 min^{-1}. Venturis with throat diameters of 0.10, 0.12, and 0.14 m are available.

    a) Choose the most suitable Venturi meter, assuming the differential pressure at maximum flow is 918 kg m^{-2}

    b) Calculate the accurate value of the differential pressure developed across the chosen Venturi at maximum flow rate.

C_d (the discharge coefficient) $= 0.990 - 0.02 \ (\frac{d}{D})^4$.

Solution:

    (a) The maximum flow rate is 2136 m^3 min^{-1} or 356 m^3 s^{-1}.

        The actual flow rate is given by

$$Q_{act} = Q_{act} = \frac{C_d \cdot A_2}{\sqrt{1 - \beta^2}} \sqrt{2 \frac{(P_1 - P_2)}{\rho}}.$$

        For this particular problem, $\rho = 1000$ kg m^{-3}, A_2 = the cross-sectional area of the Venturi throat, β is the ratio of the diameter of venture throat and that of the pipe, and P_1 and P_2 are the upstream and downstream pressure, respectively.

Here, $\rho = 1000 \text{ kg m}^{-3}$.

Case I $C_\text{d} = 0.988$ (for $d = 0.10$ m)

$$Q_1 = 340 \text{ m}^3 \text{ s}^{-1}.$$

Case II $C_\text{d} = 0.987$ (for $d = 0.12$ m)

$$Q_2 = 507.6 \text{ m}^3 \text{ s}^{-1}.$$

Case III $C_\text{d} = 0.985$ (for d $= 0.14$ m)

$$Q_3 = 737 \text{ m}^3 \text{ s}^{-1}.$$

Thus, the Venturi with the throat diameter of 0.10 m gives the most accurate reading.

(b) For this case,

$$3.56 \times 10^4 = \frac{0.988}{0.968} \times \pi \times 25 \sqrt{2 \, \Delta P g}$$

$$(\Delta P)_\text{max} = 1006.1 \text{ kg m}^{-2} \text{ (the answer).}$$

Problem 7.4. A Venturi tube is to be used to measure a maximum flow rate of water of 3.7859 kg s^{-1} or 0.0037859 m^3 s^{-1}. The Reynolds number of the flow at the throat is to be at least 10^5. Determine the size of the Venturi and the maximum range of the differential pressure.

The following data are given:
- $\rho_\text{water} = 1000 \text{ kg m}^{-3}$,
- $\mu = 0.0116 \text{ poise} = 0.0116 \times 0.1 = 0.001\ 16 \text{ Pa·s}$.

See appendix I for additional data.

Solution:
The Reynolds number should be $R_e \geqslant 10^5$. Again,

$R_e = \dfrac{\rho v d}{\mu} = \dfrac{4\rho Q}{\pi d \mu} = \dfrac{4m}{\pi d \mu}$

or $\dfrac{4m}{\pi d \mu} \geqslant 10^5$. So $d_\text{max} = \dfrac{4m}{\pi \times \mu \times 10^5}$

$= \dfrac{4 \times 3.7859}{\pi \times 0.00116 \times 10^5} = 4.15 \times 10^{-2}$ m.

From the table in appendix I, we see that $\beta = 0.5$ and that below 4.15 cm, d can have the value of 2.5 or 1.25 cm. We choose $d = 2.5$ cm and $D = 5$ cm. Thus, the Reynolds number will lie between 10^5 and 3×10^5. We find that the discharge coefficient C $= 0.976$, which is quite reasonable. For the value of $d = 2.5$, we find

$$R_e = \frac{4 \times 3.7859}{\pi \times 0.025 \times 0.001\ 16} = 1.662 \times 10^5.$$

Thus, the chosen Venturi diameter is 2.5 cm or 0.025 m (the answer).
Now,

$$Q = \frac{CA_2}{\sqrt{1 - \beta^4}} \sqrt{\frac{2g\Delta P}{\rho}}$$

$$= \frac{0.976 \times \pi \, (0.025)^2}{4\sqrt{1 - (0.5)^4}} \sqrt{\frac{2 \times 9.80 \times \Delta P}{1000}}.$$

Again, we know that the volume flow rate $Q_{\text{maximum}} = 0.003\,7859 \text{ m}^3 \text{ s}^{-1}$.
Therefore, $(\Delta P)_{\text{maximum}} = 2.92$ kPa (the answer).

Problem 7.5. A measurement of the velocity profile of a pipe is needed prior to the installation of a permanent flowmeter. The mean velocity of a high-pressure incompressible fluid through the pipe is measured using a Pitot tube. At the maximum flow rate, the mean differential pressure was found to be 310 Pa.
(a) Calculate the mean velocity of the gas at maximum flow rate.
(b) Calculate the maximum mass flow rate.
(c) Calculate the Reynolds number at the maximum flow rate.

The following data are given:
- The diameter of the pipe = 0.18 m.
- The density of the fluid = 6 kg m^{-3}.
- The viscosity of the gas = 5.0×10^{-3} Pa s^{-1}.

Solution:
Assuming energy conservation and no frictional or heat losses, the sums of the pressure, kinetic, and potential energies due to the impact with the static holes are equal. Since the kinetic energy at the impact hole is zero, we can write

$$\frac{p_1}{P} + 0 + gZ_1 = \frac{p_2}{P} + \frac{v^2}{2} + gZ_2,$$

where Z_1 and Z_2 are the elevations of the holes above the datum line. Here, $Z_1 = Z_2$. Hence, we can write

$$v = \sqrt{\frac{2(p_1 - p_2)}{\rho}}.$$

Therefore, the mean velocity is given by
$v = 10.16$ m s^{-1} (the answer).
The maximum mass flow rate is given by
$\dot{m} = (vA)\rho$, where $(vA) \rightarrow$ volume flow rate.
$\therefore \dot{m} = v \times \frac{\pi D^2}{4} \times g$, where $D = 0.18$ m.
$\dot{m} = 1.55$ kg s^{-1} (the answer).
The Reynolds number is given by
$R_e = \frac{\rho v D}{\mu}$, where $\mu =$ viscosity.

$$= \frac{6 \times 10.16 \times 0.18}{5 \times 10^{-5}}$$

$R_e = 21.94 \times 10^4$ (the answer).

Problem 7.6.   A turbine flowmeter consisting of a rotor with six blades is suspended in a fluid stream, and the rotational axis of the rotor is parallel to the direction of flow. The blades rotate at an angular velocity of ω rad s^{-1}, where $\omega = 4.2 \times 10^4$ Q.

If Q m^3 s^{-1} is the volume flow rate of the fluid, the total flux ϕ_T linked by the coil of the magnetic transducer is given by

$\phi_T = 4.637 + 0.92 \cos(6\theta)$ milliwebers.

Here, θ is the angle between the blade assembly and the transducer. The flowmeter can measure flow rates ranging between 0.22×10^{-3} m^3 s^{-1} and 3.74×10^{-3} m^3 s^{-1}.

Find the amplitude and frequency of the transducer output signals at the minimum and maximum flow rates.

Solution:

We can write

$$\phi_T(\theta) = \alpha + \beta \cos(n\theta)$$

where $\alpha =$ the mean flux, $\beta =$ the amplitude of the flux variation, and $n =$ the number of blades.

$$E = \frac{-d\phi_T}{dt} = \frac{-d\phi_T}{d\theta} \times \frac{d\theta}{dt}$$

$$\frac{d\phi_T}{d\theta} = -\beta n \sin(n\theta)$$

$$\frac{d\theta}{dt} = \omega(\text{angular velocity})$$

and $\theta = \omega t$ (assuming $\theta = 0$ at $t = 0$).

Thus,

$$e = \beta n \omega \sin(n\omega t).$$

Thus,

amplitude $= \beta \, n \, \omega$

frequency $= (n\omega)/2\pi$.

For the minimum flow rate,

$$\omega = 4.2 \times 10^4 \times 0.22 \times 10^{-3}$$
$$= 9.24 \text{ rad s}^{-1}.$$

$\therefore$ amplitude $= 0.92 \times 6 \times 9.24 = 51$ mV (the answer) frequency = $(6 \times 9.24)/2\pi = 8.82$ Hz (the answer).

For the maximum flow rate,

$$\omega = 4.2 \times 10^4 \times 3.74 \times 10^{-3}$$

$= 157.1$ rad s^{-1}.

∴ amplitude $= 867.2$ mV (the answer).

frequency $= 150$ Hz (the answer).

Problem 7.7. A Doppler shift ultrasonic flowmeter is shown below with its signal conditioning circuit. Two piezoelectric crystals, each of which has a natural frequency of 2.4 MHz, are used as the transmitter and receiver. The transmitting crystal directs an ultrasonic wave into the pipe which makes an angle of 45° with the flow. Calculate the received frequency as well as the cutoff frequency of the low-pass filter.

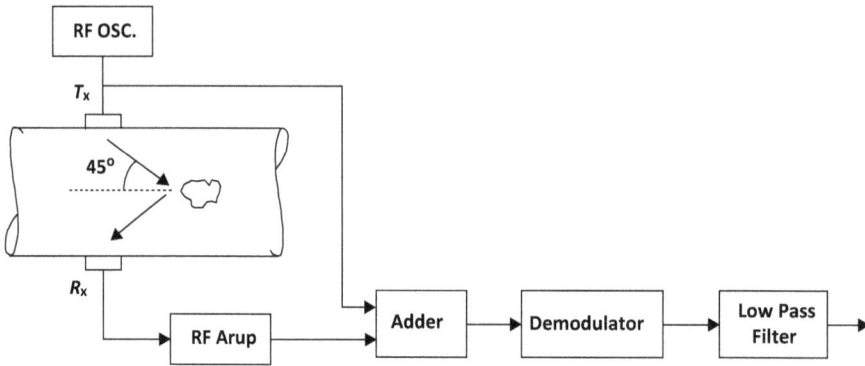

Figure 7.35.

The velocity of the flowing fluid $= 10$ m s^{-1} and the velocity of sound in the fluid $= 10^3$ m s^{-1}.

Solutions:

Here, we assume that the angle of incidence (θ_i) and angle of reflection (θ_r) are equal. Thus we can write

$$f_r = f_i \frac{\left(\dfrac{1 - v\cos\theta_i}{c}\right)}{\left(\dfrac{1 + v\cos\theta_r}{c}\right)}$$

$$= f_i \frac{\left(\dfrac{1 - v\cos\theta}{c}\right)}{\left(\dfrac{1 + v\cos\theta}{c}\right)} = f_i \left(\frac{1 - v\cos\theta}{c}\right)^2,$$

where

- f_i = the frequency of the incident beam,
- f_r = the frequency of the reflected beam,
- v = the velocity of the liquid in the pipe, and
- c = the velocity of the ultrasonic wave in the liquid when it is stationary

or $f_r \approx f_i(1 - \frac{2v \cos \theta}{c})$ ($\because v\cos\theta \langle\langle c)$

or $f_r = 2.366$ MHz (Answer).

The input signal $= v \sin 2 \pi f_i t$ and the reflected signal $= v \sin 2 \pi f_r t$.

$\therefore$ the output of the adder $= v (\sin 2 \pi f_i t + \sin 2 \pi f_r t)$

$$= 2v \cos\left[\frac{2\pi(f_r - f_i)}{2}t\right]\sin\left[\frac{2\pi(f_r + f_i)}{2}t\right]$$

$$= \left(2v \cos\left[2\pi\frac{\Delta f}{2}t\right]\right)\sin\left[2\pi\frac{(f_r + f_i)}{2}t\right].$$

The above represents an amplitude-modulated signal with a carrier frequency of $(\frac{f_r + f_i}{2})$ and a modulating frequency of $\Delta f/2$.

Let $\Delta f/2 = \omega_m$ and $\frac{f_r + f_i}{2} = \omega_c$. The signal is now passed through a demodulator. The input of the demodulator $= v[\sin(\omega_c + \omega_m)t + \sin(\omega_c - \omega_m)t]$.

The signal is demodulated with $K \sin \omega_c t$,

$\therefore$ the output of the demodulator

$$=Kv[\sin \omega_c t \; \sin(\omega_c + \omega_m)t + \sin \omega_c t \; \sin(\omega_c - \omega_m)t]$$

$$=\frac{Kv}{2}[\cos (2\omega_c + \omega_m)t - \cos \omega_m t + \cos(2\omega_c - \omega_m)t - \cos \omega_m t].$$

Thus, when it is passed through a low-pass filter, the output is

$$\frac{Kv}{2} \cos \omega_m t,$$

i.e. we only have a signal consisting of the $\Delta f/2$ term. Therefore, the cutoff frequency of the low-pass filter must be greater than $\Delta f/2$.

$$\Delta f = |(f_r - f_i)| = 33.93 \text{ kHz}$$

or $\Delta f/2 = 16.96$ kHz,

therefore the cutoff frequency of the low-pass filter $= 20$ KHz (say) (the answer).

Problem 7.8. A transit time ultrasonic flowmeter was used to measure velocity of a fluid flowing in a pipe, and it was found that the zero flow transit time was 1.2 ms; however, when there was flow, the differential transit time was 115 µs. The angle between the line connecting the transmitter to the receiver and the direction of flow of the fluid was 30°. Find the velocity of the fluid. By how much does the transit time change for a ±2% change in the velocity of sound? The velocity of sound in the fluid is 500 m s^{-1}.

Solution:

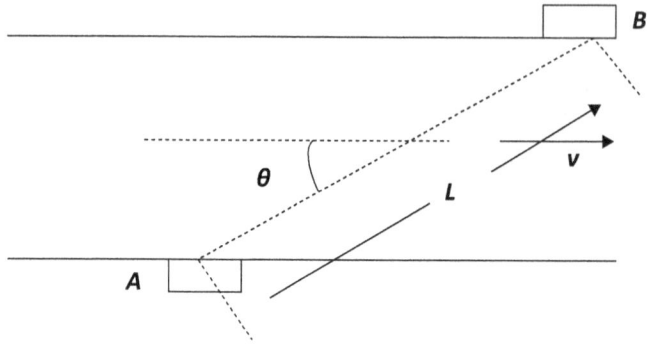

We see that $T_{BA} = \dfrac{L}{c - v \cos \theta}$.

$$T_{AB} = \dfrac{L}{c + v \cos \theta}$$

or $\Delta T = \dfrac{L 2 v \cos}{c^2}$ (as $V \cos \theta \ll c$)

$$= \dfrac{2V \cos \theta}{c} \cdot \dfrac{L}{c}.$$

Now L/c = the zero flow transit time = T_0 (say),

$$\therefore v = \left(\dfrac{\Delta T}{T_0} \right) \times \dfrac{c}{2} \times \dfrac{1}{\cos \theta}$$

where v = velocity of the gas;

$$= \dfrac{115 \times 10^{-6}}{1.2 \times 10^{-3}} \times \dfrac{500}{2} \times \dfrac{1}{\cos 30^\circ}$$

$$= 27.66 \text{ m s}^{-1} \text{ (the answer)}.$$

We know that

$$\Delta T_1 = \dfrac{2 v \cos \theta \cdot L}{c_1^2}$$

Now, $\dfrac{L}{c} = 1.2$ ms

$$\therefore L = 1.2 \times 10^{-3} \times 500$$

$$= 0.6 \text{ m}.$$

Also, $c_1 = 1.02 \times c$

or $\Delta T_1 = \dfrac{2 v \cos \theta \cdot L}{(1.02)^2 c^2}$

$$= \frac{1}{(1.02)^2} \times \Delta T$$

$$= \frac{115}{(1.02)^2} = 110.5 \ \mu\text{s}.$$

For $c_2 = 0.98 \ c$, we have $\Delta T_2 = \frac{1}{(0.98)^2} \times \Delta T = 119.7 \ \mu\text{s}.$

Therefore, the transit time variation $= \mp 3.9\%$ (the answer).

Problem 7.9. In an experiment, an electromagnetic flowmeter was used to measure the average flow rate of a liquid flowing in a cylindrical pipe. The flux density of the electric field applied had a peak value of 1.4 Tesla (Wb m^{-2}). The output from the electrodes was fed to an amplifier with a gain of ten and an input impedance of 2.2 MΩ. The internal resistance developed due to the fluid between the electrodes was found to be 200 kΩ.

(a) Determine the velocity of the liquid when the peak-to-peak output voltage of the amplifier was found to be 4 V.
(b) Find the percentage change in the reading of the amplifier for a 10% increase in the conductivity of the flowing fluid.

The diameter of pipe $= 0.1$ m.
Solution:
(a) The peak voltage at the output of the amplifier is
$E = 2$ V.
We know

$$E = e - i \times R_{\text{int}} = i \times R_{\text{amp}}$$

or $e = E(1 + \frac{R_{\text{int}}}{R_{\text{amp}}})$,
where $e = $ the voltage induced due to the magnetic field. Also, the open-circuit voltage

$$e = 2\left(1 + \frac{200 \times 10^3}{2.2 \times 10^6}\right) = 2.18 \text{ V}.$$

We also know that

$$e = B\ell v \times \text{gain},$$

where
- $B = $ the magnetic flux
- $\ell = $ the length of the inductor, i.e. the diameter of the pipe, and
- $v = $ the velocity of the liquid;

or $v = \frac{e}{B\ell \times \text{gain}}$

$$= \frac{2.18}{1.4 \times 0.1 \times 10} = 1.557 \text{ m s}^{-1} \text{ (the answer)}.$$

(b) The conductivity increases by 10%, i.e. the resistance decreases by 9.1%, or $R' = 0.91 \times 200 \text{ K}\Omega = 182 K\Omega$

$\therefore$ the output voltage $= \dfrac{2 \times 2.18}{\left[1 + \frac{182 \times 10^3}{2.2 \times 10^6} \right]}$

$$= \frac{4.36}{1.08} = 4.026 \text{ V}$$

therefore, the percentage change in the output voltage $= \dfrac{4.026 - 4}{4} \times 100$

$$= 0.67\% \text{ (the answer)}.$$

Problem 7.10. The operational principle of the vortex flowmeter is based on the natural phenomenon of vortex shedding. If a circular cylinder of diameter d meters is installed as a bluff body in a pipe of diameter D meters, then the frequency f (in Hz) of vortex shedding is given by

$$\frac{f}{Q} = \frac{4S}{\pi D^3} \frac{1}{\frac{d}{D}\left[1 - 1.4\frac{d}{D} \right]}$$

where Q = the volume flow rate of the fluid (m^3 s^{-1}) and S = the Strouhal number (0.2 in this case).

The Strouhal number is a dimensionless quantity and it is relatively constant (0.20 $\leqslant S \leqslant$ 0.21) over the range of Reynolds numbers from 300 to 150 000. Calculate the correct cylinder diameter for a 0.15 m pipe carrying water flows of between 0.1 and 1.32 m^3 s^{-1}. Find the maximum vortex shedding frequency.

Solution:

It is given that

$$f/Q = \frac{4S}{\pi D^3} \frac{1}{\frac{d}{D}\left(1 - 1.4\frac{d}{D} \right)}.$$

We know that in order to obtain the most regular and highest-amplitude shedding, we need to choose a d/D ratio that minimizes the $\left(\frac{f}{Q}\right)$ ratio.

Here, D is fixed. Hence, we differentiate the $\left(\frac{f}{Q}\right)$ ratio with respect to d. Thus, we have

$$\frac{d(f/Q)}{d(d)} = \frac{4S}{\pi D^3} \times \frac{-(D/d^2 - 2.8/d)}{[1 - 1.4\, d/D]^2}.$$

Equating the above expression to zero, we get

$$D/d^2 = 2.8/d$$

or $(d/D) = 1/2.8 = 0.357$.

Thus, $d = 0.357 \times 0.15 = 0.054$ m (the answer).

The maximum shedding frequency is obtained at the maximum flow rate.

Thus,

$$f = Q \times \frac{4S}{\pi D^3 \frac{d}{D}\left[1 - 1.4\frac{d}{D}\right]}$$

$$= 1.32 \times \frac{4 \times 0.2}{\pi \times (0.15)^3} \frac{1}{0.357[1 - 1.4 \times 0.357]}$$

$$= 557.96 \text{ Hz (the answer)}.$$

Problem 7.11. In problem 7.10, $L = 0.15$ m and $V = 0.5$ m s^{-1}. The fluctuation in frequency is up to 110 Hz. State whether the flowmeter shown in figure 7.36 is suitable for this application.

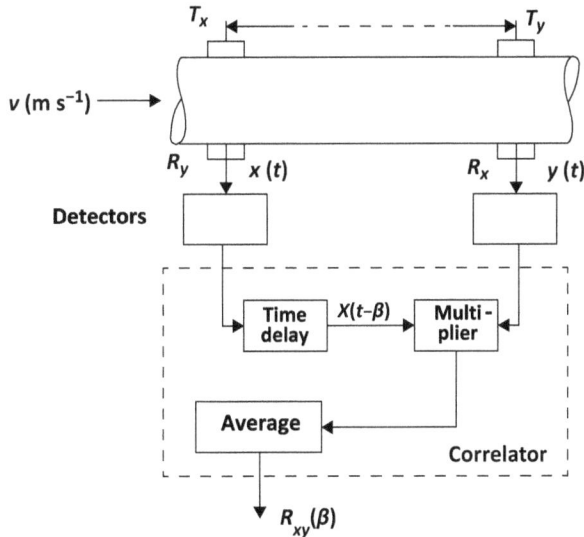

Figure 7.36.

Solution:

The output of the averaging unit is given by

$$R_{xy}(\beta) = K\frac{\sin \omega_c(\beta - \tau)}{(\beta - \tau)},$$

where $\tau = L/V$.

We see that the cross-corelation function is a sine function with a width of $(2\pi/\omega_c)$. Thus, for accurate measurements, we need to have

$$\tau \rangle\rangle 2\pi/\omega_c \quad \text{i.e.} \quad \tau \rangle\rangle 1/f_c.$$

Here, $\tau = 0.15/0.5 = 0.3$ and $1/f_c = \frac{1}{100} = 0.01$.

We find that $\tau \rangle\rangle 1/f_c$.

Therefore, the above flowmeter is suitable for the application.

Problem 7.12. A schematic of a mass flowmeter or Coriolis flowmeter is shown in figure 7.37. Assume we have a U-tube ABCD of length 1.2 m rotating with an angular velocity of 4π rad s^{-1}. Water flows in at a velocity of 1.5 m s^{-1}. Calculate the force on each of the limbs AB and CD. The diameter of the U-tube is 0.15 m.

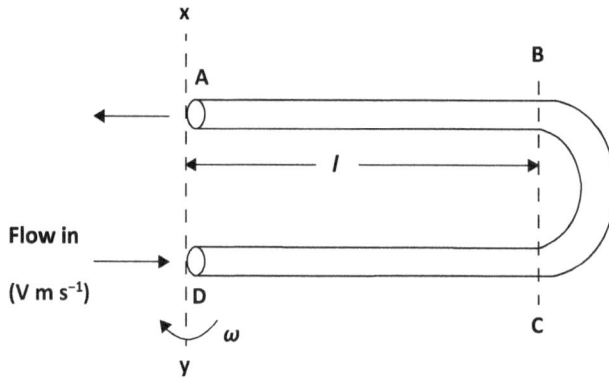

Figure 7.37.

Show that the torque developed is directly proportional to the mass flow rate.

Solution:

If we consider a small part of the limb AB as Δx, then

$$\Delta m = \rho A \Delta x.$$

This element experiences a Coriolis force

$$\Delta F = 2 \, \Delta m \omega v = 2\rho A \omega v \Delta x$$

or $F = 2\rho A \omega v \int_0^l dx = 2\rho A \omega v l.$

Now

$$A = \frac{\pi D^2}{4} = 0.0176 \text{ m}^2$$

or $F = 2 \times 1000 \times 0.0176 \times 4 \times \pi \times 1.5 \times 1.2$

$$= 799.43 \text{ kgm}^{-1}.$$

We see that both the limbs, **AB** and **CD**, experience a force which is perpendicular to the direction of the fluid velocity as well as that of the angular velocity. This results in a torque.

The torque $T = F. \, 2r$.

$$T = 4 \, \rho A \omega v l r$$

$$= 4 \omega l r (f A v)$$

$$= 4 \omega l r \dot{m} \text{ (where, } \dot{m} = \text{mass flow rate)}$$

or $T \alpha \dot{m}$.

The above derivation explains the working principle of the Coriolis flowmeter.

Problem 7.13. An electromagnetic flowmeter is used to measure the average flow rate of a liquid in a pipe of diameter 50 mm. The flux density in the liquid has a peak value of 0.1 T. The output from the flowmeter is fed to an amplifier that has a gain of 1000 and an input impedance of 1 MΩ. If the impedance of the liquid between the electrodes is 200 kΩ, then what is the peak value of the amplifier output voltage for an average flow velocity of 20 mm s^{-1}?

Solution: the output voltage from the two electrodes of the flowmeter (e_o) = Blv, where

- B = the flux density generated by the electromagnet,
- l = the length in between the electrodes = the diameter of the pipe, and
- v = the velocity of the liquid in the pipe;

$\therefore e_o = 0.1 \times 5 \times 10^{-2} \times 2 \times 10^{-2}$
$= 10^{-4}$ V.

Therefore, the output voltage from the amplifier

$V_0 = 1000 \times \left(\frac{1}{1.2}\right) \times 10^{-4}$

$= 0.08333$ V (the answer).

Problem 7.14. A tungsten constant current based hot-wire anemometer has the following specifications:

- its resistance at 0 °C is 10 Ω,
- the surface area of the wire is 10^{-4} m^2,
- the linear temperature coeifficient of resistance of the wire is 4.8×10^{-3} °C^{-1},

- the convective heat transfer coefficient is 25.2 W m^{-2} °C^{-1},
- the flowing air temperature is 30 °C,
- the hot-wire current is 100 mA, and
- the mass specific heat product is 2.5 × 10^{-5} J °C^{-1}.

(a) What is the resistance of the wire under steady-state flow conditions?
(b) Calculate the value of thermal time constant of the hot wire under steady-state flow conditions.

Solution:

(a) For a constant current hot-wire anemometer under steady-state conditions,

$$i_0^2 RT = UA(T - TF)$$

$10^{-2} \times 10(1 + 4.8 \times 10^{-3} T) = 25.2 \times 10^{-4}(T - 30).$

Therefore, $T = 86$ °C and $R = 10 (1 + 4.8 \times 10^{-3} \times 86) = 14.128 \ \Omega$ (the answer).

(b) To find the time constant, let us consider the hot-wire anemometer to be a first-order system:

$i_0^2 R_0(1 + \alpha T) - UA(T - TF) = MC\frac{dT}{dt},$

or, $T(i_0^2 R_0 \alpha - UA) + i_0^2 R0 + UAT = MC\frac{dT}{dt},$

$MC\frac{dT}{dt} + T(UA - i_0^2 R_0 \alpha) + i_0^2 R_0 + UATF = 0$

∴ the time constant $(\tau) = \dfrac{MC}{UA - i_0^2 R_0 \alpha}$

$$= \frac{2.5 \times 10^{-5}}{25.2 \times 10^{-4} - 10^{-2} \times 10 \times 4.8 \times 10^{-3}}$$

$= 0.01225$ s $= 12.25$ ms (the answer).

Chapter 9

Problem 9.1. An all-pass filter has the transfer function

$$G(s) = \frac{s^2 - 3s + 9}{s^2 + 3s + 9}.$$

Calculate the group delay (in seconds) of the filter at the frequency $\omega = 5$ rad s^{-1}.

Solution:

The delay (in seconds) of a transfer function is given by

delay $(D) = \frac{d}{d\omega} (-\phi(\omega))$,

where $\phi(\omega)$ is the phase of the transfer function.

The generalized second-order filter transfer function is $\frac{ms^2 + cs + d}{ns^2 + as + b}$.

$$\text{Delay}(D) = \frac{d}{d\omega} (-\phi(\omega)) = -\frac{c(d + m\omega^2)}{\left(d - m\omega^2\right)^2 + c^2\omega^2} + \frac{a(b + n\omega^2)}{\left(b - n\omega^2\right)^2 + a^2\omega^2}.$$

Therefore, the delay, $D = \dfrac{3(9 + \omega^2)}{\left(9 - \omega^2\right)^2 + 9\omega^2} + \dfrac{3(9 + \omega^2)}{\left(9 - \omega^2\right)^2 + 9\omega^2} = \dfrac{6(9 + \omega^2)}{\left(9 - \omega^2\right)^2 + 9\omega^2}$

$= \dfrac{6(9 + 25)}{(9 - 25)^2 + 9 \times 25} = 0.424$ s (the answer).

Problem 9.2. Find the transfer function of a second-order bandpass filter that has a centre frequency of 1000 rad s^{-1}, a selectivity of 100, and a gain of 0 dB at the centre frequency.

Solution:

The transfer function of a generalized second-order bandpass filter is

$$T(s) = \frac{Ks}{s^2 + \left(\dfrac{\omega_p}{Q_p}\right)s + \omega_p^2},$$

where ω_p Q_p and K are the central frequency, pole selectivity, and gain constant of the bandpass filter, respectively.

$\omega_p^2 = 106, \dfrac{\omega_p}{Q_p} = \dfrac{10^3}{100} = 10$

The gain at the centre frequency $\dfrac{K}{\left(\dfrac{\omega_p}{Q_p}\right)} = \dfrac{K}{10} = 0 \, dB = 1$,

$\therefore K = 10$ and the transfer function is $\dfrac{10s}{s^2 + 10s + 10^6}$ (the answer).

Problem 9.3. Synthesize the following bandpass transfer function using a Sallen and Key filter design with equal R and equal C.

$$T(s) = \frac{400s}{s^2 + 4s + 500}$$

Solution:

Let us redraw the Sallen and Key bandpass filter as shown in figure 9.6 of subsection 9.2.7.

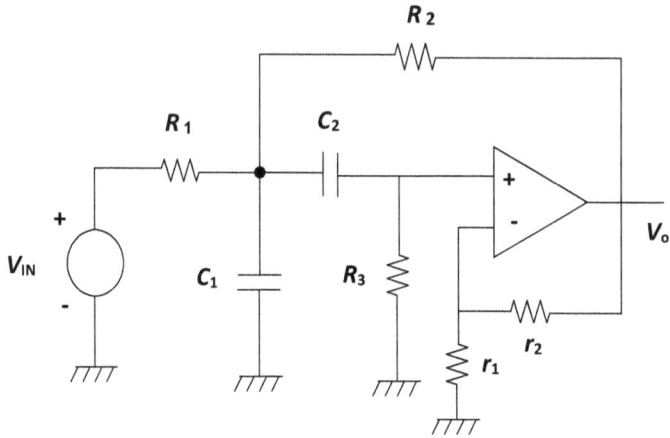

A generalized bandpass filter can be expressed as

$$\frac{Ks}{s^2 + \frac{\omega_p}{Q_p}s + \omega_p^2}.$$

Here, the pole frequency or the central frequency of the bandpass filter is $\omega_p = \sqrt{500} = 22.36$. Therefore, the pole selectivity $(Q_p) = 5.59$.

The synthesis equations for the equal R and equal C design of a single-amplifier bandpass filter circuit are as follows:

$$R1 = R2 = R3 = \frac{\sqrt{2}}{\omega_p} = 0.0632. \text{ We take } C1 = C2 = 1 \text{ F}.$$

$k = 1 + \frac{r_2}{r_1} = 4 - \frac{\sqrt{2}}{Q_p} = 4 - \frac{\sqrt{2}}{5.59} = 3.747.$ So $\frac{r_2}{r_1} = 2.747$, take $r_1 = 1$ kΩ, so $r_2 = 2.747$ kΩ.

Now, $K = \frac{k}{R_1 C_1} = \frac{3.747}{0.0632 \times 1} = 59.287.$

Gain enhancement of $\frac{400}{59.287} = 6.75$ (α) is needed. R_2 is to be replaced by $R_2' = \alpha R_2 = 0.4266$ and

$R_2'' = R_2 \frac{\alpha}{\alpha - 1} = 0.0742;$

r_2 is to be replaced by $r_2' = \alpha r_2 = 18.54$ and

$r_2'' = r_2 \frac{\alpha}{\alpha - 1} = 3.224$ Both the resistance values are in kΩ.

To find practical values for the other components, we divide the capacitor values by some scale factor and multiply the resistances by the same scale factor. The designed circuit is as follows:

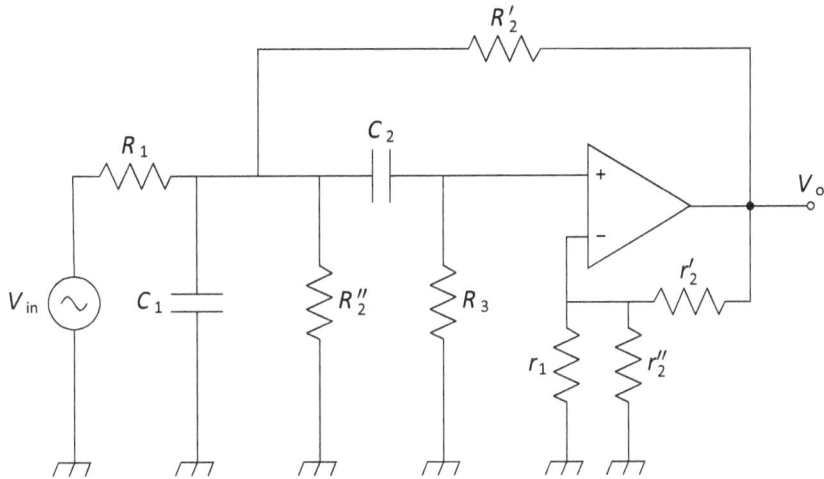

Problem 9.4. Synthesize a second-order low-pass filter that has a pole frequency (ω_p) of 2 kHz and a pole selectivity (Q_p) of 10, using the Saraga version of a Sallen and Key filter. The filter circuit of figure 9.23 is redrawn below.

The equations for the Saraga design are:

$$C_2 = 1, \quad C_1 = \sqrt{3}\, Q_p$$

$$R_1 = \frac{1}{Q_p \omega_p}, \quad R_2 = \frac{1}{\sqrt{3}\, \omega_p}$$

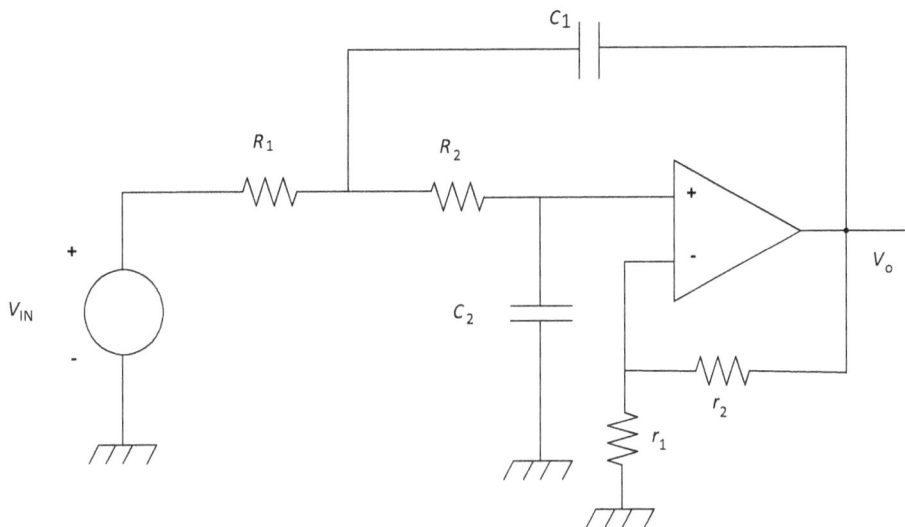

Figure 9.23.

$$k = \frac{4}{3} = \left(1 + \frac{r_2}{r_1}\right).$$

Solution:

$\omega_p = 2\pi(2000)$ rad s^{-1}, $Q_p = 10$, and the desired low-pass function is

$$\frac{K}{s^2 + 2\pi(200s) + (2\pi 2000)^2}.$$

It is given that $C2 = 1$, $C1 = 10\sqrt{3}$, $R_1 = \frac{1}{2\pi(2000 \times 10)}$, $R_2 = \frac{1}{\sqrt{3}}\frac{1}{(2\pi)(2000)}$, and
$k = \frac{4}{3} = (1 + \frac{r_2}{r_1})$, so $r_2 = 1$ kΩ, and $r_1 = 3$ kΩ.

For component scaling, if we take $C_2 = 0.001$ μF, then the other component values are $C_1 = 0.017$ μF, $R_2 = 46$ kΩ, and $R_1 = 7.96$ kΩ (the answer).

Chapter 10

Problem 10.1. A piezoelectric sensor that has a sensitivity of 2.0 pC N^{-1}, a capacitance of 1600 pF, and a leakage resistance of 10^{12} Ω is connected to a charge amplifier as shown in figure 10.18. If a force of 0.1 sin 10t N is applied to the sensor, what is the amplitude of the charge amplifier output?

Solution:
Here, $R_f = 10^8$ Ω, $Cf = 10^{-9}$ F, $R = 10^{12}$ Ω, and $C = 1600$ pF
R and C are the leakage resistance and capacitance of the crystal, respectively.

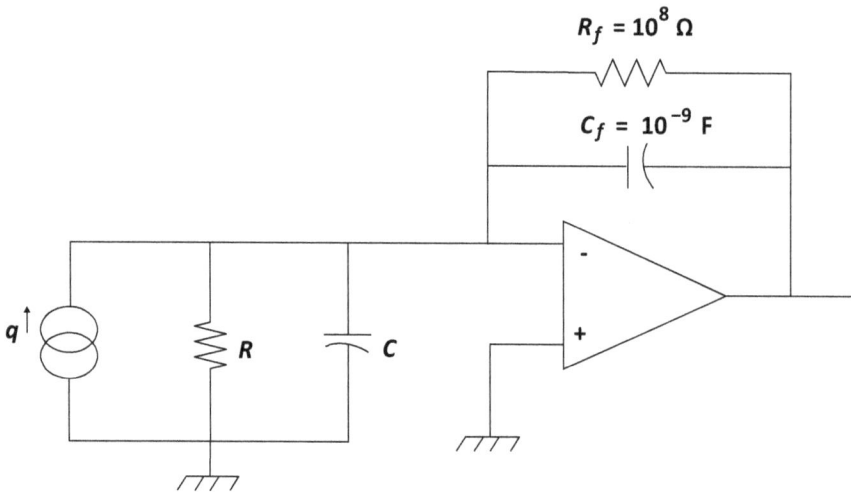

Figure 10.18. Measurement using a charge amplifier.

The inverting input of the op-amp is at virtual ground, and there is almost no flow of charge through the crystal due to its extremely low capacitance and high leakage resistance. Therefore, applying KCL at the virtual ground of the op-amp,

$$\frac{V_0}{R_f} + C_f \frac{dV_0}{dt} = - \frac{dq}{dt},$$

where V_0 is the output voltage of the charge amplifier.

Taking the Laplace transform, we get

$$V_0(s)(1 + R_f C_f s) = -R_f \times s \times q(s)$$

or $\frac{V_0}{q}(s) = \frac{-R_f \ s}{1 + \tau_f \ s}$,

where $\tau_f = R_f C_f = 10^8 \times 10^{-9} = 0.1$ s.

$$\left| \frac{V_0}{q}(j\omega) \right|_{\omega=10} = \left[\frac{R_f \omega}{\sqrt{1 + \omega^2 \tau_f^2}} \right]_{\omega=10}$$

By putting the values of ω, R_f, and τ_f into this equation, we get the output voltage amplitude as follows:

$$\frac{10^8 \times 10}{\sqrt{1 + (10 \times 0.1)^2}} \times 0.2 \times 10^{-12} = 0.141 \text{ mV (the answer).}$$

Problem 10.2. A displacement signal of $10^{-8} \sin 100t$ meter is applied to a quartz crystal of dimensions 10 mm × 10 mm × 2 mm. Find the voltage generated. The crystal data are as follows:

- charge sensitivity $(d) = 2$ pC N^{-1},
- Young's modulus $(E) = 8.6 \times 10^{10}$ N m^{-2}, and
- permittivity $(\varepsilon) = 42 \times 10^{-12}$ F m^{-1}.

Solution:

The capacitance of the crystal

$$C_{cry} = \frac{\varepsilon A}{t} = \frac{42 \times 10^{-12} \times 10 \times 10 \times 10^{-6}}{2 \times 10^{-3}} = 2.1 \times 10^{-12} \text{ F}$$

The charge generated, $q = d \times F = d \times E \times A \times \frac{x_i}{t}$, where, x_i is the input displacement in meters. A and t are the area and the thickness of the crystal, respectively. F is the force applied to the crystal. The charge sensitivity is also known as the d constant of the crystal.

Therefore, $q = \frac{2 \times 10^{-12} \times 8.6 \times 10^{10} \times 10 \times 10 \times 10^{-6}}{2 \times 10^{-3}} \ x_i$

$$= 0.0086 x_i.$$

So the output voltage generated for $x_i = 10^{-8} \sin 100t$

$$e_0 = \frac{q}{C_{cry}} = 40.95 \sin 100t \text{ V (the answer).}$$

Problem 10.3. A quartz clock employs a ceramic crystal with a nominal resonant frequency of 32.768 kHz. The clock loses 30.32 s every 23 days. What is the actual resonant frequency of the crystal?

Solution: the number of seconds in 23 days = $23 \times 24 \times 60 \times 60 = N_{23}$. The number of actual seconds recorded by the clock = N_{23}–30.32.

Therefore, $f_{\text{crystal}} = \frac{N_{23} - 30.32}{N_{23}} \times 32.768$

= 32.7675 kHz (the answer).

Chapter 11

Problem 11.1. An ultrasonic sensor is used to find cracks or gaps inside a metal. Once transmitted, the ultrasound wave is reflected back to the transmitter/receiver if there is a crack or gap in the metal. An ultrasonic transmitter uses a frequency of 330 kHz. When testing a specimen, a reflected wave is recorded 0.05 ms after the transmitted pulse. If the velocity of sound in the test object is 6.0 km s^{-1}, at what depth is the crack located?

Solution:

We know that $u.t = 2\,d$, where u is the velocity of ultrasound in the metal, t is the time of flight, and d is the depth at which the crack is located.

$\therefore d = \frac{u.t}{2} = \frac{6 \times 10^3 \times 0.05 \times 10^{-3}}{2} = 15$ cm (the answer).

Problem 11.2. An ultrasonic beam that has a frequency of 1 MHz and an intensity of 0.5 W cm^{-2} passes through a layer of soft tissue of thickness t with an attenuation coefficient of 1.18 cm^{-1}. The ratio of the output power to the input power is $\frac{1}{e^2}$. What is the thickness of the tissue?

Solution:

We know $I = I_0\, e^{-\alpha t}$. Here, $\alpha = 1.18$ cm^{-1}. Also, $\frac{I_1}{I_0} = \frac{1}{e^2}$. Therefore, $\alpha\, t = 2$, so $t = 1.695$ cm (the answer).

Chapter 12

Problem 12.1. A search coil has ten turns and a cross-sectional area of 10 cm^2. It rotates at a constant speed of 100 r.p.m. The output voltage is 80 mv. Calculate the magnetic field strength.

Solution:

The magnetic flux density is related to the rms voltage produced in a magneto-meter search coil as follows:

$$B = \frac{\sqrt{2}\, E_{\text{r.m.s.}}}{NA\omega} = \frac{\sqrt{2} \times 0.08}{10 \times 10 \times 10^{-4} \times 100 \times 2\pi/60}$$

$$= 1.08 \text{ Wb m}^{-2}.$$

Therefore, the magnetic field strength

$$H = \frac{B}{\mu} = \frac{1.08}{4\pi \times 10^{-7}} = 8.6 \times 10^5 \text{ A m}^{-1} \text{ (the answer)}.$$

Problem 12.2. A magnetometer search coil has a nominal area of 1 cm^2 and 100 turns. The rotational speed is nominally 180 r.p.m. Calculate the voltage output when the coil is placed in a magnetic field of 1 Wb m^{-2}.

Solution:

We know the r.m.s value of the voltage in a magnetometer search coil is given by

$$E_{\text{r.m.s.}} = \frac{1}{\sqrt{2}} NAB\omega$$

$$= \frac{1 \times 100 \times 1 \times 10^{-4} \times 180 \times 2\pi/60}{\sqrt{2}}$$

$$= 1.27278 \text{ V (the answer)}.$$

Problem 12.3. A germanium crystal that has dimensions of 6 × 6 mm^2 and a thickness of 3 mm is used for the measurement of flux density using a Hall-effect transducer. When the Hall field and the Lorentz force balance each other, it is observed that the current density in the crystal is 0.3 A mm^{-2} and the Hall voltage developed is −0.35 V. Find the value of the flux density and that of the electron velocity. The following data is given: the Hall coefficient for the germanium crystal = − 8 × 10^{-3} V m A^{-1} Wb m^{-2}.

Solution:

$$E_H = \frac{K_H BI}{t}$$

$$\therefore B = \frac{E_H \times t}{K_H \times I}$$

$$= \frac{- 0.35 \times 3 \times 10^{-3}}{- 8 \times 10^{-3} \times 10.8}$$

$$B = 0.012 \text{ Wb m}^{-2} \text{ (the answer)}.$$

When the Hall field = the Lorentz force,

$$BeV = eE_H/b$$

$$\therefore V = \frac{E_H}{Bb}$$

$$= \frac{0.35}{0.012 \times 3 \times 10^{-3}} = 9722 \text{ m s}^{-1} \text{ (the answer)}.$$

Problem 12.4. The distributed (self) capacitance of a low-loss coil is found to be 820 pF when measured using a Q meter. Resonance of the coil occurred at an angular frequency of 10^6 rad s^{-1} and a capacitance of 9.18 nF. What is the inductance of the coil?

Solution:

Angular frequency $\omega = \dfrac{1}{\sqrt{L(C_0 + C_1)}}$,

where C_0 is the distributed capacitance of the coil and C_1 is the tuning capacitance of the Q meter.

$$\text{Then } L = \frac{1}{\omega^2(C_0 + C_1)}$$

$$= \frac{1}{10^{12}(9.18 + 0.82) \times 10^{-9}} = 100 \text{ µH (the answer)}.$$

Chapter 14

Problem 14.1. The following design is used to measure pH: a pH electrode is connected through a shielded cable to a non-inverting amplifier as shown in figure 14.10. The input resistance of the non-inverting amplifier is given by

$$R = R_i\left(1 + A_0\frac{R_F}{R_1}\right),$$

where $A_0 =$ the open loop gain and $R_i =$ the input resistance of the op-amp.

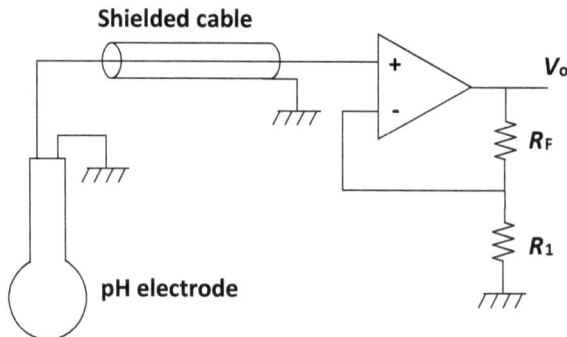

Figure 14.10.

Find the output voltage V_o of the circuit when a 225 mV signal is generated at the electrode.

The following data are given:

- The resistance of the electrode = 10^8 Ω.
- The resistance (leakage) of the shielded cable = 2×10^8 Ω.
- $A_0 = 10^5$.
- $R_i = 10^6$ Ω.
- $R_F = 2$ KΩ.
- $R_1 = 1$ KΩ.

Solution:

The op-amp given in the problem can be represented as shown below:

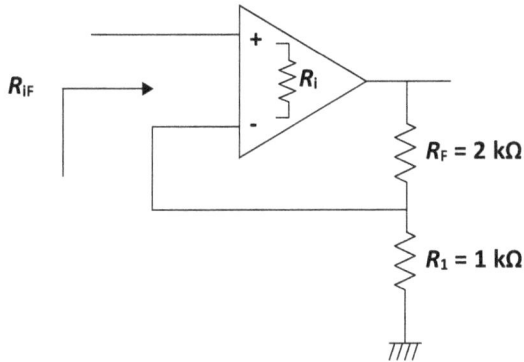

or $R_{iF} = R_i(1 + A_0\frac{R_F}{R_1})$

or $R_{iF} = 10^6(1 + 10^5\frac{2}{1})$

$= 2 \times 10^{11}\Omega$.

The equivalent circuit can be represented as shown below:-

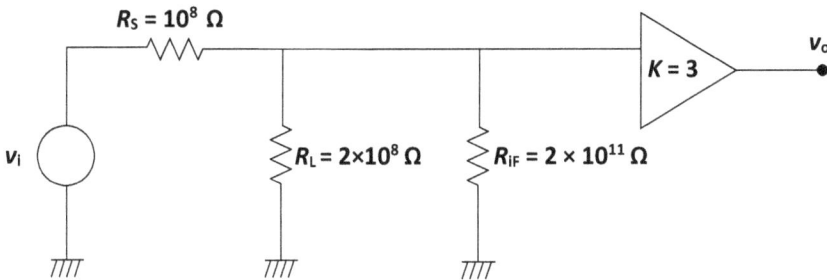

where

- R_s = the resistance of the probe,
- R_L = the leakage resistance, and
- the gain (closed loop) $K = (1 + \frac{R_F}{R_1}) = 3$.

Analyzing the above circuit, we can write

$$V_o = KV_i \frac{(R_L \| R_{iF})}{R_s + (R_L \| R_{iF})}$$

$$\approx 3 \times 225 \times \frac{R_L}{R_s + R_L} (\therefore R_{iF} = 2 \times 10^{11} >> R1 = 2 \times 10^8).$$

$$= 3 \times 225 \times \frac{2 \times 10^8}{10^8 + 2 \times 10^8}$$

$$= \frac{3 \times 225 \times 2}{3} = 450 \text{ mV (the answer).}$$

Problem 14.2. A rotating viscosity meter is shown in figure 14.11. It consists of two concentric cylinders of radii R_1 and R_2 with viscous liquid in between. When the outer cylinder is rotated by a motor at an angular velocity of ω, the suspension turns through an angle θ. Show that the viscosity of the liquid can be expressed as

$$\mu = \frac{K\theta b}{\omega L R_2 2\pi R_1^2},$$

where K is the torsional constant. Neglect the viscous effect of the liquid at the bottom of the two cylinders.

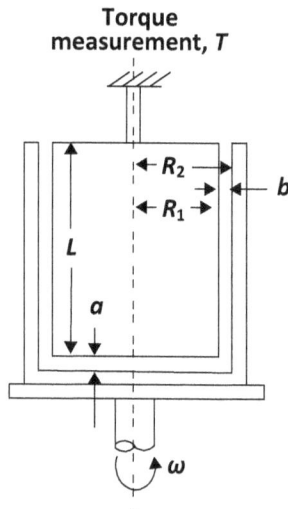

Figure 14.11. A rotating viscosity meter.

Solution:

$$\text{Viscosity } \mu = \frac{\text{Shear stress}}{\text{Velocity gradient}}.$$

The velocity gradient of the liquid on the vertical wall $= \frac{\omega R_2}{b}$. Therefore, the shear stress, $\tau = \mu \frac{\omega R_2}{b}$.

Deflecting torque = surface area × shear stress × radial distance

$$= (2 \pi R_1 L)(\tau)R_1 = 2 \pi L \tau R_1^2$$

$$= \frac{2\pi\omega R_2 \, \mu L R_1^2}{b}.$$

The restoring torque $= K\theta$

$$\therefore \; K\theta = \frac{2\pi\omega R_2 \, \mu L R_1^2}{b}$$

$$\therefore \; \text{Viscosity } (\mu) = \frac{K\theta b}{2\pi\omega R_2 L R_1^2} \quad \text{(Proved)}$$

Appendix I

Tables for the orifice meter and the Venturi meter

Table A.1. Table for the flow coefficient of the orifice meter pipe diameter 10 cm or 0.1 m, $\beta = 0.5$.

$R_e \longrightarrow$	10^4	10^5
$K \downarrow$	0.6631	0.6271

Table A.2. Table of the discharge coefficients (C_d) of the Venturi meter

Pipe × Throat (Diameter)	Reynolds No.		
	10^4	10^5	10^6
(2.54 cm × 1.27 cm)	0.946	0.972	0.9725
(5.01 cm × 2.54 cm)	0.948	0.975	0.977

Table A.3. Table of the flow coefficients of an orifice meter for different β values (Pipe diameter = 0.05 m)

β	Reynolds No.	
	10^4	10^5
0.1	0.613	0.6046
0.2	0.610	0.5987
0.3	0.615	0.6018
0.4	0.6278	0.6104
0.5	0.6521	0.6275
0.6	0.6945	0.6558
0.7	0.7630	0.7025

Flow co-efficient $K = \dfrac{Cd}{\sqrt{1 - \beta^4}}$

Fundamentals of Industrial Instrumentation (Second Edition)

Alok Barua

Appendix II

Thermocouple tables

Type E:	Chromel–constantan
Type J:	Iron–constantan
Type K:	Chromel–Alumel
Type N:	Nicrosil–nisil
Type S:	Platinum/10% rhodium–platinum
Type T:	Copper–constantan

Temperature (°C)	Type E	Type J	Type K	Type N	Type S	Type T
−270	−9.834		−6.458	−4.345		
−260	−9.795		−6.441	−4.336		
−250	−9.719		−6.404	−4.313		
−240	−9.604		−6.344	−4.277		−6.105
−230	−9.456		−6.262	−4.227		−6.003
−220	−9.274		−6.158	−4.162		−5.891
−210	−9.063	−8.096	−6.035	−4.083		−5.753
−200	−8.824	−7.890	−5.891	−3.990		−5.603
−190	−8.561	−7.659	−5.730	−3.884		−5.438
−180	−8.273	−7.402	−5.550	−3.766		−5.261
−170	−7.963	−7.122	−5.354	−3.634		−5.070
−160	−7.631	−6.821	−5.141	−3.491		−4.865
−150	−7.279	−6.499	−4.912	−3.336		−4.648
−140	−6.907	−6.159	−4.669	−3.170		−4.419
−130	−6.516	−5.801	−4.410	−2.994		−4.177

(*Continued*)

doi:10.1088/978-0-7503-3755-7ch24

(*Continued*)

Temperature (°C)	Type E	Type J	Type K	Type N	Type S	Type T
−120	−6.107	−5.426	−4.138	−2.807		−3.923
−110	−5.680	−5.036	−3.852	−2.612		−3.656
−100	−5.237	−4.632	−3.553	−2.407		−3.378
−90	−4.777	−4.215	−3.242	−2.193		−3.089
−80	−4.301	−3.785	−2.920	−1.972		−2.788
−70	−3.811	−3.344	−2.586	−1.744		−2.475
−60	−3.306	−2.892	−2.243	−1.509		−2.152
−50	−2.787	−2.431	−1.889	−1.268	−0.236	−1.819
−40	−2.254	−1.960	−1.527	−1.023	−0.194	−1.475
−30	−1.709	−1.481	−1.156	−0.772	−0.150	−1.121
−20	−1.151	−0.995	−0.777	−0.518	−0.103	−0.757
−10	−0.581	−0.501	−0.392	−0.260	−0.053	−0.383
0	0.000	0.000	0.000	0.000	0.000	0.000
10	0.591	0.507	0.397	0.261	0.055	0.391
20	1.192	1.019	0.798	0.525	0.113	0.789
30	1.801	1.536	1.203	0.793	0.173	1.196
40	2.419	2.058	1.611	1.064	0.235	1.611
50	3.047	2.585	2.022	1.339	0.299	2.035
60	3.683	3.115	2.436	1.619	0.365	2.467
70	4.329	3.649	2.850	1.902	0.432	2.908
80	4.983	4.186	3.266	2.188	0.502	3.357
90	5.646	4.725	3.681	2.479	0.573	3.813
100	6.317	5.268	4.095	2.774	0.645	4.277
110	6.996	5.812	4.508	3.072	0.719	4.749
120	7.683	6.359	4.919	3.374	0.795	5.227
130	8.377	6.907	5.327	3.679	0.872	5.712
140	9.078	7.457	5.733	3.988	0.950	6.204
150	9.787	8.008	6.137	4.301	1.029	6.702
160	10.501	8.560	6.539	4.617	1.109	7.207
170	11.222	9.113	6.939	4.936	1.190	7.718
180	11.949	9.667	7.338	5.258	1.273	8.235
190	12.681	10.222	7.737	5.584	1.356	8.757
200	13.419	10.777	8.137	5.912	1.440	9.286
210	14.161	11.332	8.537	6.243	1.525	9.820
220	14.909	11.887	8.938	6.577	1.611	10.360
230	15.661	12.442	9.341	6.914	1.698	10.905
240	16.417	12.998	9.745	7.254	1.785	11.456
250	17.178	13.553	10.151	7.596	1.873	12.011
260	17.942	14.108	10.560	7.940	1.962	12.572
270	18.710	14.663	10.969	8.287	2.051	13.137
280	19.481	15.217	11.381	8.636	2.141	13.707
290	20.256	15.771	11.793	8.987	2.232	14.281
300	21.033	16.325	12.207	9.340	2.323	14.860

(Continued)

Temperature (°C)	Type E	Type J	Type K	Type N	Type S	Type T
310	21.814	16.879	12.623	9.695	2.414	15.443
320	22.597	17.432	13.039	10.053	2.506	16.030
330	23.383	17.984	13.456	10.412	2.599	16.621
340	24.171	18.537	13.874	10.772	2.692	17.217
350	24.961	19.089	14.292	11.135	2.786	17.816
360	25.754	19.640	14.712	11.499	2.880	18.420
370	26.549	20.192	15.132	11.865	2.974	19.027
380	27.345	20.743	15.552	12.233	3.069	19.638
390	28.143	21.295	15.974	12.602	3.164	20.252
400	28.943	21.846	16.395	12.972	3.260	20.869
410	29.744	22.397	16.818	13.344	3.356	
420	30.546	22.949	17.241	13.717	3.452	
430	31.350	23.501	17.664	14.091	3.549	
440	32.155	24.054	18.088	14.467	3.645	
450	32.960	24.607	18.513	14.844	3.743	
460	33.767	25.161	18.938	15.222	3.840	
470	34.574	25.716	19.363	15.601	3.938	
480	35.382	26.272	19.788	15.981	4.036	
490	36.190	26.829	20.214	16.362	4.135	
500	36.999	27.388	20.640	16.744	4.234	
510	37.808	27.949	21.066	17.127	4.333	
520	38.617	28.511	21.493	17.511	4.432	
530	39.426	29.075	21.919	17.896	4.532	
540	40.236	29.642	22.346	18.282	4.632	
550	41.045	30.210	22.772	18.668	4.732	
560	41.853	30.782	23.198	19.055	4.832	
570	42.662	31.356	23.624	19.443	4.933	
580	43.470	31.933	24.050	19.831	5.034	
590	44.278	32.513	24.476	20.220	5.136	
600	45.085	33.096	24.902	20.609	5.237	
610	45.891	33.683	25.327	20.999	5.339	
620	46.697	34.273	25.751	21.390	5.442	
630	47.502	34.867	26.176	21.781	5.544	
640	48.306	35.464	26.599	22.172	5.648	
650	49.109	36.066	27.022	22.564	5.751	
660	49.911	36.671	27.445	22.956	5.855	
670	50.713	37.280	27.867	23.348	5.960	
680	51.513	37.893	28.288	23.740	6.064	
690	52.312	38.510	28.709	24.133	6.169	
700	53.110	39.130	29.128	24.526	6.274	
710	53.907	39.754	29.547	24.919	6.380	
720	54.703	40.382	29.965	25.312	6.486	

(Continued)

(*Continued*)

Temperature (°C)	Type E	Type J	Type K	Type N	Type S	Type T
730	55.498	41.013	30.383	25.705	6.592	
740	56.291	41.647	30.799	26.098	6.699	
750	57.083	42.283	31.214	26.491	6.805	
760	57.873	42.922	31.629	26.885	6.913	
770	58.663	43.563	32.042	27.278	7.020	
780	59.451	44.207	32.455	27.671	7.128	
790	60.237	44.852	32.866	28.063	7.236	
800	61.022	45.498	33.277	28.456	7.345	
810	61.806	46.144	33.686	28.849	7.454	
820	62.588	46.790	34.095	29.241	7.563	
830	63.368	47.434	34.502	29.633	7.672	
840	64.147	48.076	34.908	30.025	7.782	
850	64.924	48.717	35.314	30.417	7.892	
860	65.700	49.354	35.718	30.808	8.003	
870	66.473	49.989	36.121	31.199	8.114	
880	67.245	50.621	36.524	31.590	8.225	
890	68.015	51.249	36.925	31.980	8.336	
900	68.783	51.875	37.325	32.370	8.448	
910	69.549	52.496	37.724	32.760	8.560	
920	70.313	53.115	38.122	33.149	8.673	
930	71.075	53.729	38.519	33.538	8.786	
940	71.835	54.341	38.915	33.926	8.899	
950	72.593	54.949	39.310	34.315	9.012	
960	73.350	55.553	39.703	34.702	9.126	
970	74.104	56.154	40.096	35.089	9.240	
980	74.857	56.753	40.488	35.476	9.355	
990	75.608	57.349	40.879	35.862	9.470	
1000	76.357	57.942	41.269	36.248	9.585	
1010		58.533	41.657	36.633	9.700	
1020		59.121	42.045	37.018	9.816	
1030		59.708	42.432	37.402	9.932	
1040		60.293	42.817	37.786	10.048	
1050		60.877	43.202	38.169	10.165	
1060		61.458	43.585	38.552	10.282	
1070		62.040	43.968	38.934	10.400	
1080		62.619	44.349	39.315	10.517	
1090		63.199	44.729	39.696	10.635	
1100		63.777	45.108	40.076	10.754	
1110		64.355	45.486	40.456	10.872	
1120		64.933	45.863	40.835	11.991	
1130		65.510	46.238	41.213	11.110	
1140		66.087	46.612	41.590	11.229	
1150		66.664	46.985	41.966	11.348	

(Continued)

Temperature (°C)	Type E	Type J	Type K	Type N	Type S	Type T
1160		67.240	47.356	42.342	11.467	
1170		67.815	47.726	42.717	11.587	
1180		68.389	48.095	43.091	11.707	
1190		68.963	48.462	43.464	11.827	
1200		69.536	48.828	43.836	11.947	
1210			49.192	44.207	12.067	
1220			49.555	44.577	12.188	
1230			49.916	44.947	12.308	
1240			50.276	45.315	12.429	
1250			50.633	45.682	12.550	
1260			50.990	46.048	12.671	
1270			51.344	46.413	12.792	
1280			51.697	46.777	12.913	
1290			52.049	47.140	13.034	
1300			52.398	47.502	13.155	
1310			52.747		13.276	
1320			53.093		13.397	
1330			53.438		13.519	
1340			53.782		13.640	
1350			54.125		13.761	
1360			54.467		13.883	
1370			54.807		14.004	
1380					14.125	
1390					14.247	
1400					14.368	
1410					14.489	
1420					14.610	
1430					14.731	
1440					14.852	
1450					14.973	
1460					15.094	
1470					15.215	
1480					15.336	
1490					15.456	
1500					15.576	
1510					15.697	
1520					15.817	
1530					15.937	
1540					16.057	
1550					16.176	
1560					16.296	
1570					16.415	

(Continued)

(Continued)

Temperature (°C)	Type E	Type J	Type K	Type N	Type S	Type T
1580					16.534	
1590					16.653	
1600					16.771	
1610					16.890	
1620					17.008	
1630					17.125	
1640					17.243	
1650					17.360	
1660					17.477	
1670					17.594	
1680					17.711	
1690					17.826	
1700					17.942	
1710					18.056	
1720					18.170	
1730					18.282	
1740					18.394	
1750					18.504	
1760					18.612	

Bibliography

[1] Benedict R P 1984 *Fundamentals of Temperature, Pressure and Flow Measurements* 3rd edn (New York: Wiley)

[2] Doeblin E O 2004 *Measurement Systems: Application and Design* 5th edn (New Delhi: Tata McGraw Hill)

[3] Eckman D P 1975 *Industrial Instrumentation* (New Delhi: Wiley Eastern)

[4] Dally J W, Riley W F and McConnell K G 1984 *Instrumentation For Engineering Measurements* (Singapore: Wiley)

[5] Ogata K 1978 *System Dynamics* (Englewood Cliffs, NJ: Prentice Hall)

[6] Murthy D V S 2003 *Transducer and Instrumentation* (New Delhi: Prentice Hall India)

[7] Bentley J P 2007 *Principles of Measurement Systems* 3rd edn (New Delhi: Pearson Education)

[8] Barney G C 1988 *Intelligent Instrumentation* (New Delhi: Prentice Hall India)

[9] Holman J P 1989 *Experimental Methods for Engineers* 5th edn (Singapore: McGraw Hill)

[10] Johnson C D 2000 *Process Control Instrumentation Technology* 6th edn (Englewood Cliffs, NJ: Prentice Hall International)

[11] Jones B E 1988 *Instrumentation Measurement and Feedback* (New Delhi: Tata McGraw Hill)

[12] McGee T D 1988 *Principles and Methods of Temperature Measurement* (New York: Wiley-Interscience)

[13] Bell D A 1997 *Electronic Instrumentation and Measurements* 2nd edn (New Delhi: Prentice Hall India)

[14] Wobschall D 1979 *Circuit Design for Electronic Instrumentation: Analog and Digital Devices from Sensors and Display* (New York: McGraw Hill)

[15] K T V 1989 New developments in sensor technology-fibre and electro optics *Meas. Control* **22** 165–75

[16] Saha M N and Srivastava B N 1965 *A Treatise on Heat* 5th edn (Allahabad: The Indian Press)

[17] Skoog D A, James Holler F and Crouch S R 2007 *Principles of Instrumental Analysis* 6th edn (Boston, MA: Thomson Brooks)

[18] Cromwell L, Weibell F J and Pfeiffer E A 1992 *Biomedical Instrumentation and Measurement* 2nd edn (New Delhi: Prentice Hall India)

[19] Clark L C Jr 1956 Monitor and control of bood and tissue oxygen tensions *Trans. Am. Soc. Artif. Organs* **2** 41–8

[20] Lee Y H and Tsao G T 1979 Dissolved oxygen electrodes *Advances in Biochemical Engineering, Mass Transfer and Process Control* ed T K Ghose, A K Fiechter and N Blakebrough (Berlin: Springer) vol 13 pp 35–86

[21] Mancey K H, Okun D A and Reilley C N 1962 A galvanic cell oxygen analyzer *J. Electroanal. Chem.* **4** 65–92

[22] Borkowski J D and Johnson M J 1967 Long-lived steam-sterilizable membrane probes for dissolved oxygen measurement *Biotechnol. Bioeng.* **9** 635–9

[23] Young D R, Woodward D J and Rousey M J 2003 Dissolved oxygen technology is criticalcomponent in wastewater treatment *Environ. Sci. Eng. Mag.* https://esemag.com/archives/dissolved-oxygen-technology-is-critical-component-in-wastewater-treatment/

[24] Liu J, Björnsson L and Mattiasson B 2000 Immobilised activated sludge based biosensor for biochemical oxygen demand measurement *Biosens. Bioelectron.* **14** 883–93

[25] Honeywell Dissolved Oxygen Analyzer and Probes *Application: Measurement and Control of Dissolved Oxygen in Wastewater Treatment* (http://lesman.com/unleashd/catalog/analytical/analyt_hwdisox.htm)

[26] Millman J and Halkias C C 1967 *Electronic Devices and Circuits* (Tokyo: McGraw Hill Kogakusha)

[27] Millman J and Grabel A 2006 *Microelectronics* 2nd edn (New Delhi: Tata McGraw Hill)

[28] Sedra A S and Brackett P O 1978 *Filter Theory and Design: Active and Passive* (Beaverton, OR: Matrix Publishers)

[29] Barua A and Sinha S 1995 *Computer Aided Analysis Synthesis and Expertise of Active Filters* (New Delhi: Dhanpat Rai and Sons)

[30] Patranabis D 2005 *Sensors and Transducers* 2nd edn (New Delhi: Prentice Hall India)

[31] Moris A S 2001 *Measurement and Instrumentation Principles* 3rd edn (Amsterdam: Elsevier) https://shop.elsevier.com/books/measurement-and-instrumentation-principles/morris/978-0-7506-5081-6

[32] Barney G C 1988 *Intelligent Instrumentation* (New Delhi: Prentice Hall India)

[33] Breckenbridge R A and Husson C 1978 Smart sensors in spacecraft: the impact and trends *Proc. of AIAA/NASA Conf. on Smart Sensors (Hampton)* pp 1–5 https://ui.adsabs.harvard.edu/abs/1978smse.confR....B/abstract

[34] 1992 *Handbook of Intelligent Sensors for Industrial Automation* ed N Zuech (Reading, MA: Addison-Wesley)

[35] Ohba R (ed) 1992 *Intelligent Sensor Technology* (New York: Wiley)

[36] Kabisatpathy P, Barua A and Sinha S 2005 *Fault Diagnosis of Analog Integrated Circuits* (Dordrecht: Springer)

[37] Waterman D A 1986 *A Guide to Expert Systems* (Reading, MA: Addison- Wesley)

[38] Barua A, Ray S and Sinha S 1993 TRANSELEX: a knowledge-based approach for selection for transducers using dynamic database *Eng. Appl. Artif. Intell.* **6** 73–5

[39] Barua A and Sengupta S 1996 EXSENSEL: a rule based approach to selection of sensors for process variables *Chem. Eng. Technol.* **19** 443–7

[40] Schalkoff R J 1990 *Artificial Intelligence: An Engineering Approach* (Singapore: McGraw-Hill)

[41] Rich E and Knight K 1991 *Artificial Intelligence* 2nd edn (New Delhi: Tata McGraw Hill)

[42] Borland International 1986 *Turbo Prolog Owner's Handbook* 2nd edn (Scotts Valley, CA: Borland International Inc.)

[43] Schildt H 1987 *Advanced Turbo Prolog Version 1.1* (San Francisco, CA: Osborne/McGraw Hill)

[44] Lycett J E, Porter T and Maudsley D 1988 FLOSEL: expert selection of flowmeters *Eng. Appl. Artif. Intell.* **1** 37–40

[45] Barua A (ed) 2020 *Pipelined Analog to Digital Converter and Fault Diagnosis* (Bristol: IOP Publishing) https://doi.org/10.1088/978-0-7503-1732-0

Index

www.ingramcontent.com/pod-product-compliance
Lightning Source LLC
Chambersburg PA
CBHW082130210326
41599CB00031B/5928